Series Editors:

R. Bruce Lindsay — **Brown University**

Osman K. Mawardi — **Case Western Reserve University**

PUBLISHED VOLUMES AND VOLUMES IN PREPARATION

ENERGY: Historical Development of the Concept / *R. Bruce Lindsay*
APPLICATIONS OF ENERGY: Nineteenth Century / *R. Bruce Lindsay*
COAL, PART I: Social and Economic Aspects / *Mones E. Hawley*
COAL, PART II: Scientific and Technical Aspects / *Mones E. Hawley*
THE SECOND LAW OF THERMODYNAMICS / *Joseph Kestin*
IRREVERSIBLE PROCESSES / *Joseph Kestin*
ATOMIC PHYSICS AND ENERGY / *Robert Lindsay*
ENERGY STORAGE / *W. V. Hassenzahl*
DYNAMICS AND CONTROL OF POWER SYSTEMS / *A. H. El-Abiad*

Benchmark Papers
on Energy / 1

A BENCHMARK® Books Series

ENERGY:
Historical Development of the Concept

Edited by

R. BRUCE LINDSAY

Brown University

Dowden, Hutchinson
& Ross, Inc.

Stroudsburg, Pennsylvania

Distributed by

HALSTED PRESS *A Division of John Wiley & Sons, Inc.*

Benchmark Papers on Energy, Volume 1
Library of Congress Catalog Card Number: 75–30719
ISBN: 0–470–53881–3

77 76 75 1 2 3 4 5
Manufactured in the United States of America.

LIBRARY OF CONGRESS CATALOGING IN PUBLICATION DATA

Library of Congress Cataloging in Publication Data
Main entry under title:

Energy.

(Benchmark papers in energy ; 1-)
Includes index.
CONTENTS: pt. 1. Historical development of the concept.
1. Force and energy. I. Lindsay, Robert Bruce, 1900-
QC73.E52 531'.6 75-30719
ISBN 0 470-53881-3

Exclusive Distributor: **Halsted Press**
A Division of John Wiley & Sons, Inc.

PERMISSIONS

The following papers have been reprinted with the permission of the authors and copyright holders.

THE CLARENDON PRESS
The Basic Works of Aristotle
Physics
Works: The Oxford Translation
Mechanica

DOVER PUBLICATIONS, INC—*Reflections on the Motive Power of Fire and Other Papers*
The Motive Power of Heat

HARVARD UNIVERSITY PRESS
Source Book in Greek Science
The Five Simple Machines
A Source Book in Physics
Conservation of Quantity of Motion
The Impossibility of Perpetual Motion and the Problem of the Inclined Plane
The Principle of Virtual Displacements and Use of the Word Energy
The Third Law of Motion as a Forerunner of the Concept of Energy
The *Vis Viva* Controversy

NORTHWESTERN UNIVERSITY PRESS—*Dialogues Concerning Two New Sciences*
The Pendulum Experiment

PERGAMON PRESS LTD.
Men of Physics: Benjamin Thompson, Court Rumford
Source of Heat from Friction
Men of Physics: Julius Robert Mayer, Prophet of Energy
The Motions of Organisms and Their Relation to Metabolism
On the Forces of Inorganic Nature

PLENUM PUBLISHING CORP.—*Foundations of Physics*
The Concept of Energy and Its Early Historical Development

REGENTS OF THE UNIVERSITY OF WISCONSIN—*On Motion and on Mechanics*
Of the Force of Percussion
Work Involved in the Operation of Machines

SERIES EDITOR'S PREFACE

The Benchmark Papers on Energy constitute two series of volumes that make available to the reader in carefully organized form important and seminal papers in the historical development of the concept of energy and its applications in all fields of science and technology, as well as its role in civilization in general. This concept is generally admitted to be the most far-reaching idea that the human mind has developed to date, and its fundamental significance for human life and society is everywhere evident.

The first seven volumes of the series contain papers that deal primarily with the evolution of the energy concept and its current applications in the various branches of science. These will be supplemented in the future by volumes that concentrate on the technological and industrial applications of the concept and its socioeconomic implications.

Each volume has been organized and edited by an authority in the area to which it pertains and offers the editor's careful selection of the appropriate seminal papers, that is, those articles which have significantly influenced further development of that phase of the whole subject. In this way, every aspect of the concept of energy is placed in appropriate perspective, and each volume represents an introduction and guide to further work.

Each volume includes an editorial introduction that summarizes the significance of the field being covered. Every article or group of articles is accompanied by editorial commentary, with explanatory notes where necessary. An adequate index is provided for ready reference. Articles in languages other than English are either translated or abstracted in English. It is the hope of the publisher and editors that these volumes will serve as a working library of the most important scientific, technological, and social literature connected with the idea of energy.

The present volume, *ENERGY: Historical Development of the Concept,* has been prepared by the series editor. It is intended to serve as an introduction to the series as a whole and endeavors to trace the development of the concept from early times to the middle of the nineteenth century, when the concept had arrived essentially at its modern meaning. Although, in one or two cases, the reproductive quality of older papers is not as perfect as we would wish, inclusion of the original typograghy gives the flavor of the persistance of early workers on the concept of energy. This volume will be followed by a companion volume, which generalizes the concept further and stresses nineteenth-century applications. The introduction to this volume provides a more detailed picture of the basis on which the book is organized.

I am deeply indebted to Patricia Galkowski and her colleagues in the Sciences Library of Brown University for assistance in the location of source material. I am grateful to Susan Desilets Proto of the Department of Physics in Brown University for typing the translated material. Acknowledgment is made of assistance rendered by the Photo Laboratory of Brown University and the Library of the Massachusetts Institute of Technology in connection with the microfilming and photocopying of material. I owe a particular debt to my long-time friend, the late Professor Raymond S. Stites, for help in connection with the references to Leonardo da Vinci.

R. BRUCE LINDSAY

CONTENTS

PART II: THE NATURE OF HEAT

CONTENTS BY AUTHOR

INTRODUCTION: WHAT IS ENERGY?

THE BASIS OF OUR MATERIAL CIVILIZATION

To live, man must interfere with his environment. He raises food to nourish his body, builds fires to keep himself warm, and makes shelters to protect himself from the elements. He constructs vessels and other vehicles to carry him, his fellows, and his goods over sea and land. More miraculously still, he develops language to communicate meaningfully with his fellowman. All this our primitive ancestors learned how to do; if they had failed to do so, we would not be here. As man began to think more deeply about his surroundings and experience, his interference became more sophisticated and gradually developed into what we today call technology, the rational manipulation of the environment for the sake of more elaborate and comfortable living.

Parallel to the growth of technology, although somewhat later in man's development as a thinking being, came the origin of science as a method of thinking and talking about human experience, a kind of mental device to satisfy human curiosity about the way things "go" and leading to what we now call understanding of the world about us. Science operates with ideas called concepts, which have the somewhat mysterious power to pull together large pieces of experience, so that we do not have to invent new words to describe every individual element of experience we encounter in our everyday life. For example, the physical scientist concocts the concepts of force and mass to help him describe in an effective and economical manner the phenomena exhibited by the motion of bodies.

Obviously, if we can find a single word to represent an idea which applies to every element in our experience in a way that makes us feel we have a genuine grasp of it, we have achieved something economical and powerful. This is what has happened with the idea expressed by the word *energy.* No other concept has so unified our understanding of experience. Without exaggeration we can say that every single aspect of human experience, whether it be what we observe in the external world, or what we do or what is done to us, can be adequately described either as a *transfer* of energy in one form from one place to another or the *transformation* of energy from one form to another.

When you roll a barrel up an incline you are transferring the mechanical energy of your limbs and muscles to the barrel; it is represented by the mechanical energy of motion as well as by the potential energy involved in raising the barrel above the level of the ground. To accomplish this transfer of energy without undue effort on your part you employ, as your ancestors learned to do several million years ago, a device called a *machine,* in this case the inclined plane. Like every other machine, its function is to transfer mechanical energy, that is, energy of motion or position from one place to another.

When you switch on an electric light, you are taking advantage of the transfer of energy in the electrical form from the power plant to your home. That the light goes on, when you turn the switch, is an example of the ability of energy in one form to be transformed into energy of another form, in this case from electrical energy into heat energy and thence, in part, into visible light energy. This is a physical effect involving both transfer and transformation of energy, a situation encountered in most sophisticated technological operations.

Consider what happens in the running of your car. The chemical energy hidden in the explosive mixture of gasoline vapor and air is transformed by the agency of the spark into heat energy. In turn, this is transformed in part into the mechanical energy of motion of the pistons in the cylinders. The mechanical energy of the pistons is transferred to the drive shaft and from there to the wheels to move the car.

In a power plant, chemical energy stored in fossil fuel, such as coal, oil, or gas, is transformed by combustion into heat energy in the boiler. This heat energy changes water from a liquid state to steam. The heat energy of the steam is transformed in part into mechanical energy in the steam turbine. This mechanical energy is then further transformed into electrical energy in the alternating-current generator. From the latter it is transferred by electric cables to various points where it can be used for further transformation. Our whole electrically based economy is thus inextricably linked with the transfer and transformation of energy, in which indeed electrical energy plays the key role, because of the relative ease with which it is generated, transferred over long distances, and transformed into other forms. There was uncanny prescience in Michael Faraday's reported reply to a group of politicians who visited his laboratory in the early 1830s and asked him of what practical importance his discovery of electromagnetic induction could possibly be: "Some day, gentlemen, you will be able to tax it." The story may well be apocryphal. But he could have made the statement with absolute assurance of its ultimate veracity.

The detailed process of the digestion of food is a rather complicated affair, but what it amounts to is the transformation of the chemical energy locked in the food into heat energy to keep the body warm and into mechanical energy to enable the body to do work by moving its various parts or itself as a whole. There is also some transformation into electrical energy and other types of chemical energy to establish communication between various parts of the body and enable the nervous system to function. Here again transfer is involved, as well as transformation. All biological processes throughout the domain of living things can be interpreted in terms of the concept of energy.

The winds of the gale and the hurricane provide another example of the transformation of heat energy communicated to air into mechanical energy; the resulting motions are also amplified by the transfer of mechanical energy from the rotating earth. Energy also plays a characteristic role in the earthquake. When a mass of rock gives way along a fault, potential energy is transformed into kinetic energy or energy of motion, which in turn produces a large change in stress in the neighborhood. This change does not remain localized, but, since the earth's crust is to a certain extent an elastic medium, spreads out as a wave that transfers a part of the energy of the original disturbance. This energy can lead to great destruction near the surface and near the source. It can also be detected at great distances from the source by means of sensitive instruments called seismographs. Wave propagation is a very important example of energy transfer, as in both light waves and sound waves.

As an example of wave propagation as a case of energy transfer, we receive a relatively enormous amount of energy from the sun through the light waves emitted by this hot, luminous body, which is responsible for the existence and maintenance of life on the earth. What is the origin of all this energy which pours out of the sun's surface with such profusion at a rate of about 4×10^{23} kilowatts? This has been a long-standing problem of interest to astronomers and physicists. Only in relatively recent times has a plausible answer been arrived at. The source of solar energy is not a simple transformation of chemical energy into heat, as in the burning of coal, but is now considered to be the transformation of mass into energy through the building of helium nuclei out of hydrogen, a *thermonuclear* process, which is the basis of the hydrogen bomb.

In the developed parts of the world we have grown so accustomed to having something interesting and useful happen at the flick of a switch or a little pressure on the car accelerator that we are apt to forget that these happenings are nothing but manifestations of energy transfer and transformation. People in general are now growing more conscious of the role of energy in human life and activity through the frequent warnings in the mass media that our rate of energy transformation is rapidly becoming too great for the finite resources at our disposal. Modern technology is continually devising new ways of transferring and transforming energy to meet real or imagined human needs. This means faster extraction of fossil fuels and the building of power plants at a faster rate, which not only depletes the fuel supply at an accelerated rate, but also increases interference with the environment, a result now referred to with the rather unhappy term "pollution." Moreover, the rapidly increasing population of the world puts another strain on energy sources, for people have to eat to live, and food is energy, or more correctly the consumption of food is an illustration of energy transformation. In a very real sense the key problem now facing all human beings on this planet is to find new sources of energy transformation sufficient to provide a decent existence for all, without at the same time producing environmental pollution deleterious to the very life we want everyone to enjoy. It is customary to refer to this as the problem of energy supply. The alarmists refer to it as the energy crisis.

You may object that this talk about energy and its transfer and transformation, its importance in our lives, what terrible things can happen if the supply fails, and so on, is all very well, but what after all is this thing, this energy we are talking about? What *is* it really and how is it measured? Is it like water, which is measured by the gallon or the liter? If you are served by an electrical power plant, there is a device on the outside of your house with a glass cover. Inside there is a wheel and some dials with numbers. The wheel goes around as long as any of your electrical implements are operating; or, as the saying is, you are "drawing current." This device is a watt-hour meter. Its use enables the electric company to bill you for your "use" of electricity, or more correctly electrical energy. You are correct in your surmise that there is a long story back of all this. We are now going to try to tell some of the scientific details of this story. It is hoped that a better knowledge of the science of energy will give you ultimately a better appreciation of what energy really means in the life of man.

THE FUNDAMENTAL IDEA IN THE ENERGY CONCEPT: CONSTANCY IN THE MIDST OF CHANGE

We have already mentioned an obvious difficulty with the concept of energy from the standpoint of popular understanding: just what *is* it, after all? You have heard that everything in your experience is intimately connected with the transfer and transformation of energy. But what is this thing that is transferred and transformed? It seems so intangible and elusive in spite of all the very obvious happenings that are said to be explainable in terms of it. A rock or tree is *there* and seems real. But where is the reality about energy? We might consult the dictionary. After all, a dictionary exists to give us the meaning of words. Unfortunately, a look in the unabridged Webster does not help much. Like many other scientific terms, energy has attached itself to numerous aspects of ordinary life, with a wide variety of meanings. To learn that energy is "strength of expression," "internal power," "power forcibly exerted," or (as the philosophers say) "the realized state of potentialities" is not exactly helpful in attaching meaning to energy as a physical concept, although it may help us to realize what a vast circular tautology a dictionary is, since it endeavors to define every word in terms of other words in the same collection! Even in a dictionary of specifically scientific terms, we are not given much further aid. The statement "energy is the capacity to do work" conveys little unless you have an understanding of what work means. Even in mechanics this is a highly unsatisfactory definition, since it provides no idea of how to measure energy; unless this is made clear, there can be no genuine understanding of the concept. Must you then pursue a whole course of university physics with its awful array of mathematical symbolism in order to understand what energy means, how it is measured, and the reason for its overwhelmingly important role in human life? Most physicists would almost certainly answer in the affirmative. However, it seems worthwhile to try another approach—a philosophical approach that seeks to concentrate attention on what is, after all, the key idea in energy and then tries to exploit this in every possible way.

The key idea is simple: *constancy in the midst of change.* What does this mean? Probably the most obvious characteristic of human experience is *change.* Even for the sedentary person, things never stay the same for very long. We are all, of course, at the mercy of the transition from daylight to darkness and vice versa, not to mention the vagaries of the weather. The normally active person seems to encounter nothing but change as he goes about his business. He rolls along in his car and the surrounding landscape certainly does not stay put. Even if he has a monotonous job, he cannot be doing precisely the same thing with his hands or his head every moment of the working day. His wife puts water in a pan on the stove and lights the burner. Something happens; we say the water boils away, certainly an obvious change. He turns on a switch and suddenly there is light where previously there was darkness. The number of such changes happening everyday in our lives is so great it would be ridiculously tedious to enumerate them. From a scientific standpoint, indeed, we *could* describe these changes as they occur, treating each one as a separate experience without any relation to other experience. But a long time ago some intelligent people decided that to seek earnestly for something which stays constant in the midst of the observed changes might help our understanding of experience. This might be thought to be a curious attitude. Presumably, it reflected a desire in the minds of these early people to associate *order* with what otherwise seemed for the most part chaotic. Such a search might conceivably lead to the tying together of what at first sight appear to be quite disparate elements in experience. The success of this effort has developed the concept of energy to the point where it is the premier idea in the whole of science.

What has just been said may well be characterized as too much of a glittering generality. We therefore ought to give a specific example of what is meant by constancy in the midst of change. We have mentioned the machine as one of the most useful devices ever created by man. Whether it was an inclined plane, a lever, a pulley, or a wheel and axle, the machine must have appeared to our primitive ancestors as marvelous in its uncanny, miraculous ability to enable heavy loads to be lifted by the application of a very modest force. Of all machines, the simple lever is perhaps the simplest. In school textbooks it strikes most students as, of all subjects, the dullest of the dull, because its real scientific significance is never mentioned. To recapitulate the "dull" part first, you will recall that when a lever in the form of a finite solid rod is placed horizontally across a fulcrum so that its two ends are at different distances from the fulcrum, a small force applied to the end farther from the fulcrum is able to raise a large weight placed at the end closer to the fulcrum. The ratio of the weight raised (which may be called the force *exerted* by the machine) to the *applied* force is called the *mechanical advantage* of the machine. This is a measure of what the machine can accomplish by making it easier for a man to raise a heavy weight—easier, that is, than it would be to lift the weight directly off the ground with one's hands.

When a person pushes down at the end of the long arm of the lever, the weight attached to the end of the shorter arm rises. This is change, and of course a very desirable change from the standpoint of one who wants to raise the weight off the ground with the least possible effort on his part. But we have just said that this is an example of constancy in the midst of change. Where is the constancy in this process?

Unless our ancestors were very unobservant, they must have noticed that, to raise the weight in question a given distance above the ground, it was necessary to lower the end of the longer arm of the lever, at which the force is applied, a *greater* distance. In other words, the gain in force provided by the machine (as measured by the mechanical advantage) is compensated by a loss in the speed with which the weight is raised. This compensation may also be expressed by saying that, in the operation of the lever as a machine, something stays constant. This "something" is the product of the force and the distance moved by the point of application. Thus, the applied force multiplied by the distance that the point of application moves is equal to the weight raised multiplied by the distance it moves.

For the past hundred years or so it has been customary to call the product of the force and the distance moved by the end of the lever where the force is applied or exerted the "work" done by the force. The constancy implied in the action of the machine may be expressed by saying that the work done *on* the lever at one end is equal to the work done *by* the lever at the other end. Here then is constancy in the midst of change and the germ of the idea of energy. In this case, energy is said to be mechanical in form and to be measured by the work done. Thus, the lever as a machine transfers an unchanged amount of mechanical energy from one end of the lever to the other: a given *input* of mechanical energy leads to an equal output.

In actual practice it is observed that the energy at the output end is never *quite* equal to that at the input end. At first, this result might seem to blow up the whole idea of energy as constancy in the midst of change. For it certainly seems embarrassing to find that the concept does not live up to the demands laid upon it as soon as one applies it to a single example. Actually, it turns out, as we shall have abundant opportunity to observe, that it is precisely this kind of situation which has led to the development and generalization of the concept of energy to its present premier position in science.

For the moment we shall handle the difficulty in the following way. As we make the contact between the lever and the fulcrum sharper and sharper so as to approach what the engineer would call a perfect bearing, we observe that the difference between energy input and energy output becomes a smaller and smaller fraction of the measured value of either. This suggests that, in the ideal limit, there really is constancy in the energy transfer by the machine. It must be admitted that this introduction of the *ideal* case has been very commonly used by physicists and scientists in general in the development of concepts.

Let us look at what appears to be a rather different problem and see what we can make of it from the standpoint of energy as representing constancy in the midst of change. You throw a ball straight up into the air with a given velocity. What happens? The ball rises to a certain height, stops momentarily, and then falls back into your hands at the same speed with which you threw it up. Plenty of change is observed in this simple act. The ball no sooner leaves the hand than its upward speed begins to diminish and continues to do so until it becomes zero at the top of the flight. As it falls, it gradually regains the speed it had lost while rising, finally regaining it all back by the time it again reaches your hands. What then is constant in the midst of the obvious change observed in the up and down motions? When you throw the ball into the air, you are said to communicate mechanical energy of motion to it. This kind of energy is represented by one half the product of the mass of the ball and the square of the speed. The technical term is *kinetic energy.* This energy obviously does not stay constant, since it decreases steadily from the moment the ball leaves the hand until it becomes zero at the top of the path. What then does stay constant during the motion of the ball? It is clear that to the kinetic energy we must add something which steadily increases as the ball travels upward, reaching a maximum at the top of the path, and steadily decreases during the downward motion of the ball, so that the sum stays the same at every point in the ball's path. This must represent another form of energy, somehow dependent on the height of the ball above the ground. It may properly be called *positional energy*, since at each position of the ball it represents the potentiality of the ball to change its kinetic energy by moving either up or down. Actually, this energy is called *potential energy* and is measured by the product of the mass of the ball, the acceleration of gravity, and the distance above the ground. The constant in this case is the sum of the kinetic energy and the potential energy. On the upward flight, as the kinetic energy decreases, the potential energy increases in such a way that the sum remains constant. On the way down, the potential energy decreases as the kinetic energy increases, with the sum again remaining constant. The sum is called the *total mechanical energy* of the motion. Its numerical value is that of the kinetic energy at the bottom of the motion, or the equal potential energy at the top. It is easy to verify the validity of the assignment by using it to calculate the speed gained by the ball from any height above the gound when dropped from rest at the height it attains in its upward flight. This speed agrees with experience.

Here again we find a certain numerical difficulty. After the ball has been thrown into the air with a given initial speed, it is found that the speed on its return to the hand, which should be equal to the initial upward speed if the energy is really constant during the motion, does not quite live up to this requirement. It is always a bit less. Our intuition comes to the rescue, and we decide to perform the experiment with more elaborate projection equipment in a space from which the air can gradually be removed by a pump. It is then found that as the air is pumped out the return speed comes closer and closer to the initial speed, leading us to believe that in the really ideal case of motion in a vacuum the energy would indeed stay constant and satisfy our requirement of constancy in the midst of change.

But let us look at a more difficult case, also connected with the motion of a ball under gravity, that is, falling freely to the surface of the earth. We drop a ball from a certain height above the floor so that it falls into a dish rigidly fastened to the floor; the dish is full of soft putty. When the ball hits the putty, it penetrates a certain distance into it and then stops. In its final state it clearly has no energy, since it has lost its speed and obviously is in no position to resume its motion and hence has no potential energy. What then has become of the mechanical energy it had in flight? Here again constancy in the midst of change appears to be a will-of-the-wisp, and there does not seem to be any way in which we can idealize it back into existence, as we did in our previous examples. But we note that *something* has happened after the ball strikes the putty. Some of the putty has been pushed aside in order that the ball might penetrate it. We may reasonably aver that some energy of motion must be sacrificed to overcome the resistance the putty offers to the observed deformation. We can thus account for the apparent loss of the mechanical energy that the ball had at the moment of impact. The implication is clear: whenever we encounter a case in which the energy attributed to a body appears to disappear, we are entitled to look for a physical effect that was not there before and which can account for the apparent loss in energy. This restores the constancy and all is well. It is out of the apparent failure of the idea of constancy in the midst of change, as exemplified in the case just discussed, that physicists have been able to generalize the concept of energy and make it serve a more useful purpose than might at first be thought possible. It took a long time, much cogitation, and much discussion among intelligent men before this point of view became well established in the middle of the nineteenth century.

The development of the concept of energy has had a long and interesting history. It is the purpose of the present volume to trace this historical development through the papers of men who contributed to it. Our story begins in antiquity and continues to the middle of the nineteenth century, when the concept had become rather well established and scientists and technologists in various fields were preparing to apply it vigorously to a host of problems.

Part I

ENERGY: EARLY IDEAS AND DEVELOPMENT OF THE CONCEPT

Editor's Comments on Paper 1

The Concept of Energy and Its Early Historical Development
1 R. B. LINDSAY

Paper 1 is a brief summary of the historical development of the concept of energy from ancient times to the end of the eighteenth century. The aim has been to find in the writings of early scientists and philosophers vestiges of the energy idea from the standpoint of its fundamental meaning as constancy in the midst of change. Most authorities mentioned in the bibliography are represented in the reprinted material presented in this volume.

The author, who is the editor of this volume, is emeritus professor of physics in Brown University.

1

Reprinted from *Found. Phys.*, 1(4), 383–393 (1971)

The Concept of Energy and Its Early Historical Development

R. B. Lindsay
Brown University
Providence, Rhode Island

Received December 16, 1970

The concept of energy, the premier concept of physics and indeed of all science, is here investigated from the standpoint of its early historical origin and the philosophical implications thereof. The fundamental assumption is made that the root of the concept is the notion of invariance or constancy in the midst of change. Salient points in the development of this idea are presented from ancient times up to the publication of Lagrange's Mécanique Analytique *(1788).*

1. INTRODUCTION

Of all the concepts or constructs of physics, energy, by its unifying capacity, has proved by all odds to be the most significant and successful. Its domain of application has indeed by now far transcended physics and covers all branches of science. Not only has it played a major role in the logical development of physics itself, but by common consent it is the physical construct which has proved to contain the greatest meaning for all aspects of human life. Under the misnomer "power," it is the stock in trade of the engineer and that which makes the wheels of the world go round. More and more, it is recognized by economists as the real wealth of nations. The interpretation of phenomena in terms of the transfer of energy from one place to another and the transformation of energy from one form to another is the most powerful single tool in human understanding of experience.

The impact of the concept of energy on society has been enormous in the past and will be even greater in the future. What is the nature of this impact? It has both ideological and technological aspects. The ideological influence consists largely in the fact that the concept serves as a unifying element in all scientific descriptions of experience, enabling all scientists to think more effectively about their various problems and thus promoting the fundamental unity of science. As knowledge of nature becomes more specialized, this role of energy becomes of increasing significance.

The technological aspect of the impact of the concept of energy on society scarcely needs emphasis. It is necessary only to remind ourselves of the stupendous increase in the average number of energy "slaves" per head of population on the earth in the last quarter century. This has correspondingly increased the well-being and comfort of many millions. At the same time, progress along this line has not been devoid of serious sociological problems. The energy supply available for transformation has not been well distributed, and many segments of the earth's population are going without their fair share. Moreover, even in those nations in which the energy supply available for human needs has vastly increased, this has been accompanied by unpleasant by-products like water and air pollution. To solve these problems will, of course, involve further skillful application of the energy idea, so that of its impact there appears to be veritably no end.

A concept like energy obviously has had a history. One cannot hope really to understand its present state or its future implications without some appreciation of this history. Closely associated with the historical development there is, moreover, the evaluation of what may be called the philosophical significance of the concept. The two aspects are strictly speaking inseparable.

It is the aim of the present essay to take a look at the origins and early development of the energy idea. This examination will be undertaken in the light of certain assumptions which are of essentially philosophical nature, namely, that the basis of the concept of energy as we use it today is the idea of *invariance*, which here means constancy in the midst of change. We think in this connection of what we now call the mechanical energy of a system of mass particles subject only to their mutual interactions: this quantity is a function of the velocities and positions of the particles (in some inertial reference frame) that stays *constant in time*, no matter what the motions of the particles may be.

Definitely implied in our procedure is the conviction that unless we can find in earlier notions a connection with the way we look upon the concept of energy today our search will be illusory. Of course, we must face the fact that not all scientists may agree that the notion of invariance in the midst of change is the key idea in energy. An example of an opponent of the idea is Ernst Mach,[(1)] who vigorously expressed the opinion that the actual root of the energy concept is to be found in the principle of the impossibility of perpetual motion. Mach was a searching critic of the philosophy and history of science and his views are entitled to great respect. Leaving aside the fact, however, that in any case the principle of the impossibility of perpetual motion is closely associated logically with the idea of invariance, we may note that Mach's extreme view, if followed, would have prevented the generalization of the idea of energy to all physical phenomena. For example, Mach would not accept the mechanical theory of heat. His polemic against it almost rivals in intensity his attack on the atomic theory. It seems clear that his positivistic leanings prevented him from seeing any advantage in imaginative scientific theorizing. He could hardly have become a successful theoretical physicist in the sense of Maxwell, Boltzmann, Gibbs, and their twentieth century successors.

2. ROOTS OF THE CONCEPT OF ENERGY IN ANTIQUITY. THE PHILOSOPHERS

Most scientific concepts are not easy to trace historically. Energy provides no exception. One plausible source of the idea is connected with the invention of machines, an important technological development in the life of early man. People early learned the social significance of the fact that human life is impossible without somebody's labor, but rather naturally sought to reduce the terrific burden of this labor. Eventually, some clever and imaginative folk discovered the possibility of taking the sting out of human labor by the use of such devices as the lever, the inclined plane, and various forms of pulley systems. These gadgets, which we now call simple machines, must have seemed to the ancients to be endowed with almost magical powers, they made it so much easier to raise heavy weights, for example, or to give an arrow greater speed, as by the use of the bow.

The discoverers and users of such machines must have observed very early, however, that the mechanical advantage provided by them is always accompanied by a compensating disadvantage: nature is not inclined to give something for nothing. It was found, for example, that to raise a given weight by applying to a pulley system a force much less than the weight, the speed with which the pulley rope is pulled must be much greater than the speed with which the weight is raised. Alternatively, if one wishes to pull with low speed, the time needed for raising the weight is correspondingly increased. With the gain in ease of exertion in the performance of a given bit of labor provided by the machine there goes an inevitable loss of something represented in general by an increase in the time required to do the job. This fact was recognized explicitly in the writings on mechanics of Hero of Alexandria,[2] who flourished around 60 A.D. This peculiar principle of compensation, in which a certain gain in a vital effect is always balanced by a corresponding loss in an associated phenomenon, contained within itself the root of the concept of energy. The compensatory factor so evident in the behavior of machines implies that something stays constant in the midst of the obvious changes that take place in the operation of the machine. It is this constant "something" which later became quantified as energy.

At this point, we are tempted to look into Greek philosophy to see whether we can locate any reference to the general idea of constancy in the midst of change. As a matter of fact, it is there, though whether any Greek before Aristotle ever associated it with the behavior of machines is problematical. We can find what we are looking for in the alleged views of the two pre-Socratic philosophers Parmenides of Elea and Heraclitus of Ephesus (both of approximately the 6th century B.C.). Heraclitus is supposed to have taught that "all things flow" (panta rhei), or all is change. He was clearly impressed by the ever-changing flux of sensation characterizing our experience. Much of modern science is consistent with this point of view, as is shown in our concern for the changing behavior of physical systems with the passage of time. But acceptance of Heraclitus' idea in its extreme form would make all science a hopeless discipline, since we could never get a mental grip on anything before it became something else. As a matter of fact, some commentators on Heraclitus hold the view that in spite of his emphasis on the primacy of change, he also held that there is something invariant in the universe as a whole. This something he apparently took for *fire*, though he obviously did not mean fire in a modern sense, nor even in the ancient Greek practical sense. It was some ethereal essence which could be transformed into the common objects of our experience without net loss.

Parmenides comes definitely closer to the idea of constancy in the midst of change. Impressed (or possibly depressed) by the apparently chaotic sequence of events in human experience, he decided to treat change as merely an illusion. He felt that this is what men try to do when they invent names for things and so identify them continually throughout the flux of sensation. There is a strong human urge to extract from experience something that "stays put" long enough for effective observation and study; and this Parmenides emphasized. To be sure, his writings are fragmentary, and there is the obvious danger of reading too much into them. Nevertheless, the notion of invariance in the midst of change is there. If we seek an ancient patron saint of the concept of energy, it will surely be Parmenides.

Let us now return to machines and see what relevance to the concept of energy we can extract from the Greek attempts to explain their action.

It was Aristotle (384–322 B.C.) who wrote the first treatise on physics in the Western tradition. But this famous treatise *Physica*,[3] though it pays extensive attention to motion, says nothing about machines. However, there does exist a treatise attributed by some authorities to Aristotle, though others, including Marshall Clagett,[4] the well-known historian of mechanics, believe that the treatise was written by one of Aristotle's immediate successors. In the Latin version variously styled *Mechanica*, *Problemata Mechanica*, or *Quaestiones Mechanicae*, it may well be the first extant treatise on mechanics. At any rate, it contains probably the first attempt in Western science to explain how machines work. From the standpoint of the problem of the origin of the concept of energy, the importance of this treatise is that its treatment is based on a dynamical approach, in sharp contrast to the static method favored later by Euclid and Archimedes.

According to Pierre Duhem,[5] the author of *Mechanica* used the basic axiom taken from Aristotle's *Physica*: The "force" (*puissance* in French) exterted by the mover who moves a body is measured by the weight of the body and the velocity of the impressed motion. On this view, when the same "force" acts, the impressed velocity will be inversely proportional to the weight. If we represent velocity by V and weight by W, and "force" by F, Duhem expresses the content of the above axiom in the modern form:

$$F = kVW \tag{1}$$

where k is some constant. We may note in passing that the Greeks would not have used this form of expression, since they preferred always to use pure numbers in expressing mathematical relationships.

They would have expressed the content of the axiom in the form

$$V_1/V_2 = W_2/W_1 \tag{2}$$

In any case, in modern physical terminology, if F is taken as the equivalent of what we now call force, Eq. (1) makes no sense. However, it could agree with modern physics if F is interpreted as *power* or the time rate of doing work, k being set equal to unity.

In the application of Eq. (1) to the behavior of a lever with weights W_1 and W_2 suspended from the ends of the weightless lever bar at distances l_1 and l_2 respectively from the fulcrum C, the further assumption is made in *Mechanica* that when the same "force" acts, the point of the lever *further* from the fulcrum C moves with *greater* velocity. The author convinced himself of this from the geometrical properties of the circle. But this is equivalent to the relations

$$V_1 = kl_1, \qquad V_2 = kl_2 \tag{3}$$

If these are combined with (1) or (2), the result is

$$l_1W_1 = l_2W_2 \tag{4}$$

which is the law of the lever. With VW treated as power rather than "force," the above "proof" is equivalent to that based on the modern principle of virtual velocities or virtual work. Of course, this amounts to reading into the Aristotelian treatment more than is actually there. This, however, is a fairly common procedure among historians of science. That the author of *Mechanica* preferred the dynamical method of establishing the law of the lever is significant. He evidently was impressed by the fact that *something* stays the *same* at both ends of the lever, in spite of the different weights.

These considerations gain in significance with respect to the origin of the concept of energy when we reflect that the explanation of the law of the lever by Archimedes, the greatest physicist in antiquity, proceeded on quite different lines. Archimedes shunned motion in his theoretical investigations and provided a "proof" based entirely on static equilibrium considerations. His method therefore sheds no light on the idea of energy.

3. THE MIDDLES AGES

Modern scholarship has shown that during the Middle Ages in Western Europe there was a great deal of interest in the attempt to explain the behavior of machines. Most of this was in the Aristotelian tradition. We shall not discuss it here, but merely call attention to the detailed studies by Hiebert,[6] Clagett,[4] and Moody and Clagett.[7]

3.1. Stevinus and Galileo

In looking for vestiges of the concept of constancy in the midst of change during the late 16th and early 17th centuries, we are confronted by two men, both of whom devoted much attention to the behavior of machines and endeavored to understand them from different points of view. The first was the famous Flemish engineer Simon Stevin (1548–1620), better known as Stevinus, and the second his contemporary, the even greater physicist, Galileo Galilei (1564–1642).

Stevinus was definitely a disciple of Archimedes rather than Aristotle. In his two great works, *De Beghinselen der Weeghconst* (Leiden, 1586) and *Hypomnemata Mathematica* (1608),[8] he showed complete disagreement with the Aristotelian method of understanding the behavior of a machine. He says, "The reason for the equilibrium of a lever does not reside at all in the arcs of the circle which its extremities describe." We have just seen that this motion was precisely the basis of the treatment in *Mechanica*. Disagreement could not have been more complete.

It is Stevinus' handling of the inclined plane that provides his chief claim to fame in the field of the operation of machines. His method here has a definite connection with the energy concept, since it makes use of the assumption of the impossibility of perpetual motion starting from rest. His famous scheme, of which he was so proud, imagines 14 equal balls fastened together in a single loop with inextensible strings of negligible mass and length and draped over two inclined planes of the same height placed back to back. One of the planes accommodates four of the balls on its surface and the other, of half the length, permits two balls to rest on it. The other eight balls hang symmetrically below the planes. Stevinus employs the logical principle of the excluded middle class to assume that the balls either start to move or do not move. But if they move at all, they must move indefinitely and this would be perpetual motion, which Stevinus discards as impossible. Hence, he concludes (after cutting off the eight balls hanging below the planes on the ground that they contribute nothing to the problem because of symmetry) that the balls on the plane must be in equilibrium. Therefore, the weight that can be supported on any plane is directly proportional to the length of the plane. This is essentially the law of the inclined plane as a machine. Stevinus was undoubtedly lucky in his specific set up. We are more concerned here, however, with his strong adherence to the idea of the impossibility of perpetual motion. He was probably familiar with the earlier views on this subject of Leonardo da Vinci[9] (1452–1519) and Girolamo Cardano[10] (1501–1576). There is no doubt these earlier scientists were convinced that it is not possible in terrestrial phenomena to get something for nothing, which would be what would happen if motion were to start by itself and persist indefinitely. This is indeed tied in with the modern energy concept and might well serve as an epigrammatic version of the general principle of conservation of energy or the first law of thermodynamics.

It seems clear from an examination of the writings of Galileo that he fully grasped the significance of the compensatory factor in the operation of machines, which we now interpret in the light of the invariance involved in the concept of energy. By the time Galileo turned his attention to machines, the laws governing their behavior were rather well known. There is, curiously enough, no record that Galileo was familiar with the work of Stevinus, at any rate at the time when Galileo prepared his university lectures which led to the book *On Mechanics*[11] (first published in Italian in 1649, after the death of the author).

In the book just referred to, Galileo shows himself even more aware than his Aristotelian predecessor of the element of compensation involved in the action of a machine. In the very beginning, he comments on how so many mechanicians are deceived into thinking their machines can accomplish operations which are impossible. Quoting directly from the English translation of the book:

> "These deceptions appear to me to have their principal cause in the belief these craftsmen have and continue to hold of being able to raise very great weights with a small force, as if with their machines they could cheat nature whose instinct—nay, whose most firm constitution—is that no resistance may be overcome by a force that is not more powerful than it. How false such a belief is I hope to make most evident with true and rigorous demonstrations that we shall have as we go along."

This is not a completely clear and unequivocal statement, but taken in conjunction with what follows it seems to emphasize Galileo's grasp of the fundamental fact that in machines one cannot get something for nothing. A little later in the section from which the above quotation has been taken, he elucidates more extensively:

> "Now assigning any determined resistance [he means here the force to be exerted or the weight to be raised by the machine] and delimiting any force [he means here the *applied* force] there is no doubt that the given weight will be conducted by the given force to the given distance; for even though the [applied] force may be very small, by dividing the weight into many particles of which each shall not remain superior to the [applied] force and transferring them one at a time, the whole weight will finally be conducted to the appointed place; nor may it reasonably be said at the end of the operation that the great weight has been moved and translated by a force lesser than itself but rather by a force which has many times repeated that motion and space which will have been traversed only once by the whole weight. From which it appears that the speed of the force has been greater than the resistance of the weight [here the translator has followed Galileo in an illogical statement, for a speed cannot logically be compared with a resistance; what Galileo must have meant was the speed of the resistance of the weight] by as many times as that weight is greater than the force, since in the time in which the moving force has repeatedly traversed the interval between the endpoints of the motion, the thing moved has passed over this by a single time."

One is entitled to assume from this phraseology that Galileo grasped the essence of the principle of virtual velocities or virtual work. He felt so strongly the validity of this point of view that he repeated essentially the same statements on the next page of his treatise. He continually emphasized that though a machine does possess a decided mechanical advantage, it is only at the expense of the time required for it to carry out its function.

We pass over Galileo's attempts to explain the behavior of the lever and the inclined plane. In many ways he comes closest to an invariance concept in his famous pendulum experiment, devised in order to provide an experimental basis for his fundamental assumption that when a ball falls from rest at a given height from the ground, the velocity on arriving at the ground depends only on the height and is independent of the path of fall. It is not necessary to repeat the details here, as they are clearly set forth in Galileo's *Dialogues Concerning Two New Sciences.*[12] The important thing to note is Galileo's grasp of the fact that in spite of the different paths there is something which remains constant. It must have impressed the author of the ingenious experiment. Today we interpret it in terms of the invariant maximum potential energy associated with fall from a given height independently of path and time of descent.

Galileo's interest in pendulum experiments, as exemplified in the case just discussed, was undoubtedly stimulated by his very early discovery, as a young man, in the Cathedral of Pisa, of the isochronism of the small oscillations of a pendulum.

4. CONSERVATION IDEAS IN THE 17TH AND 18TH CENTURIES. DESCARTES, LEIBNIZ, AND LAGRANGE'S *MÉCANIQUE ANALYTIQUE*

After the death of Galileo and as the 17th century wore on, emphasis on the idea of conservation in physics became more marked. René Descartes (1596–1650) in France made much of it, particularly in connection with the laws of impact of bodies. His studies of these phenomena led him to what we now call the principle of the conservation of momentum or what he called conservation of quantity of motion. Descartes(13) was so impressed with this principle that he was led to the general assertion that the total momentum of the Universe is constant. He finally concluded that the proper measure of force as the entity responsible for the production of motion is the change in momentum per unit time. This view may well have had an influence on Newton when he came to systematize mechanics in his *Principia*.

Gottfried Wilhelm Leibniz (1646–1716) disagreed with the point of view of Descartes. In the year 1686, he published in the *Acta Eruditorum* (Leipzig) a brief paper(14) in which he termed the theory of Descartes a "perversion" of mechanics. He convinced himself that the "true" measure of the efficacy of a force is the product of the mass and the square of the velocity, which he termed the "*vis viva*" or "living" force, as contrasted to the "*vis mortua*" or "dead" force of statics. His argument, put in simple terms, is as follows. He imagines two masses m and $4m$. The first is assumed to be dropped from rest at the height $4h$ and the second a height h from the ground. Leibniz assumes that each mass in falling will acquire what he calls the "force" necessary to enable it to rise again to the same height. That is, the "force" involved in the fall of mass m through $4h$ will be sufficient to carry this mass up again to where it started and leave it there at rest, neglecting any friction or other resistance. But Leibniz also assumes that the same "force" is necessary to lift the mass m through the height $4h$ as to lift the mass $4m$ through the height h. We see that this is essentially treating the word "force" here as equivalent to "work" in the modern physical sense. Now, this clearly entails the result that the same "force" is involved in the fall of m through $4h$ as is involved in the fall of $4m$ through h. But the quantities of motion, in the Cartesian sense, acquired in these two falls are not the same; from the law of falling bodies, m in falling through $4h$ acquires a velocity twice as great as that which $4m$ acquires in falling through h. If we call the latter velocity v, the quantity of motion or momentum acquired by m in its fall is $m(2v)$, while that acquired by $4m$ is $4m(v)$, or twice as much. So, says Leibniz, there is no conservation of quantity of motion in this case and hence in general we should not speak of this kind of conservation. The problem remains, "What, if anything, *is* conserved here?" To Leibniz this is simple: It is the product of the mass times the square of the velocity acquired. For then, in the example under consideration $m(2v)^2 = 4mv^2$. This quantity Leibniz felt deserved a special name, and he called it the *vis viva*. It is, of course, related to what later became known in the 19th century as the *kinetic energy*, being twice the latter.

This difference in the views of Descartes and Leibniz gave rise to a celebrated controversy, which raged in scientific circles for some half a century. D'Alembert (1717–1783) felt he had finally solved it when he published his famous *Traité de Dynamique*[15] in 1743. Here, he emphasized that the apparently conflicting viewpoints are due essentially to a confusion in terminology, and that they can be readily reconciled by appropriate definitions. Descartes' concept of force involves assuming that the efficacy of a force is measured by its effect over time, or, as we should now express it, by the time integral of the force. But this is just the change in the momentum of the particle acted on by the force; to illustrate for a particle of mass m, from Newton's law of motion,

$$F = d(mv)/dt \tag{5}$$

and

$$\int_{t_0}^{t_1} F\,dt = (mv)_1 - (mv)_0 \tag{6}$$

where the right-hand side in (6) is the difference between the momentum values at the instants t_0 and t_1 between which the force is assumed to act.

On the other hand, as D'Alembert pointed out, it is perfectly possible to measure the efficacy of a force by its effect over space, and this is essentially what Leibniz had in mind. In modern notation (for the special case of the motion of a single particle along the x axis) we arrive at

$$\int_{x_1}^{x_2} F\,dx = (\tfrac{1}{2}mv^2)_1 - (\tfrac{1}{2}mv^2)_0 \tag{7}$$

or, in words, the cumulative effect of force over distance (the left side, which we now term the *work* done by the force), is equal to the change in the quantity $\frac{1}{2}mv^2$ between the two positions x_1 and x_0, brought about by the action of the force. We now call $\frac{1}{2}mv^2$ the *kinetic energy* of the particle and the equation (7) is known as the work-kinetic energy theorem.

It is of interest to note that there is now some doubt whether D'Alembert should be considered to have definitely settled the *momentum* vs. *vis viva* controversy. A recent historian of physics, Laudan,[16] has pointed out that historical evidence shows that arguments over the "true" measure of force continued long after 1743, and that many well-known writers on the subject made no mention of D'Alembert in their discussions. It seems that the 19th century writers who credited D'Alembert with the solution of the controversy did so because they were more familiar with his numerous accomplishments in mathematics and mechanics as well as his treatise on dynamics than with the works of his contemporaries and successors. The fact remains that D'Alembert did set forth the general argument that modern physics has found satisfactory.

A claim has been made in behalf of Christian Huygens (1629–1695) that he introduced the idea of *vis viva* before Leibniz. It is true that in his famous work

Horologium Oscillatorium[17] (1673), he discussed the compound pendulum and in his treatment he used effectively the product of mass times the square of the velocity for the various parts making up the pendulum. But nowhere did he single out this quantity for special attention or speak of it as a possible measure of the efficacy of a force, much less baptize it with a name to emphasize its significance in terms of invariance and conservation. It was later commentators who read the *vis viva* interpretation into Huygens' proof of the law of the compound pendulum.

D'Alembert was obviously impressed with the importance of the *vis viva* concept and devoted to it the final chapter of his *Traité de Dynamique*.[15] He entitled this chapter, "On the principle of the conservation of living force." He first states this for the perfectly elastic collisions of particles, in which it has the following form: When a number of particles collide elastically, the sum of the products of each mass times the square of its velocity remains constant. This is a true conservation law. D'Alembert does not deduce it. The modern deduction depends on the treatment of collisions by means of Newton's coefficient of restitution and the equating of this coefficient to unity to correspond to perfect elasticity. D'Alembert generalizes the principle to apply to a collection of particles held together by rigid connections, i.e., forming effectively a rigid body. If such a collection moves in such a way that no "accelerating force" (as he calls it) acts on any particle, the total *vis viva* remains constant, irrespective of the motions of the individual particles. Again, he does not demonstrate this, though he illustrates it with a number of special cases, based on the use of his well-known principle governing the motion of systems of particles subject to constraints. It corresponds of course to highly idealized and not very practical situations.

D'Alembert does indeed also discuss the case in which the masses of a system are acted on by accelerating forces and shows that then the total *vis viva* does *not* remain constant, but that the change in it is equal to the "effect" of the forces, which in modern terminology is the same as the work done by the forces. This result is equivalent to the work-kinetic energy theorem in modern mechanics. From the standpoint of our present concern, however, it is significant that nowhere deos D'Alembert interpret his result in terms of the conservation of a quantity made up of the sum of the *vis viva* and another quantity depending on the relative positions of the particles of the system. The value of introducing the notion of *potential energy* and hence the concept of the total mechanical energy was not at that time appreciated, though the germ of the idea was certainly there.

A closer approach to the energy construct as we employ it today is found in the famous treatise by Lagrange, *Mécanique Analytique*,[18] first published in 1788. One of the greatest landmarks in the history of physics, this constituted a systematic presentation of the science of mechanics from a mathematical point of view. In it, the author presented his celebrated method of generalized coordinates and derived the equations which still bear his name. In a chapter devoted to *vis viva* (or *force vive* in French), he finally showed in explicit fashion that in certain cases it is possible to set up a function of the coordinates of a system of particles which, when added to the *vis viva* of the system, yields a quantity constant in time. Of course, he does not call this the mechanical energy of the system, nor does he use the term energy anywhere in his treatise. Actually, he refers to the result as an example of the conservation of *vis viva*, for reasons which are not clear, since in the case he discusses the *vis viva* will certainly change in general as time passes. At any rate, the equation he writes corresponds to what we now call the energy equation for a dynamical system. In fact, Lagrange later recognizes this sort of equation as a first integral of the equations of motion. This certainly marks an epoch in the realization of the existence and availability of a unifying concept in the study of dynamical systems, though the time was not yet ripe for its complete exploitation. One reason for this may well have been the realization that the setting up of a first integral of the equations of motion was not always or even in general practical for terrestial dynamical systems.

From this point, the story of the evolution of the energy concept moves in the direction of other physical phenomena, notably heat.

REFERENCES

1. Ernst Mach, *History and Root of the Principle of the Conservation of Energy* (Transl. by P. E. P. Jourdain) (Open Court Publishing Company, Chicago, 1911), p. 39.
2. M. R. Cohen and I. E. Drabkin, A *Source Book of Greek Science* (McGraw-Hill, New York, 1948). See particularly the reference on p. 230 to the work of Hero of Alexandria, *Mechanics* II, 21–26.
3. Richard McKeon, *The Basic Works of Aristotle* (Random House, New York, 1941), p. 353.
4. Marshall Clagett, *The Science of Mechanics in the Middle Ages* (University of Wisconsin Press, Madison, 1959), p. 477.
5. P. Duhem, *Les Origines de la Statique* (Paris, 1905), Vol. 1, p. 177.
6. Erwin N. Hiebert, *Historical Roots of the Principle of Conservation of Energy* (University of Wisconsin Press, Madison, 1962).
7. E. A. Moody and M. Clagett, *The Medieval Science of Weights* (University of Wisconsin Press, Madison, 1952).
8. Simon Stevin, *De Beghinselen der Weeghconst* (Leiden, 1586). *Hypomnomata Mathematica* (Leiden, 1608). The latter is a Latin translation of Stevin's writings on mechanics.
9. Edward MacCurdy, *The Notebooks of Leonardo da Vinci* (Garden City Publishing Company, New York, 1941). See p. 802 for reference to impossibility of perpetual motion.
10. Girolamo Cardano, *De Subtilitate* (Milan, 1551).
11. Galileo Galilei, *On Mechanics* (ca. 1590) (Transl. by Stillman Drake) (University of Wisconsin Press, Madison, 1960). See in particular pages 138ff on the principle of virtual velocities.
12. Galileo Galilei, *Dialogues Concerning Two New Sciences* (Transl. by Henry Crew and Alfonso De Salvio) (Northwestern University Press, Evanston, Illinois, 1939). See p. 170.
13. René Dugas, *A History of Mechanics* (Transl. by J. R. Maddox) (Editions du Griffon, Neuchâtel, Switzerland and Central Book Company, New York, 1953). See p. 160.
14. Gottfried Wilhelm Leibniz, in *Acta Eruditorum* (Leipzig) (1686).
15. D'Alembert, *Traité de Dynamique* (Paris, 1743). The relevant reference to the *vis viva* controversy will be found on p. XXIII of "Discours préliminaire" in the *Traité* (2nd Ed., Paris, 1758).
16. L. L. Laudan, The vis viva controversy in a post mortem, *Isis* **59**, 131 (1968).
17. Christian Huygens, *Horologium Oscillatorium* (Paris, 1673).
18. Joseph L. Lagrange, *Mécanique Analytique* (Paris, 1788).

Editor's Comments on Paper 2

Physics: Vestiges of the Energy Idea in Greek Physics
2 ARISTOTLE

Aristotle (384–322 B.C.), the celebrated Greek philosopher and founder of the Lyceum in Athens, wrote widely on scientific subjects, although he is perhaps best known and valued for his work on logic and other branches of philosophy. Persistence and imagination are essential in the search for vestiges of the idea of energy in his writings. In his treatise *Physica* he dealt extensively with motion in a highly abstract and theoretical way, but included nothing about the nature and operation of machines. Machines were tackled intensively in a later treatise, *Mechanica*, attributed by some historians of science to a successor of Aristotle, to whom the name Pseudo-Aristotle is given. It is usually included in the collected works of Aristotle himself, indicating the uncertainty that surrounds early writings on science. Selections from *Mechanica* are presented in Paper 3.

Although the extract presented here, the first section of Book VIII of *Physica*, can hardly be considered a model of clarity and comprehensibility from the standpoint of science, it is included because there runs through it an appreciation of the significance of constancy in the midst of change, the fundamental conservation idea. On any reasonable interpretation of the passage it seems that this was what Aristotle was groping for.

From the standpoint of terminology it is worth noting that the word εηεργεια (taken over into Latin as *energia*) occurs in Aristotle's philosophical writings with the meaning, according to scholars, of the "realized state of potentialities," presumably having to do with that which has the ability to bring about something else. This has significance in light of the choice by eighteenth- and nineteenth-century scientists of the French word *énergie* and the English word *energy* to denote the physical quantity we now know by that name. However, it does not appear that Aristotle's use of the term has any relevance to the germ of the idea which ultimately evolved into our present concept of energy.

2

Reprinted from *The Basic Works of Aristotle,* R. McKeon, ed., Random House, Inc., New York, 1941, pp. 354–359; originally published in *Works: The Oxford Translation,* Vol. II, R. P. Hardie and R. K. Gaye, trans., W. D. Ross, ed., The Clarendon Press, Oxford, 1930

PHYSICS: VESTIGES OF THE ENERGY IDEA IN GREEK PHYSICS

Aristotle

BOOK VIII

11 1 It remains to consider the following question. Was there ever a becoming of motion before which it had no being, and is it perishing again so as to leave nothing in motion? Or are we to say that it never had any becoming and is not perishing, but always was and always will be? Is it in fact an immortal never-failing property of things that are, a sort of life as it were to all naturally constituted things?

Now the *existence* of motion is asserted by all who have anything 15 to say about nature, because they all concern themselves with the construction of the world and study the question of becoming and perishing, which processes could not come about without the existence of motion. But those who say that there is an infinite number of worlds, some of which are in process of becoming while others are in process of perishing, assert that there is always motion (for these processes of 20 becoming and perishing of the worlds necessarily involve motion), whereas those who hold that there is only one world, whether everlasting or not, make corresponding assumptions in regard to motion. If then it is possible that at any time nothing should be in motion, this must come about in one of two ways: either in the manner described by Anaxagoras, who says that all things were together and at 25 rest for an infinite period of time, and that then Mind introduced motion and separated them; or in the manner described by Empedocles, according to whom the universe is alternately in motion and at rest—in motion, when Love is making the one out of many, or Strife is making many out of one, and at rest in the intermediate periods of time—his account being as follows:

> 'Since One hath learned to spring from Manifold, 30
> And One disjoined makes Manifold arise,
> Thus they Become, nor stable is their life: 251[a]
> But since their motion must alternate be,
> Thus have they ever Rest upon their round':

for we must suppose that he means by this that they alternate from the one motion to the other. We must consider, then, how this matter 5 stands, for the discovery of the truth about it is of importance, not only for the study of nature, but also for the investigation of the First Principle.

Let us take our start from what we have already [1] laid down in our course on Physics. Motion, we say, is the fulfilment of the movable in so far as it is movable. Each kind of motion, therefore, necessarily 10 involves the presence of the things that are capable of that motion. In fact, even apart from the definition of motion, every one would admit that in each kind of motion it is that which is capable of that motion that is in motion: thus it is that which is capable of alteration that is altered, and that which is capable of local change that is in 15 locomotion: and so there must be something capable of being burned before there can be a process of being burned, and something capable of burning before there can be a process of burning. Moreover, these

[1] iii. 1.

things also must either have a beginning before which they had no being, or they must be eternal. Now if there was a becoming of every movable thing, it follows that before the motion in question another change or motion must have taken place in which that which was capable of being moved or of causing motion had its becoming. To
20 suppose, on the other hand, that these things were in being throughout all previous time without there being any motion appears unreasonable on a moment's thought, and still more unreasonable, we shall find, on further consideration. For if we are to say that, while there are on the one hand things that are movable, and on the other hand things that are motive, there is a time when there is a first movent and a first moved, and another time when there is no such thing but only
25 something that is at rest, then this thing that is at rest must previously have been in process of change: for there must have been some cause of its rest, rest being the privation of motion. Therefore, before this first change there will be a previous change. For some things cause motion in only one way, while others can produce either
30 of two contrary motions: thus fire causes heating but not cooling, whereas it would seem that knowledge may be directed to two contrary ends while remaining one and the same. Even in the former class, however, there seems to be something similar, for a cold thing in a sense causes heating by turning away and retiring, just as one possessed of knowledge voluntarily makes an error when he uses his
251b knowledge in the reverse way.[2] But at any rate all things that are capable respectively of affecting and being affected, or of causing motion and being moved, are capable of it not under all conditions, but only when they are in a particular condition and approach one another: so it is on the approach of one thing to another that the one causes motion and the other is moved, and when they are present under such conditions as rendered the one motive and the other mov-
5 able. So if the motion was not always in process, it is clear that they must have been in a condition not such as to render them capable respectively of being moved and of causing motion, and one or other of them must have been in process of change: for in what is relative this is a necessary consequence: e. g. if one thing is double another when before it was not so, one or other of them, if not both, must have been in process of change. It follows, then, that there will be a process of change previous to the first.

10 (Further, how can there be any 'before' and 'after' without the existence of time? Or how can there be any time without the exist-

[2] i. e. by means of his knowledge he can be sure of giving a wrong opinion and thus deceiving some one.

ence of motion? If, then, time is the number of motion or itself a kind of motion, it follows that, if there is always time, motion must also be eternal. But so far as time is concerned we see that all with one exception are in agreement in saying that it is uncreated: in fact, it is just this that enables Democritus to show that all things cannot have had 15 a becoming: for time, he says, is uncreated. Plato alone asserts the creation of time, saying [3] that it had a becoming together with the universe, the universe according to him having had a becoming. Now since time cannot exist and is unthinkable apart from the moment, and the moment is a kind of middle-point, uniting as it does in itself 20 both a beginning and an end, a beginning of future time and an end of past time, it follows that there must always be time: for the extremity of the last period of time that we take must be found in some moment, since time contains no point of contact for us except the moment. Therefore, since the moment is both a beginning and an 25 end, there must always be time on both sides of it. But if this is true of time, it is evident that it must also be true of motion, time being a kind of affection of motion.)

The same reasoning will also serve to show the imperishability of motion: just as a becoming of motion would involve, as we saw, the 30 existence of a process of change previous to the first, in the same way a perishing of motion would involve the existence of a process of change subsequent to the last: for when a thing ceases to be moved, it does not therefore at the same time cease to be movable—e. g. the cessation of the process of being burned does not involve the cessation of the capacity of being burned, since a thing may be capable of being burned without being in process of being burned—nor, when a thing ceases to be movent, does it therefore at the same time cease to be motive. Again, the destructive agent will have to be destroyed, after 252[a] what it destroys has been destroyed, and then that which has the capacity of destroying *it* will have to be destroyed afterwards, (so that there will be a process of change subsequent to the last,) for being destroyed also is a kind of change. If, then, the view which we are criticizing involves these impossible consequences, it is clear that motion is eternal and cannot have existed at one time and not at another: in fact, such a view can hardly be described as anything else than fantastic.

And much the same may be said of the view that such is the ordi- 5 nance of nature and that this must be regarded as a principle, as would seem to be the view of Empedocles when he says that the constitution of the world is of necessity such that Love and Strife alter-

[3] Aristotle is thinking of a passage in the *Timaeus* (38 B).

nately predominate and cause motion, while in the intermediate period
10 of time there is a state of rest. Probably also those who, like Anaxagoras, assert a single principle (of motion) would hold this view. But that which is produced or directed by nature can never be anything disorderly: for nature is everywhere the cause of order. Moreover, there is no ratio in the relation of the infinite to the infinite, whereas order always means ratio. But if we say that there is first a state of rest for an infinite time, and then motion is started at some moment,
15 and that the fact that it is this rather than a previous moment is of no importance, and involves no order, then we can no longer say that it is nature's work: for if anything is of a certain character *naturally,* it either is so invariably and is not sometimes of this and sometimes of another character (e. g. fire, which travels upwards naturally, does not sometimes do so and sometimes not) or there is a ratio in
20 the variation. It would be better, therefore, to say with Empedocles and any one else who may have maintained such a theory as his that the universe is alternately at rest and in motion: for in a system of this kind we have at once a certain order. But even here the holder of the theory ought not only to assert the fact: he ought also to explain the cause of it: i. e. he should not make any mere assumption or lay down any gratuitous axiom, but should employ either induc-
25 tive or demonstrative reasoning. The Love and Strife postulated by Empedocles are not in themselves causes of the fact in question, nor is it of the essence of either that it should be so, the essential function of the former being to unite, of the latter to separate. If he is to go on to explain this alternate predominance, he should adduce cases where such a state of things exists, as he points to the fact that among mankind we have something that unites men, namely Love,
30 while on the other hand enemies avoid one another: thus from the observed fact that this occurs in certain cases comes the assumption that it occurs also in the universe. Then, again, some argument is needed to explain why the predominance of each of the two forces lasts for an equal period of time. But it is a wrong assumption to suppose universally that we have an adequate first principle in virtue of the fact that something always is so or always happens so. Thus Democritus reduces the causes that explain nature to the fact that things happened in the past in the same way as they happen now:
35 but he does not think fit to seek for a first principle to explain this
252[b] 'always': so, while his theory is right in so far as it is applied to certain individual cases, he is wrong in making it of universal application. Thus, a triangle always has its angles equal to two right angles, but there is nevertheless an ulterior cause of the eternity of this truth,

whereas first principles are eternal and have no ulterior cause. Let this conclude what we have to say in support of our contention that 5 there never was a time when there was not motion, and never will be a time when there will not be motion.

Editor's Comments on Paper 3

Mechanica: The Law of the Lever
3 PSEUDO-ARISTOTLE

Psuedo-Aristotle is the name commonly given to the author of the work with the Latin title *Mechanica*, devoted to the study and explanation of the behavior of machines. It is generally agreed by classical scholars that this author was a follower of Aristotle. Although the work has an air of practicality about it that is not characteristic of Aristotle's writings on physics, the document is certainly in the Aristotelian tradition and in fact is usually included in the works of Aristotle.

We reproduce here the first four sections of the treatise as presented in an English translation by E. S. Forster in Volume VI of the *Works of Aristotle*, edited by W. D. Ross (Clarendon Press, London, 1913), pages 847a–850b.

The author seeks for an explanation of the behavior of machines in the properties of the circle, which he discusses at some length and in the course of which comes close to introducing the parallelogram law of composition of vectors. In discussing the behavior of the balance and the simple lever, he skillfully employs the properties of the circle already cited. In particular, he states the law of the lever in terms of the rotation of the lever about the fulcrum. This is interesting from the standpoint of the germ of the concept of energy, since it employs a dynamical way of looking at a problem of equilibrium in contrast to the statical method employed later by Archimedes. Some historians of mechanics, notably Pierre Duhem (1861–1916) (*Les Origines de la statique*, 2 vols., Paris, 1905, 1906, pp. 5–12) and Ernst Mach (1838–1916) (*The Science of Mechanics*, 5th English ed., Open Court Publishing Co., LaSalle, Ill., 1942, pp. 12–13, 96–99, 105–106), have expressed the feeling that for Pseudo-Aristotle to have arrived at a definite statement of the law of the lever, he must have appreciated the principle of virtual velocities; that is, the product of the force applied at once end of an unequal arm lever by the velocity at that end must for equilibrium be equal to the corresponding product at the other end. It is readily seen that this assumption, combined with Pseudo-Aristotle's knowledge that the end of the longer arm of the lever must necessarily move faster than the end of the shorter arm, leads directly to the law of the lever.

It must be admitted that it is dangerous to try to reconstruct the thoughts of ancient writers across the barriers of time and language. It is often all too tempting to read into their writings our own modern interpretations of the phenomena being discussed. Nevertheless, the dynamical attack on the problem of understanding the behavior of machines was a notable advance and paved the way for the later work of Hero of Alexandria and for the medieval commentators on the mechanics of antiquity.

3

Reprinted from *Works: The Oxford Translation,* Vol. VI, E. S. Forster, trans., W. D. Ross, ed., The Clarendon Press, Oxford, 1913, pp. 847a–850b

MECHANICA: THE LAW OF THE LEVER

Pseudo-Aristotle

OUR wonder is excited, firstly, by phenomena which occur **847ᵃ** in accordance with nature but of which we do not know the cause, and secondly by those which are produced by art despite nature for the benefit of mankind. Nature often operates contrary to human expediency; for she always follows the same course without deviation, whereas human 15 expediency is always changing. When, therefore, we have to do something contrary to nature, the difficulty of it causes us perplexity and art has to be called to our aid. The kind of art which helps us in such perplexities we call Mechanical Skill. The words of the poet Antiphon are 20 quite true:

'Mastered by Nature, we o'ercome by Art.'

Instances of this are those cases in which the less prevails over the greater, and where forces of small motive power move great weights—in fact, practically all those problems which we call Mechanical Problems. They are not quite identical nor yet entirely unconnected with Natural 25 Problems. They have something in common both with Mathematical and with Natural Speculations; for while Mathematics demonstrates *how* phenomena come to pass; Natural Science demonstrates *in what medium* they occur.

Among questions of a mechanical kind are included **847ᵇ** those which are connected with the lever. It seems strange that a great weight can be moved with but little force, and that when the addition of more weight is involved; for the very same weight, which one cannot move at all without a lever, one can move quite easily with it, in spite 15 of the additional weight of the lever.

The original cause of all such phenomena is the circle. It is quite natural that this should be so; for there is nothing strange in a lesser marvel being caused by a greater marvel, and it is a very great marvel that contraries should be present together, and the circle is made up of contraries. 20 For to begin with, it is formed by motion and rest,[1] things which are by nature opposed to one another. Hence in examining the circle we need not be much astonished at the contradictions which occur in connexion with it. Firstly, in the line which encloses the circle, being without 25 breadth, two contraries somehow appear, namely, the concave and the convex. These are as much opposed to one another as the great is to the small; the mean being in the latter case the equal, in the former the straight. Therefore just as, if they are to change into one another, the greater and smaller must become equal before they can 848a pass into the other extreme; so a line must become straight in passing from convex into concave, or on the other hand from concave into convex and curved. This, then, is one peculiarity of the circle.

Another peculiarity of the circle is that it moves in two 5 contrary directions at the same time; for it moves simultaneously to a forward and a backward position.[2] Such, too, is the nature of the radius which describes a circle. For its extremity comes back again to the same position from which it starts; for, when it moves continuously, its last position is a return to its original position, in such 10 a way that it has clearly undergone a change from that position.

Therefore, as has already been remarked, there is nothing strange in the circle being the origin of any and every marvel. The phenomena observed in the balance can be referred to the circle, and those observed in the lever to 15 the balance; while practically all the other phenomena of mechanical motion are connected with the lever. Furthermore, since no two points on one and the

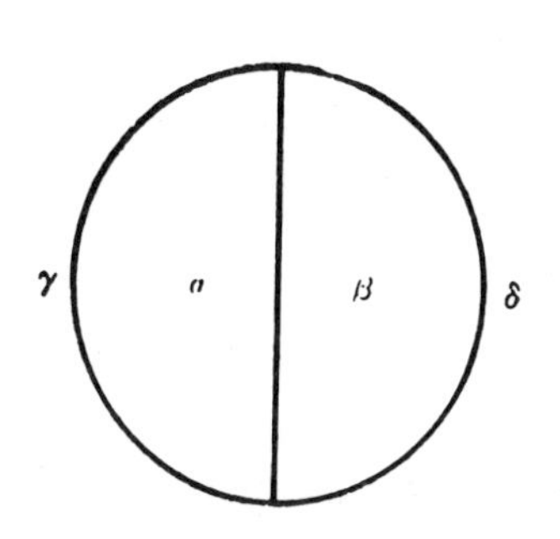

[1] 847b 20. i.e. by the motion of a line round a fixed point.

[2] 848a 5. If a circle be divided into two halves α and β, when the circle is revolved in a forward direction α will move towards δ and β towards γ.

same radius travel with the same rapidity, but of two points that which is further from the fixed centre travels more quickly, many marvellous phenomena occur in the motions of circles, which will be demonstrated in the following problems.

Because a circle moves in two contrary forms of motion 20 at the same time, and because one extremity of the diameter, A, moves forwards and the other, B, moves backwards, some people contrive so that as the result of a single movement a number of circles move simultaneously in contrary directions, like the wheels of brass and iron which they make and dedicate in the temples. Let AB be 25 a circle and ΓΔ another circle in contact with it; then if

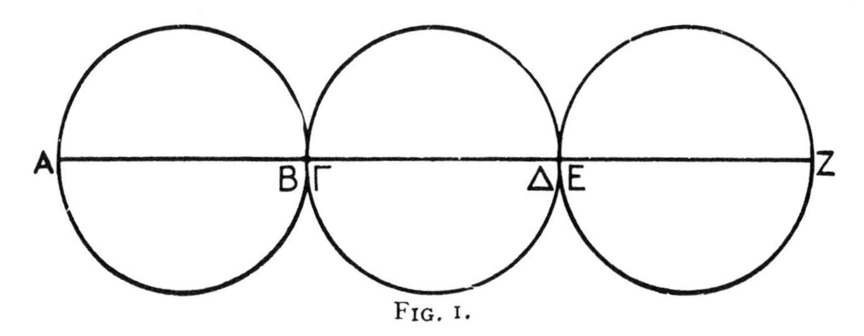

FIG. 1.

the diameter of the circle AB moves forward, the diameter ΓΔ will move in a backward direction as compared with the circle AB, as long as the diameter moves round the same point. The circle ΓΔ therefore will move in the opposite 30 direction to the circle AB. Again, the circle ΓΔ will itself make the adjoining circle EZ move in an opposite direction to itself for the same reason. The same thing will happen in the case of a larger number of circles, only one of them being set in motion. Mechanicians seizing on this inherent peculiarity of the circle, and hiding the principle, construct 35 an instrument so as to exhibit the marvellous character of the device, while they obscure the cause of it.

1 FIRST, then, a question arises as to what takes place 848[b] in the case of the balance. Why are larger balances more accurate than smaller? And the fundamental principle of this is, why is it that the radius which extends further from

the centre is displaced quicker than the smaller radius,
5 when the near radius is moved by the same force? Now we use the word 'quicker' in two senses; if an object traverses an equal distance in less time, we call it quicker, and also if it traverses a greater distance in equal time. Now the greater radius describes a greater circle in equal time; for the outer circumference is greater than the inner.
10 The reason of this is that the radius undergoes two displacements. Now if the two displacements of a body are in any fixed proportion, the resulting displacement must necessarily be a straight line, and this[1] line is the diagonal of the figure, made by the lines drawn in this proportion.
15 Let the proportion of the two displacements be as AB to

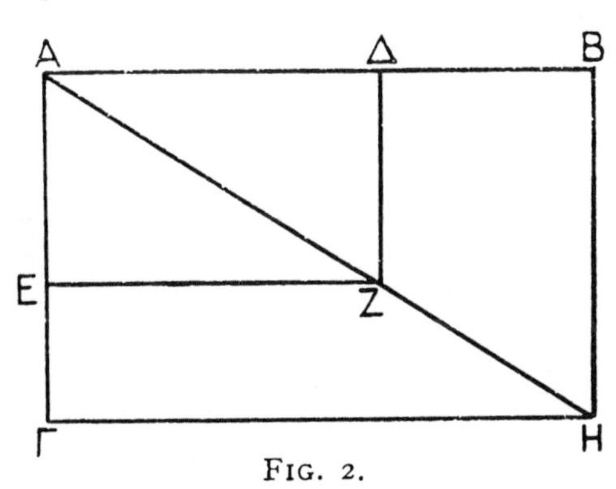

FIG. 2.

AΓ,[2] and let A be brought[3] to B, and the line AB brought down to HΓ. Again, let A be brought to Δ and the line AB to E; then if the proportion of the two displacements be maintained, AΔ must necessarily have the same proportion to AE as AB to AΓ. Therefore the small parallelo-
20 gram is similar to the greater, and their diagonal is the same, so that A will be at Z. In the same way it can be shown, at whatever points the displacement be arrested, that the point A will in all cases be on the diagonal.

And the converse is also true. It is plain that, if a point be moved along the diagonal by two displacements, it is necessarily moved according to the proportion of the sides

[1] 848b 12. Reading (with Par. A) αὕτη.
[2] 848b 15. This proposition is known as the Proof of the Parallelogram of Forces and Distances.
[3] 848b 16. Reading τὸ μὲν Α φερέσθω.

of the parallelogram; for otherwise it will not be moved along the diagonal. If it be moved in two displacements 25 in no fixed ratio for any time, its displacement cannot be in a straight line. For let it be a straight line. This then being drawn as a diagonal, and the sides of the parallelogram filled in, the point must necessarily be moved according to the proportion of the sides; for this has already been 30 proved. Therefore, if the same proportion be not maintained during any interval of time, the point will not describe a straight line; for, if the proportion were maintained during any interval, the point must necessarily describe a straight line, by the reasoning above. So that, if the two displacements do not maintain any proportion during any interval, a curve is produced. 35

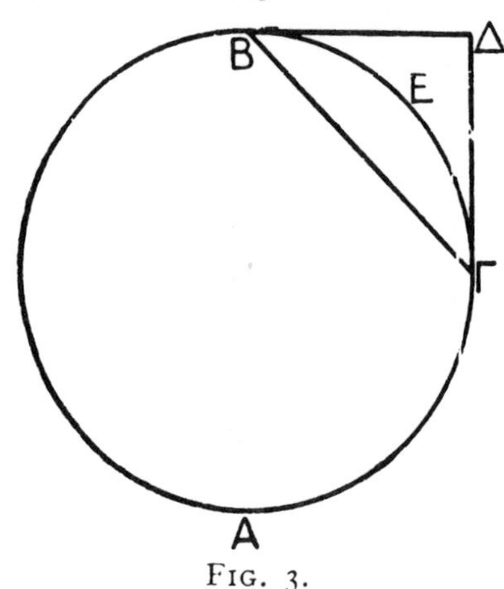

FIG. 3.

Now that the radius of a circle has two simultaneous displacements is plain from these considerations, and because the point[1] from being vertically above the 849a centre comes back to the perpendicular, so as to be again perpendicularly above the centre.

Let ABΓ be a circle, and let the point B at the summit be displaced to Δ by one force, and come eventually to Γ by the other force. If then it were moved in the proportion 5 of BΔ to ΔΓ, it would move along the diagonal BΓ. But in the present case, as it is moved in no such proportion, it moves along the curve BEΓ. And, if one of two displacements caused by the same forces is more interfered with and the other less, it is reasonable to suppose that the motion more interfered with will be slower than the motion less interfered with; which seems to happen in the case 10 of the greater and less of the radii of circles. For on account

[1] 849a 1. Omitting κατ' εὐθεῖαν which, as Capelle says, is probably corrupt. If not, it must mean moving momentarily straight, and being immediately deflected. If it continued straight, it would not come back to the original position.

of the extremity of the lesser radius being nearer the stationary centre than that of the greater, being as it were pulled in a contrary direction, towards the middle,[1] the 15 extremity of the lesser moves more slowly. This is the case with every radius, and it moves in a curve, naturally along the tangent, and unnaturally towards the centre. And the lesser radius is always moved more in respect of its unnatural motion; for being nearer to the retarding centre it is more constrained. And that the less of two radii having the same centre is moved more than the greater 20 in respect of the unnatural motion is plain from what follows.

Let ΒΓΕΔ be a circle, and ΧΝΜΞ another smaller circle within it, both having the same centre Α, and let the 25 diameters be drawn, ΓΔ and ΒΕ in the large circle, and ΜΧ and ΝΞ in the small; and let the rectangle ΔΨΡΓ be completed. If the radius ΑΒ comes back to the same position from which it started, i.e. to ΑΒ, it is plain that it moved towards itself; and likewise ΑΧ will come to 30 ΑΧ. But ΑΧ moves more slowly than ΑΒ, as has been stated, because the interference is greater and ΑΧ is more retarded.

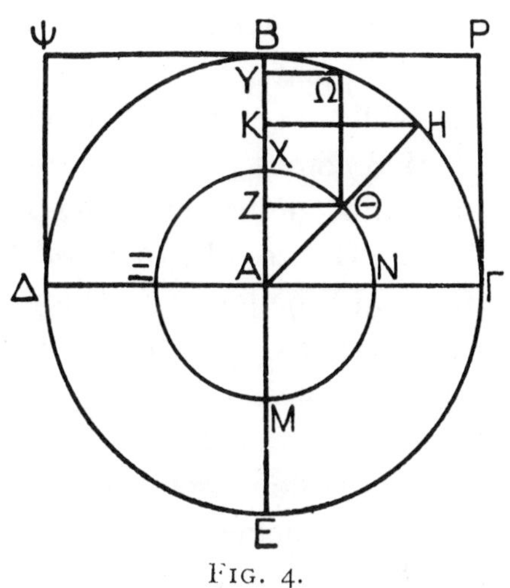

Fig. 4.

Now let ΑΘΗ be drawn, and from Θ a perpendicular upon ΑΒ within the circle, ΘΖ; and, further, from Θ let ΘΩ be drawn parallel to ΑΒ, and ΩΥ and ΗΚ perpendiculars on 35 ΑΒ; then ΩΥ and ΘΖ are equal. Therefore ΒΥ is less than ΧΖ; for in unequal circles equal straight lines drawn perpendicular to the diameter cut off smaller portions of the diameter in the greater circles; ΩΥ and ΘΖ being equal.[2]

[1] 849a 13. Punctuating εἰς τὸ ἐναντίον ἐπὶ τὸ μέσον, βραδύτερον.

[2] 849a 38. According to the parallelogram of distances, the result ought to be:—ΒΥ : ΥΩ : : ΧΖ : ΘΖ, but it is proved that ΥΩ and ΘΖ are equal, but ΒΥ and ΧΖ unequal; so that the theory of the parallelogram

Now the radius AΘ describes the arc XΘ in the same time 849b
as the extremity of the radius BA has described an arc greater than BΩ in the greater circle; for the natural displacement is equal and the unnatural less, BΥ being less than XZ. Whereas they ought to be in proportion, the two natural motions in the same ratio to each other 5
as the two unnatural motions.

Now the radius AB has described an arc BH greater than BΩ. It must necessarily have described BH in the time in which X describes XΘ; for that will be its position when in the two circles the proportion between the unnatural and natural movements holds good. If, then, the 10
natural movement is greater in the greater circle, the unnatural movement, too, would agree in being proportionally greater[1] in that case only, where B is moved along BH while X is moved along XΘ. For in that case the point B comes by its natural movement to H, and by its unnatural movement to K, HK being perpendicular from H. And as HK to BK, so is ΘZ to XZ. Which will be plain, 15
if B and X be joined to H and Θ.[2] But, if the arc described by B be less or greater than HB, the result will not be the same, nor will the natural movement be proportional to the unnatural in the two circles.

So that the reason why the point further from the centre 20
is moved quicker by the same force, and the greater radius describes the greater circle, is plain from what has been said; and hence the reason is also clear why larger balances are more accurate than smaller. For the cord by which a balance is suspended acts as the centre, for it is at rest, and the parts of the balance on either side form the radii. Therefore by the same weight the end of the balance must 25
necessarily be moved quicker in proportion as it is more distant from the cord, and some weight must be imperceptible to the senses in small balances, but perceptible in large balances; for there is nothing to prevent the 30

fails. Why is this? The answer is that the same force moves longer radii quicker than shorter.

[1] 849b 11. Reading with Capelle μεῖζον for μᾶλλον.

[2] 849a 16. For the triangles BKH and XZΘ are similar, having all their sides parallel, each to each.

movement being so small as to be invisible to the eye. Whereas in the large balance the same load makes the movement visible. In some cases the effect is clearly seen in both balances, but much more in the larger on account of the amplitude of the displacement caused by the same 35 load being much greater in the larger balance. And thus dealers in purple, in weighing it, use contrivances with intent to deceive, putting the cord out of centre and pouring lead into one arm of the balance, or using the wood towards the root of a tree for the end towards which they want it to incline, or a knot, if there be one in the wood; for the part of the wood where the root is is heavier, and a knot 850a is a kind of root.

How is it that if the cord is attached to the upper surface 2 of the beam of a balance, if one takes away the weight when the balance is depressed on one side, the beam rises again; whereas, if the cord is attached to 5 the lower surface of the beam, it does not rise but remains in the same position. Is it because, when the cord is attached above, there is more of the beam on one side of the perpendicular than on the other, the cord being the perpendicular? In that case the side on which the greater part of the beam is must necessarily sink until the line which divides the 10 beam into two equal parts reaches the actual perpendicular, since the weight now presses on the side of the beam which is elevated.

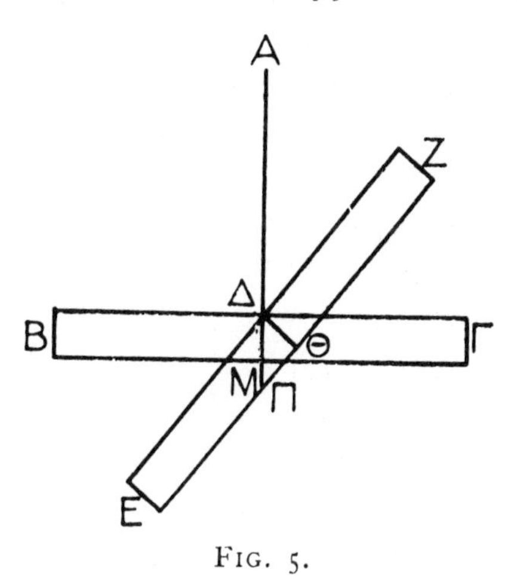

FIG. 5.

Let BΓ be a straight beam, and AΔ a cord. If AΔ be produced it will form the perpendicular AΔM. If the portion of the beam towards B be depressed, B will be displaced to E and Γ to Z; and so the line dividing the 15 beam into two halves, which was originally ΔM, part of the perpendicular, will become ΔΘ when the beam is depressed; so that the part of the beam EZ which is outside the perpendicular ΛM will be greater by ΘΠ than half the beam. If therefore the weight at E be taken away, Z must sink, because the side towards E is shorter. It has been proved then that when the cord is attached above, if the weight be removed the beam rises again. 20

But if the support be from below, the contrary takes place. For then the part which is depressed is more than half of the beam, or in other words, more than the part marked off by the original perpendicular; it does not therefore rise, when the weight is removed, for the part that is elevated is lighter. Let NΞ be the beam when horizontal, and KΛM the perpendicular dividing NΞ into two halves. 25 When the weight is placed at N, N will be displaced to O and Ξ to P, and KΛ to ΛΘ, so that KO[1] is greater than ΛP by ΘΛK. If the weight, therefore, is removed the beam must necessarily remain in the same position; for the excess of the part in which OK[2] is over half the beam acts as a weight and remains depressed.

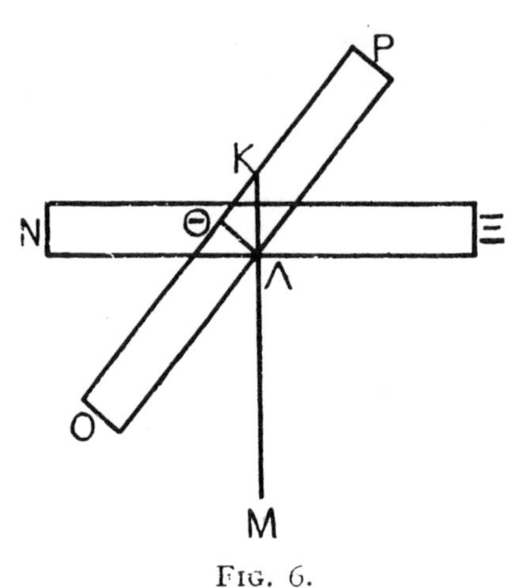

FIG. 6.

3 WHY is it, that, as has been remarked at the beginning 30 of this treatise,[3] the exercise of little force raises great weights with the help of a lever, in spite of the added weight of the lever; whereas the less heavy a weight is, the easier it is to move, and the weight is less without the lever? Does the reason lie in the fact that the lever acts like the beam of a balance with the cord attached below and

[1] 850a 27. i.e. the figure KΛOΘ is greater than the figure KPΛ by twice the triangle KΛΘ.

[2] 850a 29. Reading τὸ OK for τὸ K; Capelle apparently uses this reading in his translation, but has not altered the text.

[3] Cp. 847b 2.

35 divided into two unequal parts? The fulcrum, then, takes the place of the cord, for both remain at rest and act as the centre. Now since a longer radius moves more quickly than a shorter one under pressure of an equal weight; and since the lever requires three elements, viz. the fulcrum —corresponding to the cord of a balance and forming the centre—and two weights, that exerted by the person using the lever and the weight which is to be moved; this being so, as the weight moved is to the weight moving it, so, 850b inversely, is the length of the arm bearing the weight to the length of the arm nearer to the power. The further one is from the fulcrum, the more easily will one raise the weight; the reason being that which has already been stated,[1] namely, that a longer radius describes a larger circle. So with the exertion of 5 the same force the motive weight will change its position more than the weight which it moves, because it is further from the fulcrum.

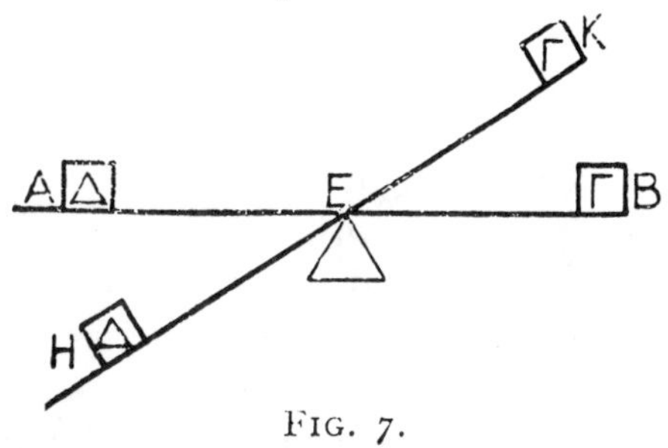

FIG. 7.

Let AB be a lever, Γ the weight to be lifted, Δ the motive weight, and E the fulcrum; the position of Δ after it has raised the weight will be H, and that of Γ, the weight raised, will be K.

[1] Ch. 1.

Editor's Comments on Paper 4

Parmenides: The Unity of Nature
4 PLATO

This dialogue of Plato is usually considered one of the most difficult of all his writings. Through a conversation between Socrates and Parmenides, Plato discusses the doctrine of ideas, to which he had throughout his professional career devoted much attention. It was perhaps appropriate that he should have made Parmenides of Elea a participant in the discussion. For although a somewhat shadowy figure, few of whose writings have come down to us, Parmenides is considered to be one of the most important of the pre-Socratic philosophers. His emphasis on the idea of unity in nature and his apparent feeling that the common emphasis on change in experience is an illusion can be interpreted as supporting the notion of constancy or conservation. Nothing in this dialogue even remotely suggests the specific properties of the concept of energy as vizualized in modern science, but the emphasis on unity and the importance of abstract ideas, of which energy is certainly a prime example, justifies the inclusion of the extract. Even the difficulties that Plato raises with respect to the existence of abstract ideas are not wholly irrelevant to the problems encountered in the historical development of the idea of energy. In this dialogue Plato shows himself the genuine forerunner of those who later stressed the role of abstract ideas in theoretical physics.

4

Reprinted from *The Works of Plato*, Vol. 4, B. Jowett, trans. and ed., Tudor Publishing Company, New York, n.d., pp. 312-328

PARMENIDES: THE UNITY OF NATURE

Plato

THE MEETING OF SOCRATES AND PARMENIDES AT ATHENS. CRITICISM OF THE IDEAS

THE Parmenides is perhaps the most difficult of all the dialogues of Plato, and one of a group which are not very attractive to the reader who has not been initiated into the "mysteries of dialectic." The subject is the Doctrine of Ideas, those metaphysical abstractions which occupied the mind of Plato more or less during the greater part of his life. They have often been supposed to be the keystone of his system, on which all his other thoughts and conceptions depend for stability.

This view, however, goes beyond the truth. It is a mistake to imagine that Plato had in view a complete scheme of philosophy, which he endeavored to draw out in a series of treatises. His genius was unsystematic and irregular; he was almost as much a poet as a philosopher; and the testimony of his own writings is sufficient to show that he fell at various periods under the influence of different teachers. The Ideas ought rather to be treated by us as an attempt to convey Plato's conviction that there was a truth unrealized beyond sense, which could only be grasped by the mind when freed from the thraldom of the body. But he was greatly perplexed by the difficulty of finding an adequate expression of his thoughts, and he was perfectly conscious of the many and serious objections which could be urged against his own doctrines.

In the Parmenides, which we may reasonably consider a work of Plato's later years, he has reached a stage at which he is able by an extraordinary effort of intellectual power to produce a criticism of the Ideas which he himself can not refute. Yet he hints by the mouth of Parmenides that he is still convinced of their reality and existence; for, without abstract ideas, thought and reasoning would be impossible. And in the Sophist he resumes the topic with more success, and clears away some of the obstacles to his theory which in the Parmenides had appeared to him to be insuperable.

We went from our home at Clazomenae to Athens, and met Adeimantus and Glaucon in the Agora. Welcome, said Adeimantus, taking me by the hand; is there anything which we can do for you in Athens?

Why, yes, I said, I am come to ask a favor of you.

What is that? he said.

I want you to tell me the name of your half-brother, which I have forgotten; he was a mere child when I last came hither from Clazomenae, but that was a long time ago; your father's name, if I remember rightly, is Pyrilampes?

Yes, he said, and the name of our brother, Antiphon; but why do you ask?

Let me introduce some countrymen of mine, I said; they are lovers of philosophy, and have heard that Antiphon was in the habit of meeting Pythodorus, the friend of Zeno, and remembers certain arguments which Socrates and Zeno and Parmenides had together, and which Pythodorus had often repeated to him.

That is true.

And could we hear them? I asked.

Nothing easier, he replied; when he was a youth he made a careful study of the pieces; at present his thoughts run in another direction; like his grandfather, Antiphon, he is devoted to horses. But, if that is what you want, let us go and look for him; he dwells at Melita, which is quite near, and he has only just left us to go home.

Accordingly we went to look for him; he was at home, and in the act of giving a bridle to a blacksmith to be fitted. When he had done with the blacksmith, his brothers told him the purpose of our visit; and he saluted me as an acquaintance whom he remembered from my former visit, and we asked him to repeat the dialogue. At first he was not very willing, and complained of the trouble, but at length he consented. He told us that Pythodorus had described to him the appearance of Parmenides and Zeno; they came to Athens, he said, at the great Panathenaea; the former was, at the time of his visit, about 65 years old, very white with age, but well favored. Zeno was nearly 40 years of age, of a noble figure and fair aspect; and in the days of his youth he was reported to have been beloved of Parmenides. He said that they lodged with Pythodorus in the Ceramicus, outside the wall, whither Socrates and others came to see them; they wanted to hear some writings of Zeno, which had been brought to Athens by them for the first time. He said that Socrates was then very young, and that Zeno read them to him in the absence of Parmenides, and had nearly finished when Pythodorus entered, and with him Parmenides and Aristoteles who was afterwards one of the Thirty; there was not much more to hear, and Pythodorus had heard Zeno repeat them before.

When the recitation was completed, Socrates requested that the first hypothesis of the first discourse might be read over again, and this having been done, he said: What do you mean, Zeno? Is your argument that the existence of many necessarily involves like and unlike, and that this is impossible, for neither can the like be unlike, nor the unlike like; is that your position? Just that, said Zeno. And if the unlike can not be like, or the like unlike, then neither can the many exist, for that would involve an impossibility. Is the design of your argument throughout to disprove the existence of the many? and is each of your treatises intended to furnish a separate proof of this, there being as many proofs in all as you have composed arguments, of the non-existence of the many? Is that your meaning, or have I misunderstood you?

No, said Zeno; you have quite understood the general drift of the treatise.

I see, Parmenides, said Socrates, that Zeno is your second self in his writings too; he puts what you say in another way, and half deceives us into believing that he is saying what is new. For you, in your compositions, say that the all is one, and of this you adduce excellent proofs; and he, on the other hand, says that the many is naught, and gives many great and convincing evidences of this. To deceive the world, as you have done, by saying the same thing in different ways, one of you affirming and the other denying the many, is a strain of art beyond the reach of most of us.

Yes, Socrates, said Zeno. But although you are as keen as a Spartan hound in pursuing the track, you do not quite apprehend the true motive of the performance, which is not really such an artificial piece of work as you imagine; there was no intention of concealment effecting any grand result — that was a mere accident. For the truth is, that these writings of mine were meant to protect the arguments of Parmenides against those who ridicule him, and urge the many ridiculous and contradictory results which were supposed to follow from the assertion of the one. My answer is addressed to the partisans of the many, and intended to show that greater or more ridiculous consequences follow from their hypothesis of the existence of the many if carried out, than from the hypothesis of the existence of the one. A love of controversy led me to write the book in the days of my youth, and some one stole the writings, and I had therefore no choice about the publication of them; the motive, however, of writing, was not the ambition of an old man, but the pugnacity of a young one. This you do not seem to see, Socrates; though in other respects, as I was saying, your notion is a very just one.

That I understand, said Socrates, and quite accept your account. But tell me, Zeno, do you not further think that there is an idea of likeness in the abstract, and another idea of unlikeness, which is the opposite of likeness, and that in these two, you and I and all other things to which we apply the term many, participate; and that the things which participate in likeness are in that degree and manner like; and that those which participate in unlikeness are in that degree unlike, or both like and unlike in the degree in which they participate in both? And all things may partake of both opposites, and be like and unlike to themselves, by reason of this participation. Even in that there is nothing wonderful. But if a person could prove the absolute like to become unlike, or the absolute unlike to become like, that, in my opinion, would be a real wonder; not, however, if the things which partake of the ideas experience likeness and unlikeness — there is nothing extraordinary in this. Nor, again, if a person were to show that all is one by partaking of one, and that the same is many by partaking of many, would that be very wonderful? But if he were to show me that the absolute many was one, or the absolute one many, I should be truly amazed. And I should say the same of other things. I should be surprised to hear that the genera and species had opposite qualities in themselves; but if a person wanted to prove of me that I was many and also one, there would be no marvel in that. When he wanted to show that I was many he would say that I have a right and a left side, and a front and a back, and an upper and a lower half, for I can not deny that I partake of multitude; when, on the other hand, he wants to prove that I am one, he will say, that we who are here assembled are seven, and that I am one and partake of the one, and in saying both he speaks truly. Or if a person shows that the same wood and stones and the like, being many are also one, we admit that he shows the existence of the one and many, but he does not show that the many are one or the one many; he is uttering not a wonder but a truism. If, however, as I was suggesting just now, we were to make an abstraction, I mean of like, unlike, one, many, rest, motion, and similar ideas, and then to show that these in their abstract form admit of admixture and separation, I should greatly wonder at that. This part of the argument appears to be treated by you, Zeno, in a very spirited manner; nevertheless, as I was saying, I should be far more amazed if any one found in the ideas themselves which are conceptions, the same puzzle and entanglement which you have shown to exist in visible objects.

While Socrates was saying this, Pythodorus thought that Parmenides and Zeno were not altogether pleased at the successive steps of the argument; but still they gave the closest attention, and often looked at one another, and smiled as if in admiration of him. When he had finished, Parmenides expressed these feelings in the following words: —

Socrates, he said, I admire the bent of your mind towards philosophy; tell me now, was this your own distinction between abstract ideas and the things which partake of them? and do you think that there is an idea of likeness apart from the likeness which we possess, or of the one and many, or of the other notions of which Zeno has been speaking?

I think that there are such abstract ideas, said Socrates.

Parmenides proceeded. And would you also make abstract ideas of the just and the beautiful and the good, and of all that class of notions?

Yes, he said, I should.

And would you make an abstract idea of man distinct from us and from all other human creatures, or of fire and water?

I am often undecided, Parmenides, as to whether I ought to include them or not.

And would you feel equally undecided, Socrates, about things the mention of which may provoke a smile? — I mean such things as hair, mud, dirt, or anything else that is foul and base; would you suppose that each of these has an idea distinct from the phenomena with which we come into contact, or not?

Certainly not, said Socrates; visible things like these are such as they appear to us, and I am afraid that there would be an absurdity in assuming any idea of them, although I sometimes get disturbed, and begin to think that there is nothing without an idea; but then again, when I have taken up this position, I run away, because I am afraid that I may fall into a bottomless pit of nonsense, and perish; and I return to the ideas of which I was just now speaking, and busy myself with them.

Yes, Socrates, said Parmenides; that is because you are still young; the time will come when philosophy will have a firmer grasp of you, if I am not mistaken, and then you will not despise even the meanest things; at your age, you are too much disposed to look to the opinions of men. But I should like to know whether you mean that there are certain forms or ideas of which all other things partake, and from which they are named; that similars, for example, become similar, because they partake of similarity; and great things become great, because they partake of greatness; and that just and beautiful things become just and beautiful, because they partake of justice and beauty?

Yes, certainly, said Socrates, that is my meaning.

And does not each individual partake either of the whole of the idea or of a part of the idea? Is any third way possible?

Impossible, he said.

Then do you think that the whole idea is one, and yet being one, exists in each one of many?

Why not, Parmenides? said Socrates.

Because one and the same existing as a whole in many separate individuals, will thus be in a state of separation from itself.

Nay, replied the other; the idea may be like the day which is one and the same in many places, and yet continuous with itself; in this way each idea may be one and the same in all.

I like your way, Socrates, of dividing one into many; and if I were to spread out a sail and cover a number of men, that, as I suppose, in your way of speaking, would be one and a whole in or on many — that will be the sort of thing which you mean?

I am not sure.

And would you say that the whole sail is over each man, or a part only?

A part only.

Then, Socrates, the ideas themselves will be divisible, and the individuals will have a part only and not the whole existing in them?

That seems to be true.

Then would you like to say, Socrates, that the one idea is really divisible and yet remains one?

Certainly not, he said.

Suppose that you divide greatness, and that of many great things each one is great by having a portion of greatness less than absolute greatness — is that conceivable?

No.

Or will each equal part, by taking some portion of equality less than absolute equality, be equal to some other?

Impossible.

Or suppose one of us to have a portion of smallness; this is but a part of the small, and therefore the small is greater; and while the absolute small is greater, that to which the part of the small is added, will be smaller and not greater than before.

That is impossible, he said.

Then in what way, Socrates, will all things participate in the ideas, if they are unable to participate in them either as parts or wholes?

Indeed, he said, that is a question which is not easily determined.

Well, said Parmenides, and what do you say of another question?

What is that?

I imagine that the way in which you are led to assume the existence of ideas is as follows: — You see a number of great objects, and there seems to you to be one and the same idea of greatness pervading them all; and hence you conceive of a single greatness.

That is true, said Socrates.

And if you go on and allow your mind in like manner to contemplate the idea of greatness and these other greatnesses, and to compare them, will not another idea of greatness arise, which will appear to be the source of them all?

That is true.

Then another abstraction of greatness will appear over and above absolute greatness, and the individuals which partake of it; and then another, which will be the source of that, and then others, and so on; and there will be no longer a single idea of each kind, but an infinite number of them.

But may not the ideas, asked Socrates, be cognitions only, and have no proper existence except in our minds, Parmenides? For in that case there may be single ideas, which do not involve the consequences which were just now mentioned.

And can there be individual cognitions which are cognitions of nothing?

That is impossible, he said.

The cognition must be of something?

Yes.

Of something that is or is not?

Of something that is.

Must it not be of the unity, or single nature, which the cognition recognizes as attaching to all?

Yes.

And will not this unity, which is always the same in all, be the idea?

From that, again, there is no escape.

Then, said Parmenides, if you say that other things participate in the ideas, must you not say that everything is made up of thoughts or cognitions, and that all things think; or will you say that being thoughts they are without thought?

But that, said Socrates, is irrational. The more probable view, Parmenides, of these ideas is, that they are patterns fixed in nature, and that other things are like them, and resemblances of them; and that what is meant by the participation of other things in the ideas, is really assimilation to them.

But if, said he, the individual is like the idea, must not the idea also be like the individual, in as far as the individual is a resemblance of the idea? That which is like, can not be conceived of as other than the like of like.

Impossible.

And when two things are alike, must they not partake of the same idea?

They must.

And will not that of which the two partake, and which makes them alike, be the absolute idea [of likeness]?

Certainly.

Then the idea can not be like the individual, or the individual like the idea; for if they are alike, some further idea of likeness will always arise, and if that be like anything else, another and another; and new ideas will never cease being created, if the idea resembles that which partakes of it?

Quite true.

The theory, then, that other things participate in

the ideas by resemblance, has to be given up, and some other mode of participation devised?

That is true.

Do you see then, Socrates, how great is the difficulty of affirming self-existent ideas?

Yes, indeed.

And, further, let me say that as yet you only understand a small part of the difficulty which is involved in your assumption, that there are ideas of all things, which are distinct from them.

What difficulty? he said.

There are many, but the greatest of all is this: — If an opponent argues that these self-existent ideas, as we term them, can not be known, no one can prove to him that he is wrong, unless he who is disputing their existence be a man of great genius and cultivation, and is willing to follow a long and laborious demonstration — he will remain unconvinced, and still insist that they can not be known.

How is that, Parmenides? said Socrates.

In the first place, I think, Socrates, that you, or any one who maintains the existence of absolute ideas, will admit that they can not exist in us.

Why, then they would be no longer absolute, said Socrates.

That is true, he said; and any relation in the absolute ideas, is a relation which is among themselves only, and has nothing to do with the resemblances, or whatever they are to be termed, which are in our sphere, and the participation in which gives us this or that name. And the subjective notions in our mind, which have the same name with them, are likewise only relative to one another, and not to the ideas which have the same name with them, and belong to themselves, and not to the ideas.

How do you mean? said Socrates.

I may illustrate my meaning in this way, said Parmenides: — A master has a slave; now there is nothing absolute in the relation between them; they are both relations of some man to another man; but there is also an idea of mastership in the abstract, which is relative to the idea of slavery in the abstract; and this abstract nature has nothing to do with us, nor we with the abstract nature; abstract natures have to do with themselves alone, and we with ourselves. Do you see my meaning?

Yes, said Socrates, I quite see your meaning.

And does not knowledge, I mean absolute knowledge, he said, answer to very and absolute truth?

Certainly.

And each kind of absolute knowledge answers to each kind of absolute being?

Yes.

And the knowledge which we have, will answer to the truth which we have; and again, each kind of knowledge which we have, will be a knowledge of each kind of being which we have?

Certainly.

But the ideas themselves, as you admit, we have not, and can not have?

No, we can not.

And the absolute ideas or species, are known by the absolute idea of knowledge?

Yes.

And that is an idea which we have not got?

No.

Then none of the ideas are known to us, because we have no share in absolute knowledge?

They are not.

Then the ideas of the beautiful, and of the good, and the like, which we imagine to be absolute ideas, are unknown to us?

That appears to be the case.

I think that there is a worse consequence still.

What is that?

Would you, or would you not, say, that if there is such a thing as absolute knowledge, that must be a far more accurate knowledge than our knowledge, and the same of beauty and other things?

Yes.

And if there be anything that has absolute knowledge, there is nothing more likely than God to have this most exact knowledge?

Certainly.

But then, will God, having this absolute knowledge, have a knowledge of human things?

And why not?

Because, Socrates, said Parmenides, we have admitted that the ideas have no relation to human notions, nor human notions to them; the relations of either are in their respective spheres.

Yes, that has been admitted.

And if God has this truest authority, and this most exact knowledge, that authority can not rule us, nor that knowledge know us, or any human thing; and in like manner, as our authority does not extend to the gods, nor our knowledge know anything which is divine, so by parity of reason they, being gods, are not our masters; neither do they know the things of men.

Yet, surely, said Socrates, to deprive God of knowledge is monstrous.

These, Socrates, said Parmenides, are a few, and only a few, of the difficulties which are necessarily involved in the hypothesis of the existence of ideas, and the attempt to prove the absoluteness of each of them; he who hears of them will doubt or deny their existence, and will maintain that even if they do exist, they must necessarily be unknown to man, and he will think that there is reason in what he says, and as we were remarking just now, will be wonderfully hard of being convinced; a man must be a man of real ability before he can understand that everything has a class and an absolute essence; and still more remarkable will he be who makes out all these things for himself, and can teach another to analyze them satisfactorily.

I agree with you, Parmenides, said Socrates; and what you say is very much to my mind.

And yet, Socrates, said Parmenides, if a man, fixing his mind on these and the like difficulties, refuses to acknowledge ideas or species of existences, and will not define particular species, he will be at his wit's end; in this way he will utterly destroy the power of reasoning; and that is what you seem to me to have particularly noted.

Very true, he said.

But, then, what is to become of philosophy? What resource is there, if the ideas are unknown?

I certainly do not see my way at present.

Yes, said Parmenides; and I think that this arises, Socrates, out of your attempting to define the beautiful, the just, the good, and the ideas generally, without sufficient previous training. I noticed your deficiency, when I heard you talking here with your friend Aristoteles, the day before yesterday. The impulse that carries you towards philosophy is noble and divine — never doubt that — but there is an art which often seems to be useless, and is called by the vulgar idle talking; in that you must train and exercise yourself, now that you are young, or truth will elude your grasp.

And what is the nature of this exercise, Parmenides, which you would recommend?

That which you heard Zeno practising; at the same time, I give you credit for saying to him that you did not care to solve the perplexity in reference to visible objects, or to consider the question in that way; but only in reference to the conceptions of the mind, and to what may be called ideas.

Why, yes, he said, there appears to me to be no difficulty in showing that visible things experience likeness or unlikeness or anything else.

Quite true, he said; but I think that you should go a step further, and consider not only the consequences which flow from a given hypothesis, but the consequences which flow from denying the hypothesis; and the exercise will be still better.

What do you mean? he said.

I mean, for example, that in the case of this very hypothesis of Zeno's about the many, you should inquire not only what will follow either to the many in relation to themselves and to the one, or to the one in relation to itself and the many, on the hypothesis of the existence of the many, but also what will follow to the one and many in their relation to themselves or to one another, on the opposite hypothesis. Or if likeness does or does not exist — what will follow on either of these hypotheses to that which is supposed, and to other things in relation to themselves and to one another, and the same of unlikeness; and you may argue in a similar way about motion and rest, about generation and destruction, and even about existence and non-existence; and in a word, whatever you like to suppose as existing or non-existing, or experiencing any sort of affection. You must look at what follows in relation to the thing supposed, and to any other things which you choose, — to the greater number, and to all in like manner; and you must also look at other things in relation to themselves and to anything else which you choose, whether you suppose that they do or do not exist, if you would train yourself perfectly and see the real truth.

Editor's Comments on Paper 5

Excerpts from *The Dynamics of Machinery: The Five Simple Machines*
5 HERO OF ALEXANDRIA

Hero of Alexandria has been one of the enigmas of the history of science. At one time he was only a name given to a collection of writings on physical science that came down to the Middle Ages in Arabic and Latin translations, although presumably written originally in Greek. For a long time great uncertainty existed about his dates, but it is now considered that this matter was settled through historical-astronomical analysis by Otto Neugebauer in 1938. The latter concluded that Hero must have flourished around A.D. 62 although he was unable to fix the dates of birth and death.

There have been widely varying opinions about the real character of Hero. Some have held him to be a mere ignoramus who copied the writings of others without really understanding them. Others have been convinced from his writings that he was primarily an artisan, who was lucky enough to understand how various physical devices work in practice and may indeed have been clever enough to have invented some of these devices. Still others consider Hero a mathematician of ability, with a complete tehcnical grasp of what he was writing about.

Hero and his contemporaries were familiar with the nature and operation of machines. Hero wrote a treatise on the five simple machines, which has survived in the *Mathematical Collection* of Pappus of Alexandria (who was active about A.D. 300). We present here extracts from this treatise, concentrating only on the wheel and axle, the lever, and pulley systems. It is clear from the remarks at the end of the article that Hero fully understood the fundamental compensation involved in the operation of machines (that is, what is gained in force is lost in speed). This is our principal reason for including extracts from his work.

5

Reprinted from *A Source Book in Greek Science,* M. R. Cohen and I. E. Drabkin, eds., McGraw-Hill Book Company, New York, 1948, pp. 224-227, 230-232, 234-235

THE DYNAMICS OF MACHINERY: THE FIVE SIMPLE MACHINES

Hero of Alexandria

1. We shall now give, from the works of Hero, an abridged account of the five aforesaid machines. We shall also set forth, for the benefit of students, the essential facts with regard to cranes of one, two, three, and four masts on the chance that one who seeks the books in which these subjects have been treated may not find them available. For we have met with many copies which have been considerably mutilated, lacking both beginning and end.[2]

There are five machines by the use of which a given weight is moved by a given force, and it will be our task to give the forms, the applications, and the names of these machines. Now both Hero and Philo[3] have shown that these machines, though they differ considerably from one another in their external form, are reducible to a single principle. The names are as follows: wheel and axle, lever, system of pulleys, wedge, and, finally, the so-called endless screw.

[2] The reference seems to be to copies of Hero's works.

[3] I.e., Philo of Byzantium (see p. 255).

The wheel and axle is constructed in the following way. One must take a strong log squared off like a beam, make its ends round by planing,[1] and fit bronze end-pivots to the axle, so that when inserted in round openings in the immovable framework, they turn easily. For these openings also have a bronze lining upon which the end-pivots rest. Now this beam is called the axle. At the middle of this axle is placed a wheel having a square opening to fit the axle so that the axle and wheel turn together.

Having indicated the construction we must now speak of the use of the machine. If we wish to move a large weight with a smaller force, we attach a cable to the weight and fasten this cable around the curved portion of the axle. We then insert spokes in holes bored in the wheel and turn the wheel by pressing on the spokes. In this way the weight will readily be lifted by a smaller force, as the cable is wound around the axle, or else bunched to keep the whole cable from covering the axle.[2] The size of this machine must be adapted to the weight which one intends to move; and the ratio of the parts of the machine[3] must be adapted to the ratio of the weight moved to the moving force, as will be proved below.[4]

2. The second machine is the lever, which may well have been man's first discovery in connection with the moving of large weights. For, wishing to move large weights, men found it necessary first to lift them from the ground. But they had no means of grasping the weight because all parts of its base rested on the ground. They therefore dug a small groove under the weight, inserted here the end of a long wooden pole, and placed a stone, called hypomochlion (fulcrum), under the pole near the weight. They then pressed down on the other end. When they saw that the motion was quite easy they realized that large weights could be moved in this way. Now the pole is called a lever and is either squared or rounded. The nearer the fulcrum is placed to the weight, the more easily is the weight moved, as will be shown below.

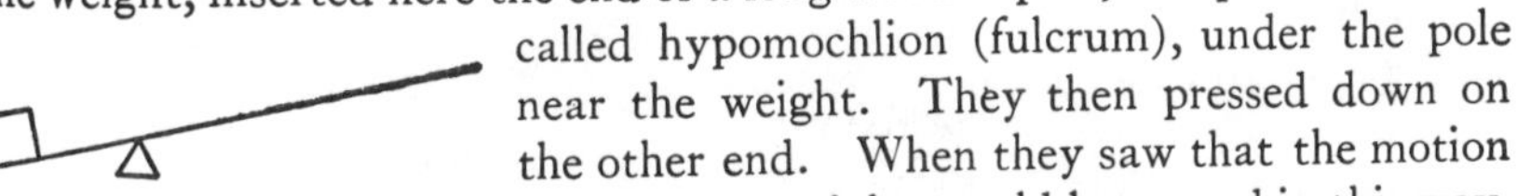

[1] The Greek is obscure. As the sequel shows, the entire part about which the cable turned was also to be rounded.

[2] This last clause, though bracketed by Hultsch, is present in the Arabic version.

[3] I.e., the ratio of the diameter of the wheel to that of the axle.

[4] Hero, *Mechanics* II.22, not quoted by Pappus but extant in the Arabic version (p. 232, below).

3. The third machine is the system of pulleys. If we wish to lift a given weight we may attach a rope to it and pull with a force equal to the weight in question. If, however, we untie[1] the rope from the weight, attach one end to an immovable beam, pass the other end around a pulley-wheel made fast to the weight, and pull this free end we shall move the weight more easily. Again, if to the fixed beam we attach another pulley-wheel, pass the free end of the rope over it, and then pull, we shall move the weight still more easily. And still further, if we attach another pulley-wheel to the weight, pass the free end of the rope over it and pull, we shall move the weight still more easily. Thus, as we attach more pulley-wheels both to the fixed beam and to the weight, and pass the free end of the rope successively around the wheels, we shall move the weight more and more easily. The greater the number of parts into which the rope is divided by the successive turns, the more easily will the weight be moved. But that end of the rope which is not pulled must be attached to the fixed beam.[2] However, to avoid the necessity of attaching each separate pulley-wheel to the fixed beam or to the weight, those pulleys which were described as attached to the fixed beam are instead inserted into a single wooden frame, called a pulley-block (manganon), and permitted to rotate about their axes. This pulley-block is itself attached by another cable to the fixed beam. Again, those pulley-wheels connected with the weight are inserted into a second pulley-block which is like the first and is, in turn, connected only with the weight. The pulley-wheels must be so arranged in the blocks that the cord lengths do not become entangled with one another and hard

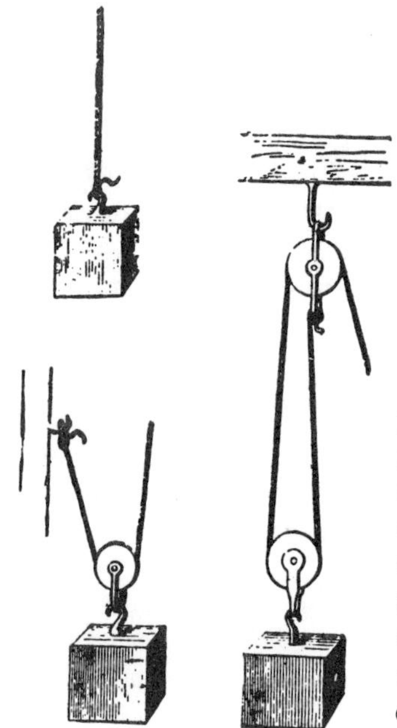

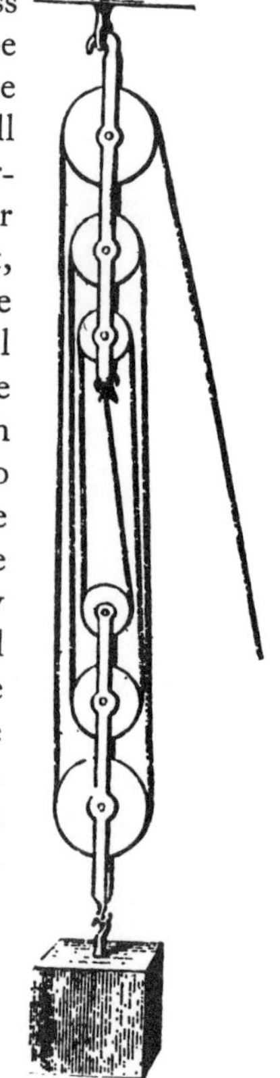

[1] Reading ἐκλύσαντες instead of ἑλκύσαντες (Hultsch).

[2] I.e., directly or indirectly (see the figures). The last two sentences, though bracketed by Hultsch, are present in the Arabic version.

to move. We shall show below[1] why an increase in the number of turns of the rope makes the lifting of the weight easier, and why one end of the rope is connected with the fixed beam.

THE RELATION OF FORCE AND VELOCITY IN SIMPLE MACHINES

Hero of Alexandria, *Mechanics* II. 21–26

21. We assert that of all figures the circle possesses the greatest and easiest mobility, whether it revolves about its center or on a plane to which it is perpendicular. The same holds true of the figures related to the circle, that is, spheres and cylinders, for their motion is circular, as we have shown in the preceding book.

Suppose we wish, in the first place, to move a large weight with a small force by the use of the wheel and axle, without the difficulty [discussed in the preceding chapter].[3] Let the weight that we desire to move be 1,000 talents and the force with which we wish to move it be five talents. Now we must, in the first place, bring the force and the weight into equilibrium, since, after equilibrium is attained, we can make the force predominate over the weight by the addition of a slight force to the machine. We place at *A* the axle on which turns the cable attached to the weight, and we place at *B* the wheel connected with that axle. To facilitate the construction of the apparatus we make the diameter of the wheel five times that of the axle. In that case the force required to move wheel *B*, which balances the weight of 1,000 talents, will be 200 talents. But the force we have available, is, as we assumed, only five talents. With this force, therefore, we cannot move the given weight merely with wheel *B*. Hence we construct a geared axle, *C*, which fits the teeth of wheel *B* in such a way that if axle *C* moves, wheel *B* is also set in motion together with the first axle, *A*, with the result that the weight moves if axle *C* turns. This axle is set in motion by the force which moves wheel *B*, for we have shown that all objects revolving about fixed centers may be moved by a small force. Hence it makes no difference whether the weight is moved by wheel *B* or by axle *C*. Now let a wheel, *D*, be attached to axle *C* and let the diameter of *D* be, say, five times that of axle *C*. In that case a force of 40 talents is required at wheel *D* to hold the weight in equilibrium. Again, if we take another axle, *E*, fitting into wheel *D*, the moving force at *E* will likewise be 40 talents. Now let there be still another wheel, *Z*, at-

---

[1] Hero, *Mechanics* II.24, not quoted by Pappus but extant in the Arabic version (p. 233, below).

[3] It had been indicated that if a single wheel and axle is used, and if it is required, for example, to lift a 1,000-talent weight with a force of five talents, the diameter of the wheel must be more than 200 times as large as that of the axle.

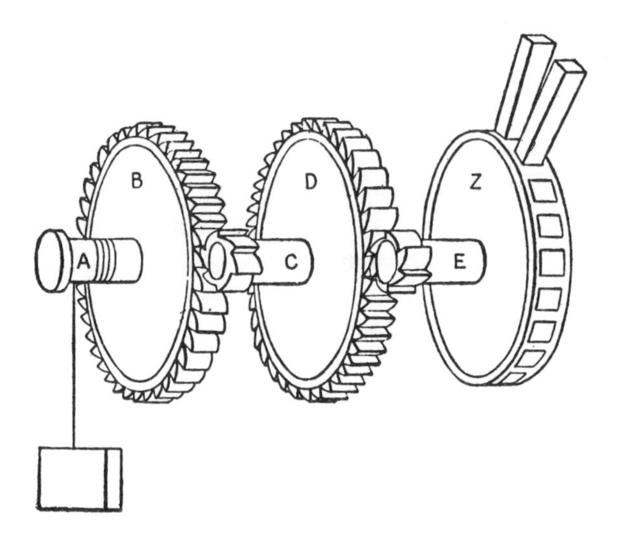

tached to axle E, of diameter eight times that of axle E, since the force of 40 talents is eight times the force of five talents. The force required at Z to hold the weight in equilibrium will then be five talents, which was the given available force. In order, however, that the force overcome the weight, we must either make wheel Z a little larger or axle E a little smaller. If we do this the force will overcome the weight. If we wish to use several wheels and axles in this process we must employ the same ratio, since the combination of all the ratios must correspond to the weight if we wish to bring force and weight into equilibrium.[1] But if we want the force to overbalance the weight we must apply an amount of force just in excess of that required, according to the combined ratios, to preserve the equilibrium of the weight.[2]

Thus a given weight may be moved by an axle passing through a wheel. If, however, we do not wish to have geared wheels, we may wind cables around the wheels and axles. In that case the same work may be performed, since the first axle, that one which draws the load, will be moved by the wheel which is ultimately moved.[3] Wheels and axles of this type must, if they are to be used properly, have strong supports with holes into which the ends of the axles fit. If the weight is to be lifted these supports must be set on a firm and secure foundation.

22. In this and similar machines in which great force is developed there

---

[1] Though it is not put very clearly, the idea seems to be that no matter how many or how few wheels and axles are used, the same rules for computing what we call the "mechanical advantage" obtain and determine whether our force is sufficient to balance the weight.

[2] Note that the efficiency of the machine is not discussed. Any force just larger than the amount *theoretically* required to keep the weight in equilibrium is presumed to be sufficient to lift the weight.

[3] We might expect the designations of "first" and "ultimate" to be interchanged.

is a retardation since we must use more time in proportion as the moving force is smaller than the weight to be moved. *Force is to force as time is to time, inversely.* For example, when the force at wheel *B* is 200 talents and is sufficient to move the weight, a single revolution [of *B*] is needed to effect a sufficient winding of the cable around *A* that, by the movement of wheel *B* the weight may be lifted a distance equal to the circumference of *A*. But if the weight is moved through a movement of wheel *D*, the circumference of *C* must make five revolutions in order that axle *A* make one revolution, since the diameter of *B* is five times that of axle *C*. Hence five revolutions of *C* are equal to one revolution of *B* if we bring the axles and wheels back to their original position. If not, we will have a corresponding ratio.[4] Wheel *D* moves according as *B* moves, and five revolutions of *D* require five times as much time as a single revolution of *B*. Now 200 talents are five times 40 talents. Thus the ratio of moving force to moving force is equal inversely to the ratio of time to time. This is also the case with several axles and wheels. The proof is the same.

25. Again, the same weight [1,000 talents] may be moved by the same force [five talents] by means of the lever, according to a similar procedure.

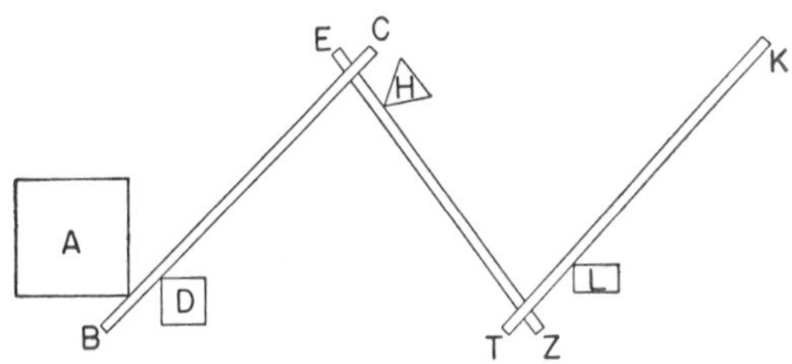

Let the weight be at point *A*, the lever at *BC*, and the fulcrum at point *D*. Let us move the weight by means of the lever parallel to the ground, making *CD* five times *BD*. Thus the force at *C* sufficient to balance the 1,000 talents will be 200 talents. Now let there be another lever, *EZ*, and let *E*, the end of this lever, be joined with point *C* so that the latter will be set in motion by the motion of *E*. Let the fulcrum be at point *H* and let arm *E* of the lever move in the direction of *D*. If *ZH* be made five times *HE*, the force at *Z* will amount to 40 talents. Now let there be still another lever, *TK*, end *T* of which is connected with *Z* and moves in a direction opposite to that in which *E* moves. If the fulcrum of this lever be at point *L*, and if this lever move in a direction opposite to that in which *E* moves, and *KL* be eight times *LT*, the force at *K* will be five talents and will balance the weight. If, however, we wish the force to overbalance the weight we must make *KL* larger than eight times *LT*. Or, if *KL* is eight times *LT*, *ZH* five times *HE*, and *CD* more than five times *DB*, the force will lift the weight.

26. Also in this case there is a retardation in the same ratio [as in the other machines], since there is no difference between these levers and the

[4] The meaning seems to be that not merely in the case of complete revolutions but in general the arc traveled by a point on the circumference of *C* is, in circular measure, five times that traveled by a point on the circumference of *B*.

combinations of wheel and axle moving about a center. For the levers, like axles, move about points D, H, and L, the fulcra round which the levers turn. Corresponding to the circles described by axles are the circles which points B, E, and T describe; corresponding to the circles described by wheels are those which points C, Z, and K describe. In the same way as we proved in the case of axles that the ratio of force to force is equal, inversely, to the ratio of time to time, so we may prove it also in the case of levers.

Editor's Comments on Papers 6, 7, and 8

The Impossibility of Perpetual Motion
6 L. DA VINCI

Comments on Perpetual Motion
7 J. CARDAN

The Impossibility of Perpetual Motion and the Problem of the Inclined Plane
8 S. STEVIN

PERPETUAL MOTION

Leonardo da Vinci (1452–1519), the great Italian artist, scientist, engineer and inventor, was probably the most versatile intellect of fifteenth-century Europe. His achievements in art and anatomical science are so well known as to need no commentary. His restless and fertile mind led him to the design of many interesting devices based on a rather sound knowledge of mechanics and often of a strikingly practical character. His technical ideas were not organized in the form of elaborate treatises but were for the most part jotted down as notes; the manuscripts of many have been preserved.

Among other mechanical problems, Leonardo directed his attention to the possibility of constructing a perpetual motion machine. This has an obvious bearing on the development of the concept of energy. For if such a machine could be constructed, the idea of energy would have no value whatever.

In the note reproduced here, Leonardo visualizes two types of alleged perpetual-motion machines, both based on the rotating wheel idea. He gives reasons for believing that these devices will not work as alleged. It must be confessed that the extracts presented are by no means models of clarity, but they do bring out Leonardo's conviction of the impossibility of perpetual motion. As we shall note in a later extract from the work of Cardan and Stevinus, his view undoubtedly had an influence on the development of effective and successful principles of mechanics.

Jerome Cardan (in Italian, Girolamo Cardano) (1501–1576) was an Italian physician, mathematician, and generally versatile scholar who spent most of his professional career in Milan and other cities of northern Italy. He wrote voluminously on many subjects, including medicine, physical science, and mathematics. His treatise on algebra is considered by competent authorities to be the best treatise on the subject up to his time. He also wrote successfully on probability. Cardan's excursions into physics were not so successful. However, in his encyclopedic work *De Subtilitate* (commonly translated *On Subtlety*, although this provides no genuine idea of its contents), he devoted much attention to motion and said several things that one could consider reasonable today. Among other subjects, he discussed the possibility of perpetual motion and argued against it. Although his remarks are by no means conclusive, he does show a glimmering of the notion that terrestrial motions cannot go on indefinitely without renewal from some source. He showed a keen awareness of the unlikelihood that we ever get something for nothing. To what extent this entitles Cardan to be considered one of the forerunners of the energy concept is debatable. However, what he has to say is of some interest, and hence we include here a brief selection from Book 17 of *De Subtilitate.*

Simon Stevin (1548–1620), known more commonly as Stevinus, the Latinized version of his name, was a Flemish-Dutch mathematician, physicist, and engineer, most of whose life was lived during the struggle of the Low Countries to throw off the yoke of Spain. As an engineer he served Prince Maurice of Nassau (1567–1625) in connection with both civil and military operations. His contributions to physics were principally concerned with statics and the properties of fluids, although he also studied terrestrial magnetism. Perhaps Stevin's most famous achievement was his derivation of the law of the inclined plane. Here he departed entirely from the Aristotelian dynamical method, which had guided practically all medieval investigators of the behavior of machines. He had no use for Aristotle's emphasis on the properties of the circle as important in the understanding of machines. In fact, he derided it. At heart, Stevin was really a follower of Archimedes in his views on statics. He could not swallow the idea that static equilibrium is a special case of dynamics and hence in general refused to employ a dynamical method in his treatment of equilibrium. In spite of this, his handling of the problem of the inclined plane deserves an important place in the evolution of the energy concept. The reason lies in his use of the denial of the possibility of perpetual motion in the terrestrial sphere.

We reproduce here Stevin's derivation in English translation. It is an elegant demonstration from a geometrical point of view, although its validity rests entirely on the hypothesis of the impossibility of perpetual motion. It is interesting that in his derivation all that Stevin says about such motion is that it is absurd to believe it possible (or, in a more literal translation, the assumption of it is "false"). No further support for the principle is given. Competent scholars believe that Stevin was familiar with the views of Leonardo da Vinci and Cardan on perpetual motion, although he mentions neither in his demonstration. That he does not take the trouble to do this nor to emphasize the basis for the belief of these two predecessors is perhaps not too surprising when one considers the medieval tendency to ignore specific reference to previous work unless the author was in serious disagreement with it. It may well be that Stevin felt that in his time the absurdity of the possibility of perpetual motion was so well known as to require no comment. Still, one could wish to confront Stevin with the specific question, "Why do you believe in the impossibility of perpetual motion?" to see what answer he would give. If the whole idea was absurd in the sixteenth century, why have so many later hopefuls attempted to construct perpetual-motion machines?

Ernst Mach (1838–1916), the distinguished Austrian physicist, psychologist, and philosopher of science, in his book *History and Root of the Principle of Conservation of Energy* (original German edition, Prague, 1872; English translation by P. E. B. Jourdain, Open Court Publishing Co., Chicago, 1911, p. 22ff), presents Stevin's discussion of the inclined plane with complete approval, for Mach felt that the impossibility of perpetual motion is the cornerstone of the concept of energy. But even he does not indicate why Stevin had the right to be so sure about his hypothesis. Mach was a phenomenalist and laid great stress on the basic importance of what one can simply observe. Physics, in its development since the seventeenth century, has achieved greatest success by looking behind phenomena and constructing theories. Today it is fair to say that most students of the theory of mechanics are inclined to view the impossibility of perpetual motion as a direct consequence of the principles of thermodynamics, which in turn rest on the concept of energy based on deeper grounds. However, no one can question the intimate connection between the assumption Stevin used in his demonstration of the law of the inclined plane and the idea of energy.

Although Stevin proclaimed himself a vehement anti-Aristotelian in mechanics, in his treatment of other machines, like the lever, he effectively used the principle of virtual displacements as it came down from the Greeks.

6

Reprinted from *The Notebooks of Leonardo da Vinci*, E. MacCurdy, ed., Garden City Publishing Co., New York, 1941, pp. 802–803

THE IMPOSSIBILITY OF PERPETUAL MOTION

Leonardo da Vinci

METHOD OF THE SMALL COMPARTMENTS OF THE ROUND MACHINE GIVEN BELOW

[*Drawing*]

Make it so that the buckets which are plunging with the mouth downwards have such an opening that the air cannot escape; it will also be a good thing that the covered exits to the buckets should be of terracotta so that they may be better able to pass beneath the water; and of copper would be best of all. Forster I 50 v.

Moreover you might set yourself to prove that by equipping such a wheel with many balances, every part however small which turned over as the result of percussion would suddenly cause another balance to fall, and by this the wheel would stand in perpetual movement. But by this you would be deceiving yourself; for as there are these twelve pieces and only one moves to the percussion, and by this percussion the wheel may make such a movement as may be one twentieth part of its circle, if then you give it twenty-four balances the weight would be doubled and the proportion of the percussion of the descending weight diminished by half, and by this the half of the movement would be lessened; consequently if the first was one twentieth of the circle this second would be one fortieth, and it would always go in proportion, continuing to infinity.

Forster II 89 v.

Whatever weight shall be fastened to the wheel, which weight may

[*Ed. note:* The two figures presented with Leonardo's note were kindly provided by the Elmer Belt Library of Vinciana in the Dickson Art Center of the University of California at Los Angeles (Kate T. Steinetz, Curator) and are used with the permission of the curator.]

be the cause of the movement of this wheel, without any doubt the centre of such weight will remain under the centre of its axis.

And no instrument which turns on its axis that can be constructed by human ingenuity will be able to avoid this result.

O speculators about perpetual motion, how many vain chimeras have you created in the like quest? Go and take your place with the seekers after gold. Forster II 92 v.

[*Diagram*]

To try again the wheel which continually revolves.

I have many weights attached to a wheel at various places: I ask you the centre of the whole sum of the weight.

I take a wheel revolving on its axis, upon which are attached at various places weights of equal gravity, and I would wish to know which of these weights will remain lower than any of the others and at what stage it will stop. I will do as you see above, employing this rule for four sides of the circle, and that where you will see greater difference upon the arms of the balance, that is that experiment which will throw you the sum of one of the gravities more distant from the pole of the balance, that will go on and become stationary below; and if you want all the details repeat the experiment as many times as there are weights attached to the wheel. Forster II 104 v.

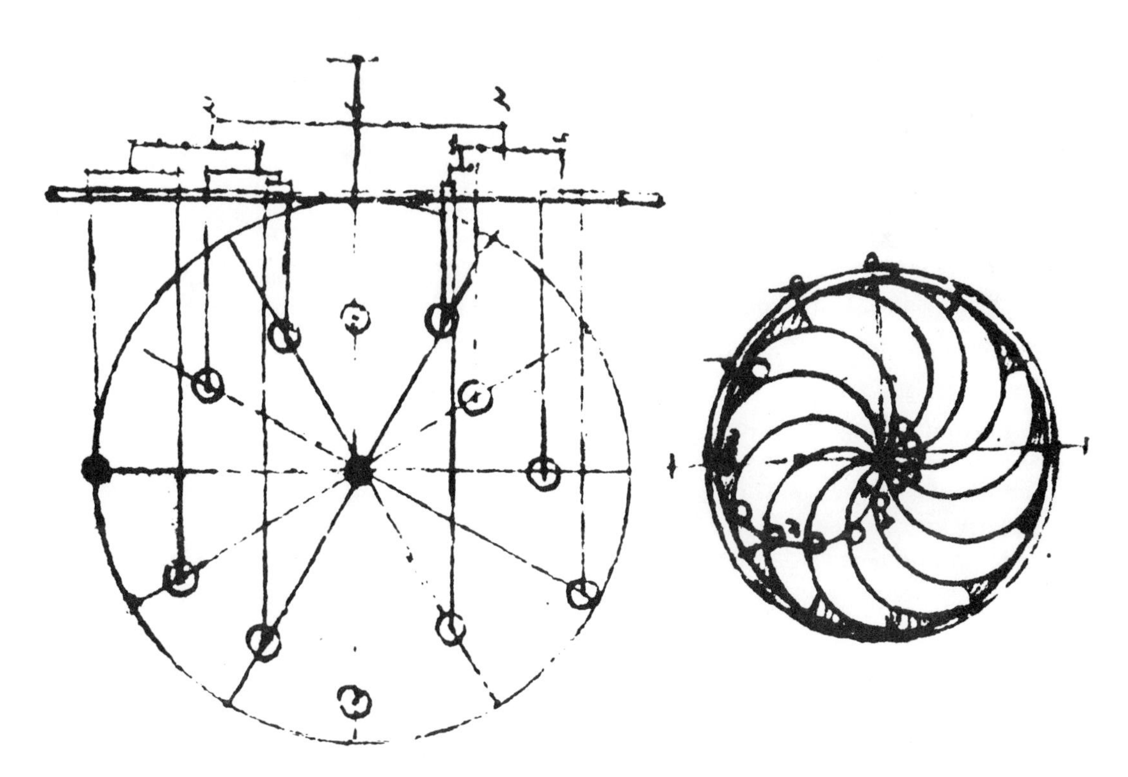

7

This article was translated expressly for this Benchmark volume by R. Bruce Lindsay, Brown University, from De Subtilitate, *Book 17,* Opera Omnia, *Vol. 3, Leiden, 1663, p. 625*

COMMENTS ON PERPETUAL MOTION

Jerome Cardan

We must now see whether perpetual motion is possible among artificially produced motions, a matter of current argument. There is no doubt that natural motion in the heavens is perpetual. It is the same with rivers; as long as water is provided by the source, they continually flow down the slope of their beds. The existence of perpetual motion is not to be sought in motions that are really perpetual. For all natural bodies, the more they are moved, the more over a long period of time are their motions attenuated and consumed. It is therefore proper to inquire whether any motion can be found that can continue by itself without some generating cause. For example, in clocks, in which the motion is indicated by the strokes of the hour, weights must be raised and there is activity involved in keeping them running. There can be three kinds of natural motions of heavy bodies: either to the center per se (that is, center of the universe); or not directly toward the center, as in the flowing of rivers; or, finally, motion of a special nature like that of iron toward a magnet. Obviously, it is necessary that perpetual motion should be looked for in the first two kinds of motion. When something is pulled out too much or contracted too much, the resulting motion is, to be sure, natural, yet not without a violent aspect. These two cases are seen in the weights of clocks. In every such case, however, there is a beginning and an end to the motion. Those things that move in a circle never stay constantly so except in the heavens or the air, and they have their origin in things that move along a straight line. For water itself, as I have said, moves along a straight line. For perpetual motion to exist, a body that has reached the end of its path must be brought back to the beginning. But it is not possible that it be brought back save by the using up of something else. Therefore, continuity of motion comes from that which is according to nature, or it is not uniform. That which always diminishes unless it is continually renewed cannot be perpetual.

8

Reprinted from *A Source Book in Physics,* W. F. Magie, ed.,
McGraw-Hill Book Company, New York, 1935, pp. 23-27

THE IMPOSSIBILITY OF PERPETUAL MOTION AND THE PROBLEM OF THE INCLINED PLANE

Simon Stevin

The extract from Stevinus' work contains an interesting demonstration of the law of equilibrium of a body on an inclined plane and a partial if not a complete statement of the parallelogram of forces. It was first published in Leyden in 1586 and has been translated from the French as it appears in his collected works published at Leyden in 1634.

The Inclined Plane

Up to this point there have been considered the properties of weights acting directly downward; in what follows will be treated the properties and qualities of oblique forces, of which the general foundation is contained in the following theory.

Theorem XI, *Proposition* XIX

If a triangle has its plane perpendicular to the horizon and its base parallel to it; and if on each one of the two other sides there is placed a spherical body of equal weight; then as the right side of the triangle is to the left so is the force of the left hand body to that of the right hand body.

Given. Let *ABC* (Fig. 9) be a triangle with its plane perpendicular to the horizon and its base *AC* parallel to the horizon; and let there be placed on the side *AB* (which is twice the length of *BC*) a globular body *D*, and on *BC* another, *E*, equal in weight and in size.

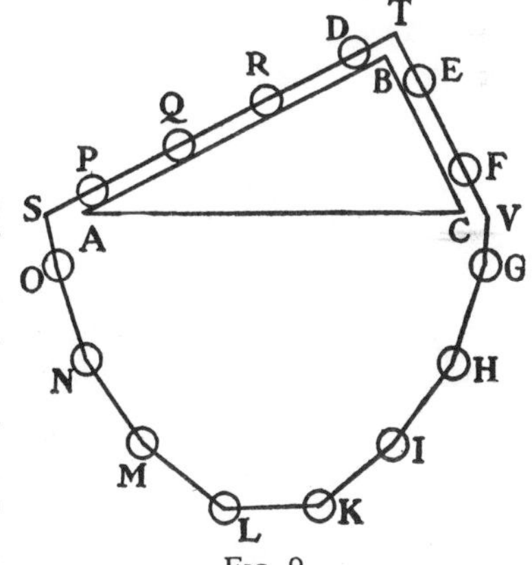

Fig. 9.

Required. It is to be demonstrated that as the side *AB*2 is to the side *BC*1 so the force or power of the weight *E* is to that of *D*.

Preparation. Let there be arranged about the triangle a set of fourteen spheres, equal in weight and in size, and set at equal distances from each other, such as *D, E, F, G, H, I, K, L, M, N, O, P, Q, R*, strung on a line passing through their centers, so that they can turn about their respective centers, and so placed that there are two spheres on the side *BC* and four on *BA*, then as the one line is to the other line so is the number of the spheres on the one to the number of the spheres on the other; also in *S, T, V*, let there be three rigidly fixed points, over which the line or the thread can slide, so that the two parts of it above the triangle are parallel to the sides *AB*, *BC*; in such a way that the whole string can run freely and without catching over the sides *AB*, *BC*.

Demonstration

If the power of the spheres *D*, *R*, *Q*, *P* were not equal to the power of the two spheres *E*, *F*, one side would be heavier than the other; thus, if it were possible, the four bodies *D*, *R*, *Q*, *P*, will be heavier than the two bodies *E*, *F*; but the four bodies *O*, *M*, *N*, *L*, are equal to the four bodies *G*, *H*, *I*, *K*; wherefore, the set of eight bodies *D*, *R*, *Q*, *P*, *O*, *N*, *M*, *L*, will be heavier, as they are placed, than the six bodies *E*, *F*, *G*, *H*, *I*, *K*, and since the heavier part overpowers the lighter part, the eight spheres will descend and the other six will rise; let this be so, and let *D* come where *O* is at present, and so for the rest; so that *E*, *F*, *G*, *H*, will come where *P*, *Q*, *R*, *D* are at present and *I*, *K*, where *E*, *F* are at present. Nevertheless, the set of spheres will have the same arrangement as before, and by the same reasoning the eight spheres will exceed the others in weight, and in falling will make eight others come into their places, and so this motion will have no end, which is absurd. The demonstration will be the same on the other side; therefore the part of the arrangement *D*, *R*, *Q*, *P*, *O*, *N*, *M*, *L*, will be in equilibrium with the part *E*, *F*, *G*, *H*, *I*, *K*; so that if we take away from the two sides the equal weights which are similarly arranged, as are the four spheres *O*, *N*, *M*, *L*, on the one side and the four *G*, *H*, *I*, *K*, on the other side; the four remaining spheres *D*, *R*, *Q*, *P*, will be and will remain in equilibrium with the two *E*, *F*; wherefore *E* will have a power double of the power of *D*; therefore as the side *BA*2 is to the side *BC*1, so is the power of *E* to the power of *D*.

Conclusion. Therefore if a triangle has its plane, etc.

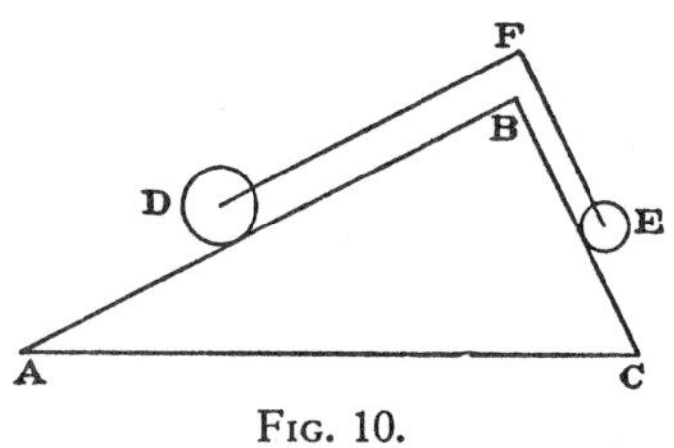

Fig. 10.

Corollary 1.

Let *ABC* (Fig. 10) be a triangle as before, and *AB* twice as great as *BC*, and let *D* be a sphere on *AB* twice as massive as *E*, which is on *BC*; at *F* let there be a rigidly fixed point over which the line *DFE* can slide freely, and so that *DF*, *FE*, are parallel to the sides of the triangle *ABC*, drawn from the centers of the spheres; it is plain that *D*, *E*, will still be in equilibrium, just as before *P*, *Q*, *R*, *D*, were in equilibrium with *E*, *F*, because as *AB* is to *BC* so is the sphere *D* to the sphere *E*.

Corollary 2.

Further, let one of the sides of the triangle, like *BC* (Fig. 11) (which is half the length of the other side *AB*), be perpendicular to *AC*; the sphere *D* which is twice the mass *E* will still be in equilibrium with *E*, for as the side *AB* is to *BC* so is the sphere *D* to the sphere *E*.

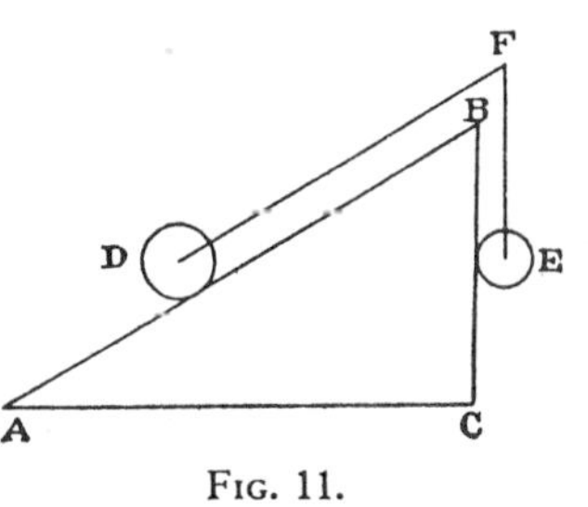

FIG. 11.

In Corollary 3 a pulley is substituted for the fixed point at the top of the plane, and in Corollary 4 the sphere on the plane is replaced by a column or cylinder with its axis perpendicular to the slant face of the plane.

Corollary 5.

Let a perpendicular be drawn from the center of the column *D* (Fig. 12), such as *DK*, cutting the side of it at *L*; then the triangle *LDI* will be similar to the triangle *ABC*, for the angles *ACB*

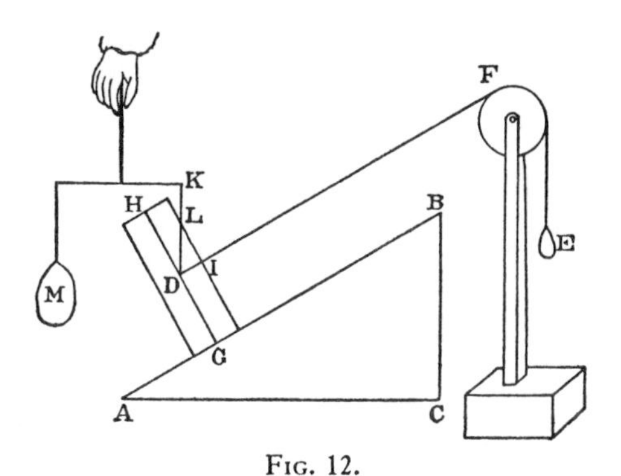

FIG. 12.

and *LID* are right-angles, *LD* is parallel to *BC*, and *DI* to *AB*; wherefore, as *AB* is to *BC* so is *LD* to *DI*; but as *AB* is to *BC* so is the weight of the column to the weight *E*, by the fourth corollary: therefore as *LD* is to *DI* so is the weight of the column to *E*: so that if in *KD* we apply a lifting force *M* in equilibrium with the column, it will be equal in weight to it by the fourteenth proposition: and finally as *LD* is to *DI* so is *M* to *E*.

Corollary 6.

Let us draw the line *BN* (Fig. 13) cutting *AC* produced in *N*, and also *DO*, cutting *LI* produced in *O*, in such a way that the angle

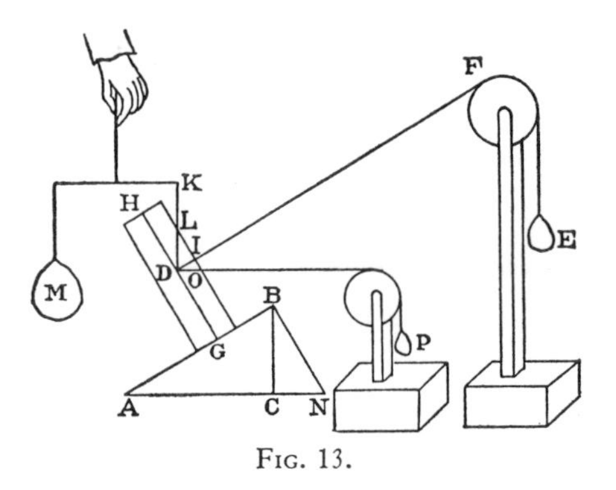

Fig. 13.

IDO is equal to the angle *CBN*, and then let there be applied the direct force *P* along *DO*, holding the column in position (having removed the weights *M*, *E*); then, in as much as *LD* is homologous to *BA* in the triangle *BAC*, and *DI* to *BC*, it follows, since *BA* is to *BC* as the weight on *BA* is to the weight on *BC*, by the second Corollary, that also *DL* is to *DI* as the weight belonging to *DL* is to the weight belonging to *DI*, as *M* to *E*; similarly, the three lines *LD*, *DI*, *DO* being homologous to the three lines *AB*, *BC*, *BN*; then *BA* being to *BN* as the weights pertaining to them so also *LD* will be to *DO* as the weights pertaining to them, that is to say, as *M* to *P*: and the same would be true if *BN* were drawn on the other side of the perpendicular *BC*, that is, between *AB* and *BC*, and similarly *DO* between *DL* and *DI*, and this proportion holds not only when the elevation *DI* is perpendicular to the axis but for every angle.

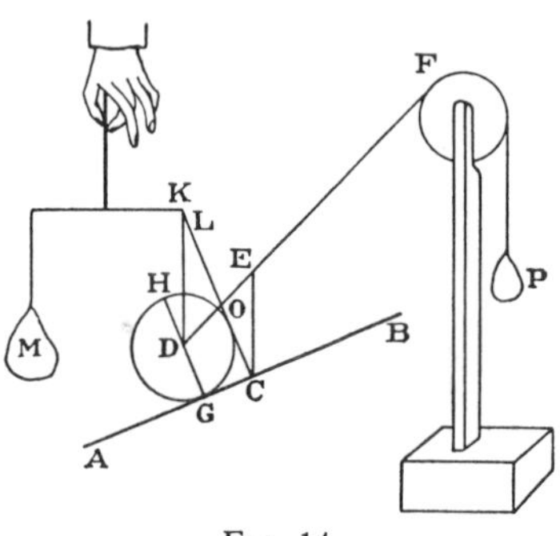

Fig. 14.

The above also may be applied to a sphere on the line *AB* (Fig. 14) as here represented, where we shall say as before: that as *LD* is to *DO* so is *M* to *P* (provided that *CL* is at right-angles to *AB*, that is to say, parallel to the axis *HG* of the sphere *D*) and so the weight of the sphere is to *P* as *LD* is to *DO*; but since *LD*, *DO* cannot be easily drawn inside the sphere let the perpendicular *CE* be drawn; then we shall have a triangle *CEO* outside the solid sphere, similar to *LDO*: of which the homologous sides are *LD*, *CE*, also *DO*, *EO*; wherefore as *LD* is to *DO*, that is to say as *CE* is to *EO*, so the weight of the sphere is to *P*.

There follows a diagram in which all the lines have been removed except those of the triangle *COE*, and the result is announced that as *CE* is to *EO* so is the weight of the sphere *D* to the weight *P*. The author adds here that this relation holds true not only for spheres but for all sorts of solid bodies.

Editor's Comments on Papers 9A, 9B, and 9C

Work Involved in the Operation of Machines
9A GALILEO

Of the Force of Percussion
9B GALILEO

The Pendulum Experiment
9C GALILEO

FORESHADOWINGS OF THE IDEA OF ENERGY AS CONSTANCY WITHIN CHANGE

Galileo Galilei (1564–1642), usually known in English-speaking countries by his first name, was one of the great natural philosophers of the sixteenth and early seventeenth centuries and one of the founders of modern physics. His achievements in the study of astronomy, mechanics, acoustics, heat, and light are well known. His derivation of the law of falling bodies was a classic in what came to be called theoretical physics. Although he could not claim complete originality in all his discoveries, his codification of earlier work was done with great skill and unusually keen insight and generally carried conviction.

Obviously, we should look into the writings of Galileo for vestiges and foreshadowings of the idea of energy, and they are not difficult to find. In his early work *On Mechanics,* prepared about 1600 and recently translated with a commentary by Stillman Drake in the book *On Motion and on Mechanics* under the authorship of I. E. Drabkin and Stillman Drake (University of Wisconsin Press, Madison, Wisc., 1960), Galileo makes clear his complete grasp of the essential feature of any machine—what is gained in force is lost in speed, which is essentially the principle of virtual velocities. We reproduce here from Drake's translation not only the general comments of Galileo about machines, but also his specific treatment of the lever.

Further on in the same treatise Galileo examines the problem of percussion, for example, the driving of a nail with a hammer into a piece of wood. He seems to realize that we need a new concept beyond that of static force to account for what a hammer blow can accomplish. Of course he does not introduce this concept (what we now call the kinetic energy of the motion of the hammer), but his very groping is a significant advance. We reproduce his whole observation on this phenomenon.

One of the most beautiful and striking examples of Galileo's grasp of the significance of constancy in the midst of change is found in his famous pendulum experiment. He carried this out to provide an experimental basis for his assumption that when a ball falls freely under gravity the velocity it gains on reaching the ground depnds only on the height from which it falls and not at all on the path of fall. This assumption was necessary to provide cogency to his experimental verification of his law of fall. We reproduce here his account of the pendulum experiment, given in his famous work *Dialogues Concerning Two New Sciences* (1638). Galileo convinced himself that, even though the pendulum string is shortened during flight, as long as the bob falls from the same height, something stays constant in all cases. This constant we now recognize as the total mechanical energy of the pendulum bob. In this experiment Galileo came very close to the introduction of the concept of energy.

9A

Reprinted from *On Motion and On Mechanics,* S. Drake, trans., I. E. Drabkin and S. Drake, eds., University of Wisconsin Press, Madison, Wis., 1960, pp. 147-150, 157-159

WORK INVOLVED IN THE OPERATION OF MACHINES

Galileo Galilei

On the Utilities That Are Derived from the Mechanical Science and from Its Instruments

It has seemed well worthwhile to me, before we descend to the theory of mechanical instruments, to consider in general and to place before our eyes, as it were, just what the advantages are that are drawn from those instruments. This I have judged the more necessary to be done, the more I have seen (unless I am much mistaken) the general run of mechanicians deceived in trying to apply machines to many operations impossible by their nature, with the result that they have remained in error while others have likewise been defrauded of the hope conceived from their promises.[1] These deceptions appear to me to have their principal cause in the belief which these craftsmen have, and continue to hold, in being able to raise very great weights with a small force, as if with their machines they could cheat nature, whose instinct—nay, whose most firm constitution—is that no resistance may be overcome by a force that is not more powerful than it. How false such a belief is, I hope to make most evident with true and rigorous demonstrations that we shall have as we go along.

Meanwhile, since it has been mentioned that the utility which is drawn from machines is not the ability to move with a small force, by means of a machine, those weights which without it could not be moved by the same force, it will not be amiss to

1. According to stories narrated by two of Galileo's biographers who had undoubtedly heard them from his own lips, his move from Pisa to Padua was at least partly the result of having offended a "principal personage" of Tuscany (probably Giovanni de' Medici) by giving an unfavorable opinion on some mechanical contrivance devised by that person and approved by others who had been consulted. If such an event took place, the incident was fresh in Galileo's mind when the *Mechanics* was composed during his early years at Padua.

declare what are the advantages brought to us by this study; for if nothing useful were to be expected from it, all the work employed in its acquisition would be vain.

Taking our start, then, from this consideration, there lie before us at first four things to be considered; the first is the weight to be transferred from one place to another; second is the force or power that must move it; third is the distance between the beginning and the end of the motion; and fourth is the time in which the change must be made—which time comes to the same thing as the swiftness and speed of the motion, that motion being determined to be speedier than another which passes an equal distance in less time. Now assigning any determined resistance, and delimiting any force, and noting any distance, there is no doubt whatever that the given weight will be conducted by the given force to the given distance; for even though the force be very small, by dividing the weight into many particles of which each shall not remain superior to the force, and transferring them one at a time, the whole weight will finally be conducted to the appointed place; nor may it reasonably be said at the end of this operation that this great weight has been moved and translated by a force lesser than itself, but rather by a force which has many times repeated that motion and space which will have been traversed only once by the whole weight. From which it appears that the speed of the force has been greater than the resistance of the weight by as many times as this weight is greater than the force, since in the time in which the moving force has repeatedly traversed the interval between the endpoints of the motion, the thing moved has passed over this but a single time; nor may it therefore be said that a greater resistance has been overcome by a smaller force, against the constitution of nature. One could say that the natural arrangement had been overcome only if the lesser force should transfer the greater resistance with its speed of motion equal to that with which the latter travels—which we absolutely affirm to be impossible to accomplish with any machine imagined or imaginable.[2]

But since it may sometimes happen that, having but a small force, we need to move a great weight all at once without dividing it into pieces, on such an occasion it will be necessary to have

2. The repudiation of the possibility of perpetual motion which, according to Ernst Mach at least, is a basic principle in the formulation of any sound theoretical mechanics, was to Stevin a static principle. With Galileo it took this dynamic form from which the idea of equivalence readily emerged, as seen in the ensuing sentence.

recourse to the machine, by means of which the given weight will be transferred through the assigned space by the given force; yet this does not remove the necessity for that same force to travel and measure that same (or an equal) space as many times as it is exceeded by the said weight. So that at the end of the action we will find that the only profit we have gained from the machine is to have transported the given weight in one piece with the given force to the given end; which weight, divided into pieces, would have been transported without any machine by the same force in the same time through the same distance. And this must be counted as one of the utilities that are derived by the mechanic; for indeed it often happens that, with a paucity of force but not of time, we must move great weights as units. But whoever hopes and attempts by means of machines to gain the same effect without slowing down the movable body will surely be mistaken, and will demonstrate that he does not understand the nature of mechanical instruments and the reasons for their effects.

Another utility is derived from mechanical instruments, which depends upon the place in which the operation must be carried out, for all the instruments are not adapted to all places with equal convenience. Thus we see (to explain by means of an example) that to draw water from a well we make use of a simple cord, with a container suited to receiving and containing water. With this we procure a determined quantity of water in a certain time with our limited force, and whoever might think himself able by machines of any sort to draw with the same force in the same time a greater quantity of water would be seriously in error; and he would be the more mistaken the more varied and complicated the inventions he might go about contriving.[3] Yet we see water extracted with other instruments, as with pumps to empty the holds of ships. But here it must be noted that pumps were not introduced for this purpose because of their carrying a larger quantity of water in the same time by means of the same force as that required for a simple bucket, but only because in such a place the use of a bucket or similar vessel could not accomplish the desired effect, which is to keep the hold dry of even a small quantity of water; and this the bucket cannot do, not being capable of plunging and submerging where there is no appreciable depth

3. Here Galileo probably referred to the fact, well known to practical artisans, that loss of power attends the multiplication of moving parts. But he may have meant only to stress the increase of self-delusion in seeking very roundabout means to fool and defraud nature, a common characteristic of the inventors of perpetual motion machines.

of water. And thus we see winecellars dried with the same instrument, where the water can be removed only obliquely, which would not be done by the ordinary use of a bucket that is raised and lowered perpendicularly by its rope.

The third and perhaps the greatest advantage brought to us by mechanical instruments is with regard to the mover, either some inanimate force like the flow of a river being utilized, or an animate force of much less expense than would be necessary to maintain human power—as when we make use of the flow of a river to turn mills, or the strength (*forza*) of a horse to effect that for which the power of several men would not suffice. In this way we can gain an advantage also in the raising of water or making other strong exertions which doubtless could be carried out by men in the absence of other devices. For men can take water in a simple container, and raise it, and empty it where it is needed; but since a horse or other such mover lacks reason as well as those instruments which are required in order to grasp the container and empty it at the proper time, returning then to refill it, and is endowed only with strength (*forza*), it is necessary for the mechanic to remedy the natural deficiencies of such a mover by supplying artifices and inventions such that with the mere application of its strength it can carry out the desired effect. And in this there is very great utility, not because those wheels or other machines accomplish the transportation of the same weight with less force or greater speed, or through a larger interval, than could be done without such instruments by an equal but judicious and well-organized force, but rather because the fall of a river costs little or nothing, while the maintenance of a horse or similar animal whose power exceeds that of eight or more men is far less expensive than it would be to sustain and maintain so many men.

These, then, are the utilities that are drawn from mechanical instruments, and not those which, to the deception of so many princes and to their own shame, engineers of little understanding go dreaming about when they apply themselves to impossible undertakings. Of this we shall be assured both by the little that has been hinted thus far, and by much more which will be demonstrated in the course of this treatise, if we but attentively heed what is to be said.

Of the Steelyard and of the Lever

Now that we have understood through a sure proof one of the prime principles from which, as from a most fertile source, many of the mechanical instruments derive, we will be able to acquire a knowledge of the nature of these without any difficulty whatever.

And first, speaking of the steelyard,[15] a very widely used instrument with which various kinds of merchandise are weighed, even though extremely heavy, by the weight of a small counterpoise (commonly called the *romano*), we shall prove that in such operations nothing more is done than to reduce to a practical act precisely that on which we have theorized above. For let us suppose *AB* to be a steelyard whose support (called *trutina*) is at the point *C*, near which at a small distance *CA* hangs the heavy weight *D*; and along the greater distance *CB* (called the *ago* of the steelyard) there runs back and forth the *romano E*, of small weight in comparison with the heavy body *D*, yet nevertheless capable of moving far enough from the *trutina C* so that the proportion existing between the two weights *D* and *E* may exist between the distances *FC* and *CA*; and then equilibrium will be made, unequal weights being found hanging at distances inversely proportional to them.

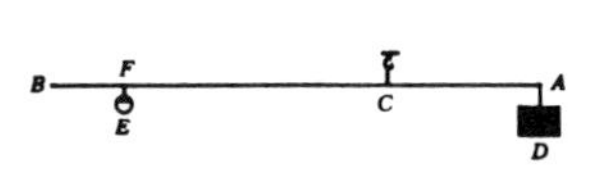

Nor is this instrument different from that other called the *vette*, or commonly the lever, with which very large stones and other weights are moved with a small force. Its application is according to the next diagram where the lever is denoted by the bar *BCD*, of wood or other solid material; the heavy weight to be raised is *A*, and a firm support or fulcrum upon which the lever presses and moves is designated *E*. Placing one end of the lever under the weight *A*, as is seen at the point *B*, the force weighing down at the other end *D*, though small, will be able to raise the weight *A*, so long as the proportion of the distance *BC* to *CD* exists between the force placed at *D* and the resistance made by the heavy body *A* upon the point *B*. From which it is made clear that the more closely the support *E* approaches the extremity *B*, increasing the proportion of the distance *DC* to the distance *CB*, the more the force at *D* may be diminished in raising the weight *A*.

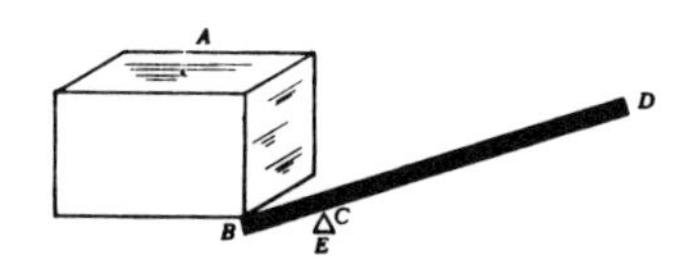

15. *Stadera*, also called the Roman balance; Galileo's terms for its parts are used untranslated below. Guido's translator called the counterpoise *marco* and the arm of the balance *fusto*.

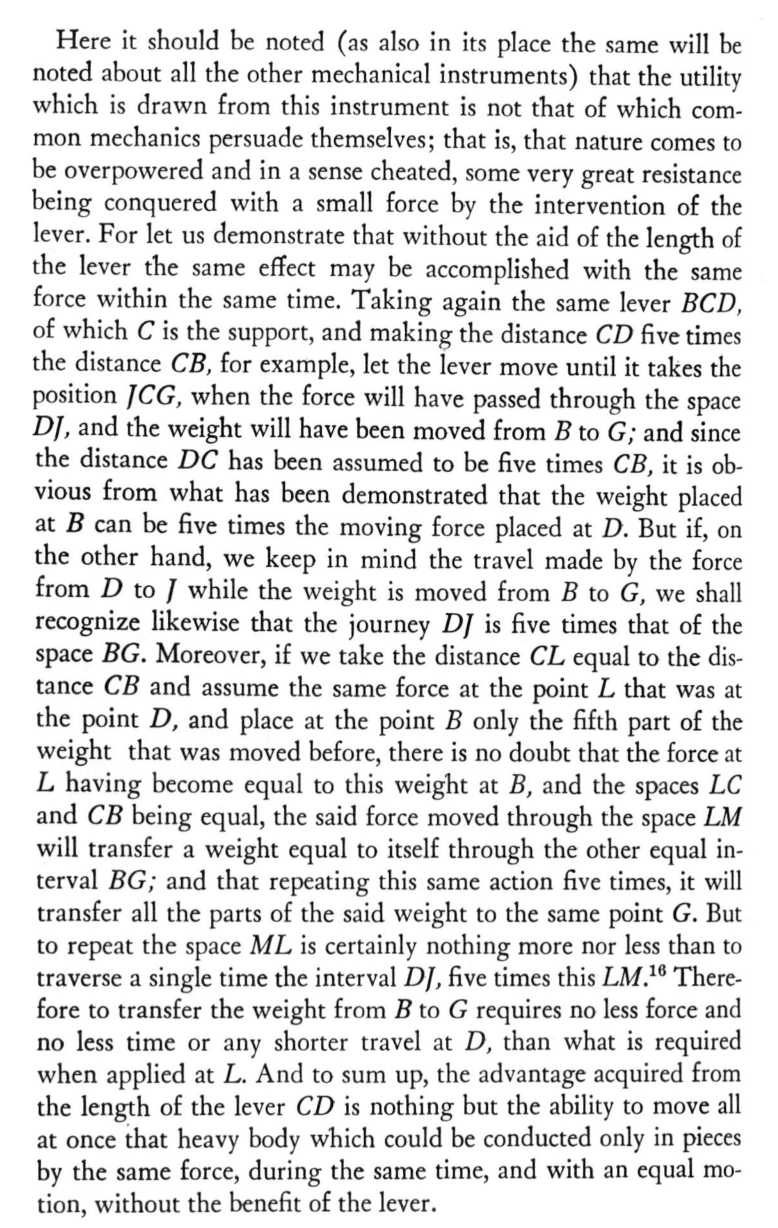

Here it should be noted (as also in its place the same will be noted about all the other mechanical instruments) that the utility which is drawn from this instrument is not that of which common mechanics persuade themselves; that is, that nature comes to be overpowered and in a sense cheated, some very great resistance being conquered with a small force by the intervention of the lever. For let us demonstrate that without the aid of the length of the lever the same effect may be accomplished with the same force within the same time. Taking again the same lever *BCD*, of which *C* is the support, and making the distance *CD* five times the distance *CB*, for example, let the lever move until it takes the position *JCG*, when the force will have passed through the space *DJ*, and the weight will have been moved from *B* to *G;* and since the distance *DC* has been assumed to be five times *CB*, it is obvious from what has been demonstrated that the weight placed at *B* can be five times the moving force placed at *D*. But if, on the other hand, we keep in mind the travel made by the force from *D* to *J* while the weight is moved from *B* to *G*, we shall recognize likewise that the journey *DJ* is five times that of the space *BG*. Moreover, if we take the distance *CL* equal to the distance *CB* and assume the same force at the point *L* that was at the point *D*, and place at the point *B* only the fifth part of the weight that was moved before, there is no doubt that the force at *L* having become equal to this weight at *B*, and the spaces *LC* and *CB* being equal, the said force moved through the space *LM* will transfer a weight equal to itself through the other equal interval *BG;* and that repeating this same action five times, it will transfer all the parts of the said weight to the same point *G*. But to repeat the space *ML* is certainly nothing more nor less than to traverse a single time the interval *DJ*, five times this *LM*.[16] Therefore to transfer the weight from *B* to *G* requires no less force and no less time or any shorter travel at *D*, than what is required when applied at *L*. And to sum up, the advantage acquired from the length of the lever *CD* is nothing but the ability to move all at once that heavy body which could be conducted only in pieces by the same force, during the same time, and with an equal motion, without the benefit of the lever.

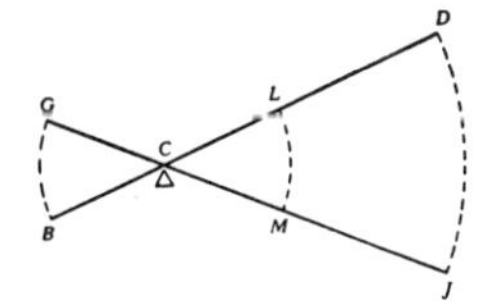

9B

Reprinted from *On Motion and On Mechanics,* S. Drake, trans., I. E. Drabkin and S. Drake, eds., University of Wisconsin Press, Madison, Wis., 1960, pp. 179–182

OF THE FORCE OF PERCUSSION

Galileo Galilei

To investigate what is the cause of the force of percussion is most necessary for many reasons. First, because in this there appears to be something much more marvelous than that which is perceived in any other mechanical instrument; for when a nail is struck to fix it in some very hard wood, or a stick that must penetrate into very firm ground, each is driven forward by the force of the percussion alone; without which, if we were to place the hammer on either, it would not move, nor would it do so even if a much heavier weight than the hammer were so placed.[33] This effect is truly marvelous, and is the more worthy of speculation in my opinion because, so far as I know, none of those who have philosophized about it before us has hit the mark, which we may take as a certain and sure sign of the obscurity and difficulty of such speculation. For as to Aristotle and others who have tried to reduce the cause of this remarkable effect to the length of the handle of the hammer, it seems to me that without any long argument one can expose the weakness of their reasoning by the effect of those instruments which, without having any handle, give percussion either by falling from on high or by being driven speedily from the side.[34] Therefore we must have recourse to

33. *Mechanical Problems,* sec. 19: "Why is it that if one puts a large axe on a block of wood and a heavy weight on top of it, it does not cut the wood to any extent; but if one raises the axe and strikes with it, it splits it in half, even if the striker has far less weight than one placed on it and pressing it down? Is it because all work is produced by movement, and a heavy object produces the movement of weight more when it is moving than when it is at rest?"

34. It is not quite fair to attribute this remark to Aristotle, or even to the Aristotelian writer of the *Mechanical Problems,* whose reference to long handles was made in quite another connection (sec. 13). It was Guido who introduced

some other principle if we want to find the truth of this matter. And though the cause is by its nature somewhat abstruse and difficult to explain, yet we shall attempt with as much lucidity as we can command to render it clear and palpable, showing in the end that the basis and origin of the effect derives from no other source than that from which the causes of other mechanical effects have sprung.

Now this will be done by keeping before our eyes that which has been seen to happen in all other mechanical operations, which is that the force, the resistance, and the space through which the motion is made respectively follow that proportion and obey those laws by which a resistance equal to the force will be moved by this force through an equal space and with equal velocity to that of the mover. Likewise, a force that is one-half of a resistance will be able to move it, provided the former moves with double the velocity, or let us say through twice as great a distance, as that passed through by the resistance moved. And in brief, it is seen in all other instruments that any great resistance may be moved by any given little force, provided that the space through which this force is moved shall have to the space through which the resistance shall be moved that proportion which exists between this large resistance and the small force; and this is by the necessary constitution of nature. Whence, turning the argument about and arguing by the converse, where will be the marvel if that power which would move a small resistance through a large interval should drive one a hundred times greater through one-hundredth of the said interval? No wonder, certainly; and for things to be otherwise would be not only absurd, but impossible.[35]

We shall consider, therefore, the resistance to being moved that exists in the hammer at the point where it goes to strike, and how far it would be driven by the received force if it did not strike; and moreover what is the resistance to being moved of that which it strikes, and how much it is moved by a single

---

this idea explicitly, deducing it from a general rule used in the Aristotelian work (*Le Mechaniche,* ff. 113v.–114r.).

35. This bold application of a general principle of physics obtained by induction led Galileo close to a solution of the problem; but in his attempt to measure the moving force he became ensnared in difficulties, as the next paragraph discloses. In later years he came to regard the force of percussion as infinite, but he never completed and published his researches, which in their unfinished form make up the so-called sixth day intended for the *Two New Sciences* (*Opere,* VIII, 321–46).

stroke. And having found how far this great resistance goes forward under one stroke, by however much this is less than the distance through which the hammer would go by the impetus (*empito*) of what moves it, by so much is this resistance greater than that of the hammer; and so the wonder of the effect ceases in us, since it stems only from the terms of natural arrangements and from what has already been said.[36]

Let there be added, for better comprehension, an example in particular terms. There is a hammer which, having four of resistance, is moved by a force such that if it were set free from this at that point where the stroke is made, it would go ten paces beyond if it met with no obstruction. But let it rather be opposed there by a great beam whose resistance to motion is as four thousand; that is, one thousand times as great as that of the hammer (and yet not immovable, so that it would overcome the resistance of the hammer beyond all proportion). Now the percussion being made on this, it will indeed be driven forward, but by the thousandth part of the ten paces in which the hammer would be moved. And thus, reflecting with a converse method that which has been theorized on other mechanical effects, we may investigate the cause of the force of percussion.[37]

I know that this will give rise in some to difficulties and objections, which however there are means to remove with little trouble, and these we willingly leave among the mechanical problems which will be appended at the end of this discourse.[38]

36. The "resistance to being moved" is certainly a primitive mass concept, which Galileo avoids measuring in terms of weight because he is contemplating horizontal motion. The next concept, however, presents real difficulties. How far would the hammer go if set free? Galileo seems to have been thinking only of the empirical fact that a thrown hammer will fall to the ground at a distance somehow proportional to the force given it by the thrower. It is curious that he did not revert to the idea of velocity here, as in the discussion of the steelyard, especially after having remarked that no force is required to carry a body horizontally.

37. The attempt to supply a model calculation only makes matters worse. But considering that Galileo was not given to circulating in writing conjectures that were devoid of internal consistency in his mind, let us attempt to reconstruct a possible line of thought he may have followed.

The hammer is given four units of resistance; that is, of some property which also exists and can be measured in the other object that is to be moved. Hence in saying that the hammer will go ten paces if unobstructed, he ought to have considered it as moving this hypothetical distance under conditions similar to those governing the thing struck. If he did so, we may forgive him his utter incapacity to specify such conditions, as he seems to have been considering a large beam resting on the ground. And for theoretical purposes he now had a legitimate work concept. But it is unfortunate that he did not take both his beam and his hammer off the ground and suspend them in trusses, or place them on his frictionless frozen lake, in which case he would have been forced to remove the ambiguities and contradictions in his exposition; and if it then did not result in what he set out to accomplish, we might have had from him instead a discussion of the laws of impact.

38. The *Problems of Mechanics* here promised by Galileo have never been found.

9C

Reprinted from *Dialogues Concerning Two New Sciences,* H. Crew and A. De Salvio, trans., Northwestern University Press, Evanston, Ill., 1939, pp. 169–172

THE PENDULUM EXPERIMENT

Galileo Galilei

SALV. This definition established, the Author makes a single assumption, namely,

> The speeds acquired by one and the same body moving down planes of different inclinations are equal when the heights of these planes are equal.

By the height of an inclined plane we mean the perpendicular let fall from the upper end of the plane upon the horizontal line drawn through the lower end of the same plane. Thus, to illustrate, let the line AB be horizontal, and let the planes CA and CD be inclined to it; then the Author calls the perpendicular CB the "height" of the planes CA and CD; he supposes that the speeds acquired by one and the same body, descending along the planes CA and CD to the terminal points A and D are equal since the heights of these planes are the same, CB; and also it must be understood that this speed is that which would be acquired by the same body falling from C to B.

Sagr.

SAGR. Your assumption appears to me so reasonable that it ought to be conceded without question, provided of course there are no chance or outside resistances, and that the planes are hard and smooth, and that the figure of the moving body is perfectly round, so that neither plane nor moving body is rough. All resistance and opposition having been removed, my reason tells me at once that a heavy and perfectly round ball descending along the lines CA, CD, CB would reach the terminal points A, D, B, with equal momenta [*impeti eguali*].

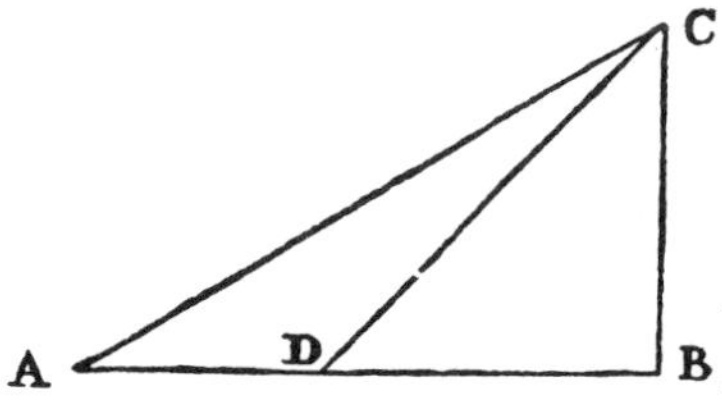

Fig. 45

SALV. Your words are very plausible; but I hope by experiment to increase the probability to an extent which shall be little short of a rigid demonstration.

[206]

Imagine this page to represent a vertical wall, with a nail driven into it; and from the nail let there be suspended a lead bullet of one or two ounces by means of a fine vertical thread, AB, say from four to six feet long, on this wall draw a horizontal line DC, at right angles to the vertical thread AB, which hangs about two finger-breadths in front of the wall. Now bring the thread AB with the attached ball into the position AC and set it free; first it will be observed to descend along the arc CBD, to pass the point B, and to travel along the arc BD, till it almost reaches the horizontal CD, a slight shortage being caused by the resistance of the air and the string; from this we may rightly infer that the ball in its descent through the arc CB acquired a momentum [*impeto*] on reaching B, which was just sufficient to carry it through a similar arc BD to the same height. Having repeated this experiment many times, let us now drive a nail into the wall close to the perpendicular AB, say at E or F, so that it projects out some five or six finger-breadths in order that the thread, again carrying the bullet through the arc CB, may strike upon the nail E when the bullet reaches B, and thus compel it to traverse the arc BG, described about E as center. From this we we can see what can be done by the same momentum [*impeto*] which previously starting at the same point B carried the same body through the arc BD to the horizontal CD. Now, gentlemen, you will observe with pleasure that the ball swings to the point G in the horizontal, and you would see the same thing happen if the obstacle were placed at some lower point, say at F, about which the ball would describe the arc BI, the rise of the

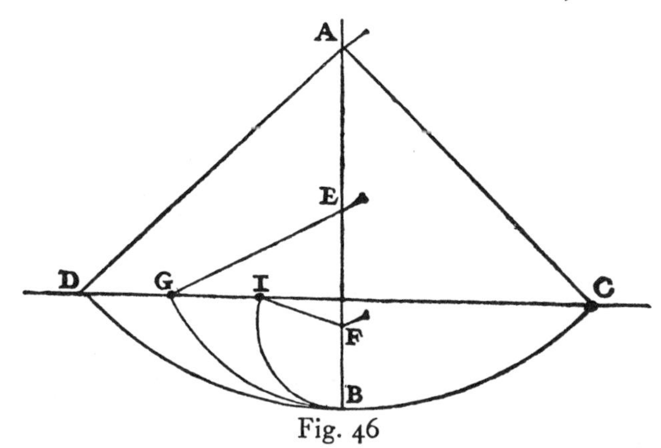

Fig. 46

ball always terminating exactly on the line CD. But when the nail is placed so low that the remainder of the thread below it will not reach to the height CD (which would happen if the nail were placed nearer B than to the intersection of AB with the [207] horizontal CD) then the thread leaps over the nail and twists itself about it.

This experiment leaves no room for doubt as to the truth of our supposition; for since the two arcs CB and DB are equal and similarly placed, the momentum [*momento*] acquired by the fall through the arc CB is the same as that gained by fall through the arc DB; but the momentum [*momento*] acquired at B, owing to fall through CB, is able to lift the same body [*mobile*] through the arc BD; therefore, the momentum acquired in the fall BD is equal to that which lifts the same body through the same arc from B to D; so, in general, every momentum acquired by fall through

through an arc is equal to that which can lift the same body through the same arc. But all these momenta [*momenti*] which cause a rise through the arcs BD, BG, and BI are equal, since they are produced by the same momentum, gained by fall through CB, as experiment shows. Therefore all the momenta gained by fall through the arcs DB, GB, IB are equal.

SAGR. The argument seems to me so conclusive and the experiment so well adapted to establish the hypothesis that we may, indeed, consider it as demonstrated.

SALV. I do not wish, Sagredo, that we trouble ourselves too much about this matter, since we are going to apply this principle mainly in motions which occur on plane surfaces, and not upon curved, along which acceleration varies in a manner greatly different from that which we have assumed for planes.

So that, although the above experiment shows us that the descent of the moving body through the arc CB confers upon it momentum [*momento*] just sufficient to carry it to the same height through any of the arcs BD, BG, BI, we are not able, by similar means, to show that the event would be identical in the case of a perfectly round ball descending along planes whose inclinations are respectively the same as the chords of these arcs. It seems likely, on the other hand, that, since these planes form angles at the point B, they will present an obstacle to the ball which has descended along the chord CB, and starts to rise along the chord BD, BG, BI.

In striking these planes some of its momentum [*impeto*] will be lost and it will not be able to rise to the height of the line CD; but this obstacle, which interferes with the experiment, once removed, it is clear that the momentum [*impeto*] (which gains

[208]

in strength with descent) will be able to carry the body to the same height. Let us then, for the present, take this as a postulate, the absolute truth of which will be established when we find that the inferences from it correspond to and agree perfectly with experiment.

Editor's Comments on Paper 10

Conservation of Quantity of Motion
10 R. DESCARTES

René Descartes (1596–1650), the distinguished French mathematician, scientist, and philosopher of the first part of the seventeenth century, made perhaps his most important contribution in the invention of analytic geometry. But he was also deeply interested in physics and, in particular, in the principles of mechanics. He devoted particular attention to the collisions of bodies and was aware of the conservation of momentum (mass times velocity, called by him "quantity of motion") exhibited in such collisions. This undoubtedly stimulated his interest in the general notion of conservation. Looking for something that might stay constant in the midst of the multifarious motions encountered in human experience, he decided quite plausibly that quantity of motion is the thing conserved. In his elaborate work, *Principles of Philosophy*, from which we reproduce a short extract from Part II, Section 36, Descartes asserts his belief that God had so arranged the universe that the total quantity of motion would forever remain constant.

From this it was but a step for Descartes and his followers to conclude that force as a cause of motion and change in motion should properly be measured by change (or rate of change) of quantity of motion or momentum. This may well have had some influence on Newton when he came to write his *Mathematical Principles of Natural Philosophy*. But it certainly stirred up considerable controversy when in 1686 Leibniz concluded that he had proved that it is really *vis viva*, or the product of mass times the square of the velocity, which is conserved in motion and which should therefore better serve as the measure of force. We present Leibniz's views in Papers 12A and 12B.

The essential importance of Descartes' view was his emphasis on conservation, on constancy in the midst of change. Logically, there would seem to be no reason why this type of constancy should not have become the fundamental one in physics. The reason why the constancy involved in the energy concept ultimately triumphed will become evident in succeeding articles.

10

Reprinted from *A Source Book in Physics,* W. F. Magie, ed.,
McGraw-Hill Book Company, New York, 1935, pp. 50–51

CONSERVATION OF QUANTITY OF MOTION

René Descartes

René Descartes was born in La Haye on March 31, 1596. He was of a family of wealth and position in the country. He received his first education in the school of La Flèche, where he remained eight years. His studies covered a wide range, though he was especially interested in mathematics. After leaving school in 1612 he spent some years in Paris, preparing himself for military life. He there met some of the distinguished mathematicians of his time and was led gradually to withdraw from social life and to devote himself to intensive study. He served for a while in Breda as a volunteer in the army of the Statthalder, Maurice of Orange, and later in the Bavarian army and in other military services. He abandoned the military life and returned home at the age of twenty-five. After various journeys and adventures he went to Holland with the intention of devoting himself to the study of philosophy. The system of philosophy which he developed in the twenty years of his life in Holland had an immense influence upon the thought of the time. In so far as it dealt with the system of the universe it was finally overthrown by the Newtonian system. In 1649 Descartes accepted an invitation from the Queen of Sweden to visit Stockholm and instruct her in his philosophy. The climate was unfavorable for his health and he died in Stockholm on February 11, 1650.

For many years there was a discussion of the question whether the quantity of motion was to be measured by the product of mass and velocity or by the product of mass and the square of velocity. The first view was that of Descartes, the second of Leibnitz. The short extract from Descartes' *Principles of Philosophy*, part II, Section 36, presents Descartes' view that motion in the sense in which he defined it always remains the same or is conserved.

Quantity of Motion

§ 36. Now that we have examined the nature of motion, we come to consider its cause, and since the question may be taken in two ways, we shall commence by the first and more universal way, which produces generally all the motions which are in the world; we shall consider afterwards the other cause, which makes each portion of matter acquire motion which it did not have before. As for the first cause, it seems to me evident that it is nothing other than God, Who by His Almighty power created matter with motion and rest in its parts, and Who thereafter conserves in the universe by His ordinary operations as much of motion and of rest as He put in it in the first creation. For while it is true that motion is only the behavior of matter which is moved, there is, for all that, a quantity of it which never increases nor diminishes, although there is sometimes more and sometimes less of it in some of its parts; it is for this reason that when a part of matter moves twice as rapidly as another part, and this other part is twice as great as the first part, we have a right to think that there is as much motion in the smaller body as in the larger, and that every time and by as much as the motion of one part diminishes that of some other part increases in proportion. We also know that it is one of God's perfections not only to be immutable in His nature but also to act in a way which never changes: to such a degree that besides the changes that we see in the world, and those that we believe in because God has revealed them, and that we know have happened in nature without any change on the part of the Creator, we ought not to attribute to Him in His works any other changes for fear of attributing to Him inconstancy; from which it follows that, since He set in motion in many different ways the parts of matter when He created them and since He maintained them with the same behavior and with the same laws as He laid upon them in their creation, He conserves continually in this matter an equal quantity of motion.

Editor's Comments on Paper 11

The Third Law of Motion As a Forerunner of the Concept of Energy
11 I. NEWTON

Isaac Newton (1642–1727) needs no introduction since his place in the pantheon of science is secure. His contributions to mathematics, mechanics, astronomy, heat, optics, and acoustics mark him as one of the greatest natural philosophers who ever lived. Since Newton's famous *Mathematical Principles of Natural Philosophy* (1686) deals extensively with the motion of bodies, one would expect to find in it some vestige of the concept of energy. It is rather strange that although Newton stated the laws of motion he never wrote down the second law as a differential equation governing the motion of a particular system. If he had done so, it seems almost certain that he would have been led to what we now call the energy equation of the system, as was his great successor Lagrange 100 years later. On the other hand, Newton did not seem to be very much interested in the properties of machines, evidently considering them a concern for artisans. However, he does mention machines briefly in a scholium at the end of the introductory section on the laws of motion. From his statements it is clear that he realizes the role of the principle of virtual velocities in their operation.

We reproduce the relevant extract in the translation of Andrew Motte. In their famous *Treatise on Natural Philosophy* (Vol. 1, Cambridge University Press, 1879, p. 247ff), Kelvin and Tait felt that in the final two sentences of the scholium Newton provided the firm basis for both the dynamical principle of d'Alembert and the energy equation of Lagrange.

11

Reprinted from *A Source Book in Physics,* W. F. Magie, ed.,
McGraw-Hill Book Company, New York, 1935, pp. 39–46

THE THIRD LAW OF MOTION AS A FORERUNNER OF THE CONCEPT OF ENERGY

Isaac Newton

For action and its opposite re-action are equal, by Law III, and therefore, by Law II, they produce in the motions equal changes towards opposite parts. Therefore if the motions are directed towards the same parts, whatever is added to the motion of the preceding body will be subducted from the motion of that which follows; so that the sum will be the same as before. If the bodies meet, with contrary motions, there will be an equal deduction from the motions of both; and therefore the difference of the motions directed towards opposite parts will remain the same.

. . . .

Corollary IV

The common centre of gravity of two or more bodies does not alter its state of motion or rest by the actions of the bodies among themselves; and therefore the common centre of gravity of all bodies acting upon each other (excluding outward actions and impediments) is either at rest, or moves uniformly in a right line.

. . . .

Corollary V

The motions of bodies included in a given space are the same among themselves, whether that space is at rest, or moves uniformly forwards in a right line without any circular motion.

. . . .

Corollary VI

If bodies, any how moved among themselves, are urged in the direction of parallel lines by equal accelerative forces, they will all continue to move among themselves, after the same manner as if they had been urged by no such forces.

. . . .

Scholium

Hitherto I have laid down such principles as have been received by mathematicians, and are confirmed by abundance of experiments. By the first two Laws and the first two Corollaries, Galileo discovered that the descent of bodies observed the duplicate ratio of the time, and that the motion of projectiles was in the curve of a parabola; experience agreeing with both, unless so far as these motions are a little retarded by the resistance of the air. When a body is falling, the uniform force of its gravity acting equally, impresses, in equal particles of time, equal forces upon that body, and therefore generates equal velocities; and in the whole time impresses a whole force, and generates a whole velocity proportional to the time. And the spaces described in proportional times are as the velocities and the times conjunctly; that is, in a duplicate ratio of the times. And when a body is thrown upwards, its uniform gravity impresses forces and takes off velocities proportional to the times; and the times of ascending to the greatest heights are as the velocities to be taken off, and those heights are as the velocities and the times conjunctly, or in the duplicate ratio of the velocities. And if a body be projected in any direction, the motion arising from its projection is compounded with the motion arising from its gravity. As if the body *A* (Fig. 16) by its motion of projection alone could describe in a given time the right line *AB*, and with its motion of falling alone would describe in the same time the altitude *AC*; complete the parallelogram *ABCD*, and the body by that compounded motion will at the end of the time be found in the place *D*; and the curve line *AED*, which that body describes, will be a parabola, to which the right line *AB* will be a tangent in *A*; and whose ordinate *BD* will be as the square of the line *AB*. On the same Laws and Corollaries depend those things which have been demonstrated concerning the times of the vibration of pendulums, and are confirmed by the daily experiments of pendulum clocks. By the same, together with the third Law, Sir Christ. Wren, Dr. Wallis, and Mr. Huygens, the greatest geometers of our times, did severally determine the rules of the congress and reflexion of hard bodies, and much about the same time communicated their discoveries to the Royal Society, exactly agreeing among themselves as to those rules. Dr. Wallis, indeed, was something more early in the publication; then followed Sir Christopher Wren, and, lastly, Mr. Huygens. But Sir Christopher Wren confirmed the truth of the thing before the Royal Society by the experiment of pendulums, which Mr. Mariotte soon after thought fit to explain in a treatise entirely upon that subject. But to bring this experiment to an accurate agreement with the theory, we are to have a due regard as well to the resistance of the air as to the elastic force of the concurring

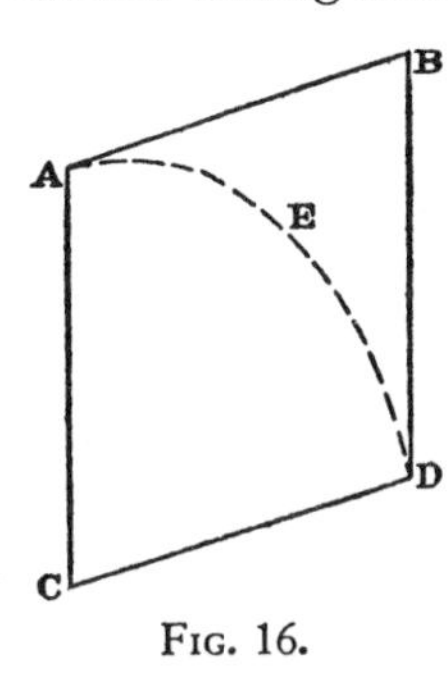

Fig. 16.

bodies. Let the spherical bodies *A*, *B* (Fig. 17) be suspended by the parallel and equal strings *AC*, *BD*, from the centres *C*, *D*. About these centres, with those intervals, describe the semicircles *EAF*, *GBH*, bisected by the radii *CA*, *DB*. Bring the body *A* to any point *R* of the arc *EAF*, and (withdrawing the body *B*) let it go from thence, and after one oscillation suppose it to return to the point *V*: then *RV* will be the retardation arising from the resistance of the air. Of this *RV* let *ST* be a fourth part, situated in the middle, to wit, so as *RS* and *TV* may be equal, and *RS* may be to *ST* as 3 to 2: then will *ST* represent very nearly the retardation during the descent from *S* to *A*. Restore the body *B* to its

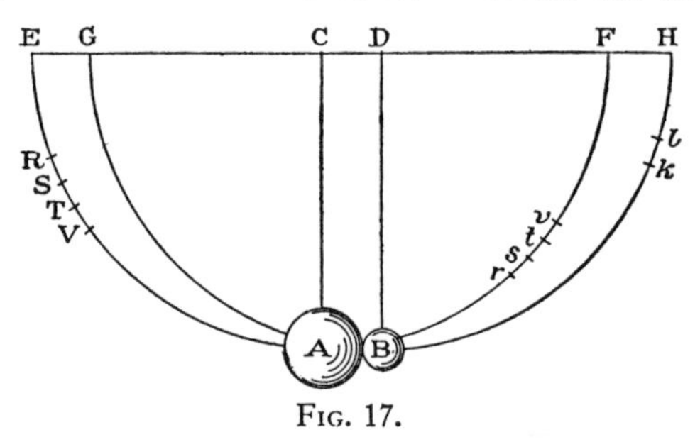

Fig. 17.

place: and, supposing the body *A* to be let fall from the point *S*, the velocity thereof in the place of reflexion *A*, without sensible error, will be the same as if it had descended *in vacuo* from the point *T*. Upon which account this velocity may be represented by the chord of the arc *TA*. For it is a proposition well known to geometers, that the velocity of a pendulous body in the lowest point is as the chord of the arc which it has described in its descent. After reflexion, suppose the body *A* comes to the place *s*, and the body *B* to the place *k*. Withdraw the body *B*, and find the place *v*, from which if the body *A*, being let go, should after one oscillation return to the place *r*, *st* may be a fourth part of *rv*, so placed in the middle thereof as to leave *rs* equal to *tv*, and let the chord of the arc *tA* represent the velocity which the body *A* had in the place *A*. immediately after reflexion. For *t* will be the true and correct place to which the body *A* should have ascended, if the resistance of the air had been taken off. In the same way we are to correct the place *k* to which the body *B* ascends, by finding the place *l* to which it should have ascended *in vacuo*. And thus everything may be subjected to experiment, in the same manner as if we were really placed *in vacuo*. These things being done, we are to take the product (if I may so say) of the body *A*, by the chord of the arc *TA* (which represents its velocity), that we may have its motion in the place *A* immediately before reflexion; and then by the chord of the arc *tA*, that we may have its motion in the place *A* immediately after reflexion. And so we are to take the product of the body *B* by the chord of the arc *Bl*, that we may have the motion of the same immediately after reflexion. And in like manner, when two bodies are let go together from different places, we are to find the motion of each, as well before as after reflexion; and then we may compare the motions between themselves, and collect the effects of the reflexion. Thus trying the thing with pendulums of ten feet, in unequal as well as equal bodies, and making the bodies to concur after a descent through large spaces, as of 8, 12, or 16 feet, I found always, without an error of 3 inches, that when the bodies concurred together directly, equal changes towards the contrary parts were produced in their motions, and, of consequence, that the action and reaction were always equal. As if the body *A* impinged upon the body *B* at rest with 9 parts of motion, and losing 7, proceeded after reflexion with 2, the body *B* was carried backwards with those 7 parts. If the bodies concurred with contrary motions, *A* with twelve parts of motion, and *B* with six, then if *A* receded with 2, *B* receded with 8; to wit, with a deduction of 14 parts of motion on each side. For from the motion of *A* subducting twelve parts, nothing will remain; but subducting 2 parts more, a motion will be generated of 2 parts towards the contrary way; and so, from the motion of the body *B* of 6 parts, subducting 14 parts, a motion is generated of 8 parts towards the contrary way. But if the bodies were made both to move towards the same way, *A*, the swifter, with 14 parts of motion, *B*, the slower, with 5, and after reflexion *A* went on with 5, *B* likewise went on with 14 parts; 9 parts being transferred from *A* to *B*. And so in other cases. By the congress and collision of bodies, the quantity of motion, collected from the sum of the motions directed towards the same way, or from the difference of those that were directed towards contrary ways, was never changed. For the error of an inch or two in measures may be easily ascribed to the difficulty of executing everything with accuracy. It was not easy to let go the two pendulums so exactly together that the bodies should impinge one upon the other in the lowermost place *AB*; nor to mark the places *s*, and *k*, to which the bodies ascended after congress. Nay, and some errors, too, might have happened from the unequal density of the parts of the pendulous bodies themselves, and from the irregularity of the texture proceeding from other causes.

But to prevent an objection that may perhaps be alledged against the rule, for the proof of which this experiment was made, as if this rule did suppose that the bodies were either absolutely hard, or at least perfectly elastic (whereas no such bodies are to be found in nature), I must add, that the experiments we have been describing, by no means depending upon that quality of hardness, do succeed as well in soft as in hard bodies. For if the rule is to be tried in bodies not perfectly hard, we are only to diminish the reflexion in such a certain proportion as the quantity of the elastic force requires. By the theory of Wren and Huygens, bodies absolutely hard return one from another with the same velocity with which they meet. But this may be affirmed with more certainty of bodies perfectly elastic. In bodies imperfectly elastic the velocity of the return is to be diminished together with the elastic force; because that force (except when the parts of bodies are bruised by their congress, or suffer some such extension as happens under the strokes of a hammer) is (as far as I can perceive) certain and determined, and makes the bodies to return one from the other with a relative velocity, which is in a given ratio to that relative velocity with which they met. This I tried in balls of wool, made up tightly, and strongly compressed. For, first, by letting go the pendulous bodies, and measuring their reflexion, I determined the quantity of their elastic force; and then, according to this force, estimated the reflexions that ought to happen in other cases of congress. And with this computation other experiments made afterwards did accordingly agree; the balls always receding one from the other with a relative velocity, which was to the relative velocity with which they met as about 5 to 9. Balls of steel returned with almost the same velocity: those of cork with a velocity something less; but in balls of glass the proportion was as about 15 to 16. And thus the third Law, so far as it regards percussions and reflexions, is proved by a theory exactly agreeing with experience.

In attractions, I briefly demonstrate the thing after this manner. Suppose an obstacle is interposed to hinder the congress of any two bodies A, B, mutually attracting one the other: then if either body, as A, is more attracted towards the other body B, than that other body B is towards the first body A, the obstacle will be more strongly urged by the pressure of the body *A* than by the pressure of the body *B*, and therefore will not remain in equilibrio: but the stronger pressure will prevail, and will make the system of the two bodies, together with the obstacle, to move directly towards the parts on which *B* lies; and in free spaces, to go forward *in infinitum* with a motion perpetually accelerated; which is absurd and contrary to the first Law. For, by the first Law, the system ought to persevere in its state of rest, or of moving uniformly forward in a right line; and therefore the bodies must equally press the obstacle, and be equally attracted one by the other. I made the experiment on the loadstone and iron. If these, placed apart in proper vessels, are made to float by one another in standing water, neither of them will propel the other; but, by being equally attracted, they will sustain each other's pressure, and rest at last in an equilibrium.

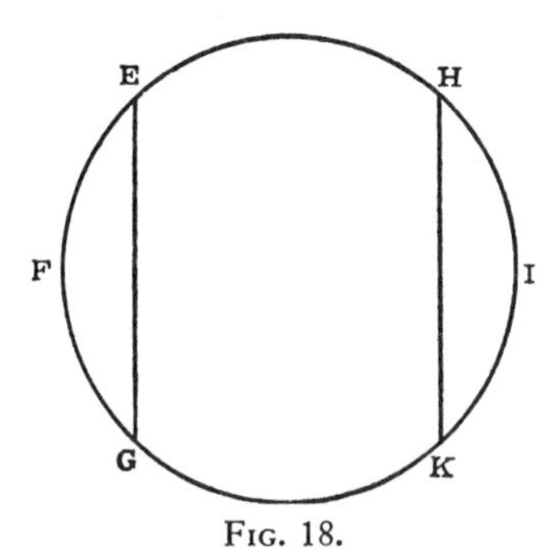

FIG. 18.

So the gravitation betwixt the earth and its parts is mutual. Let the earth *FI* (Fig. 18) be cut by any plane *EG* into two parts *EGF* and *EGI*, and their weights one towards the other will be mutually equal. For if by another plane *HK*, parallel to the former *EG*, the greater part *EGI* is cut into two parts *EGKH* and *HKI*, whereof *HKI* is equal to the part *EFG*, first cut off, it is evident that the middle part *EGKH*, will have no propension by its proper weight towards either side, but will hang as it were, and rest in an equilibrium betwixt both. But the one extreme part *HKI* will with its whole weight bear upon and press the middle part towards the other extreme part *EGF*; and therefore the force with which *EGI*, the sum of the parts *HKI* and *EGKH*, tends towards the third part *EGF*, is equal to the weight of the part *HKI*, that is, to the weight of the third part *EGF*. And therefore the weights of the two parts *EGI* and *EGF*, one towards the other, are equal, as I was to prove. And indeed if those weights were not equal, the whole earth floating in the non-resisting aether would give way to the greater weight, and, retiring from it, would be carried off *in infinitum*.

And as those bodies are equipollent in the congress and reflexion, whose velocities are reciprocally as their innate forces, so in the use of mechanic instruments those agents are equipollent, and mutually sustain each the contrary pressure of the other, whose velocities, estimated according to the determination of the forces, are reciprocally as the forces.

So those weights are of equal force to move the arms of a balance; which during the play of the balance are reciprocally as their velocities upwards and downwards; that is, if the ascent or descent is direct, those weights are of equal force, which are reciprocally as the distances of the points at which they are suspended from the axis of the balance; but if they are turned aside by the interposition of oblique planes, or other obstacles, and made to ascend or descend obliquely, those bodies will be equipollent, which are reciprocally as the heights of their ascent and descent taken according to the perpendicular; and that on account of the determination of gravity downwards.

And in like manner in the pully, or in a combination of pullies, the force of a hand drawing the rope directly, which is to the weight, whether ascending directly or obliquely, as the velocity of the perpendicular ascent of the weight to the velocity of the hand that draws the rope, will sustain the weight.

In clocks and such like instruments, made up from a combination of wheels, the contrary forces that promote and impede the motion of the wheels, if they are reciprocally as the velocities of the parts of the wheel on which they are impressed, will mutually sustain the one the other.

The force of tne screw to press a body is to the force of the hand that turns the handles by which it is moved as the circular velocity of the handle in that part where it is impelled by the hand is to the progressive velocity of the screw towards the pressed body.

The forces by which the wedge presses or drives the two parts of the wood it cleaves are to the force of the mallet upon the wedge as the progress of the wedge in the direction of the force impressed upon it by the mallet is to the velocity with which the parts of the wood yield to the wedge, in the direction of lines perpendicular to the sides of the wedge. And the like account is to be given of all machines.

The power and use of machines consist only in this, that by diminishing the velocity we may augment the force, and the contrary: from whence, in all sorts of proper machines, we have the Solution of this problem: *To move a given weight with a given power,* or with a given force to overcome any other given resistance.

For if machines are so contrived that the velocities of the agent and resistant are reciprocally as their forces, the agent will just sustain the resistant, but with a greater disparity of velocity will overcome it. So that if the disparity of velocities is so great as to overcome all that resistance which commonly arises either from the attrition of contiguous bodies as they slide by one another, or from the cohesion of continuous bodies that are to be separated, or from the weights of bodies to be raised, the excess of the force remaining, after all those resistances are overcome, will produce an acceleration of motion proportional thereto, as well in the parts of the machine as in the resisting body. But to treat of mechanics is not my present business. I was only willing to show by those examples the great extent and certainty of the third Law of motion. For if we estimate the action of the agent from its force and velocity conjunctly, and likewise the reaction of the impediment conjunctly from the velocities of its several parts, and from the forces of resistance arising from the attrition, cohesion, weight, and acceleration of those parts, the action and reaction in the use of all sorts of machines will be found always equal to one another. And so far as the action is propagated by the intervening instruments, and at last impressed upon the resisting body, the ultimate determination of the action will be always contrary to the determination of the reaction.

Editor's Comments on Papers 12 Through 19

THE INTRODUCTION OF *VIS VIVA* AND RELATED CONCEPTS

Gottfried Wilhelm von Leibniz (1646–1716) was a German mathematician, philosopher, and man of affairs who busied himself with a host of problems of both theoretical and practical interest. He invented the differential and integral calculus independently of Newton and at about the same time. His notation for derivatives and integrals has survived to our time. He made contributions to practically every branch of science, but was also active in metaphysical research, theology, law, and engineering. As a diplomat he served many German rulers and encouraged the founding of scientific societies. He must have had an optimistic disposition since he convinced himself that all is for the best in this best of all possible worlds, a view that Voltaire satirized in his novel *Candide.*

Leibniz's restless mind soon made itself evident with respect to mathematics and mechanics during his sojourn in Paris from 1672 to 1676. Here he naturally heard much about the philosophy of Descartes, which, although its author had been dead for over 25 years, still dominated scientific thinking in France. Leibniz found much to disagree with in Descartes' ideas. It was indeed in the process of extending the domain of Cartesian analytical geometry that Leibniz invented his brand of the calculus. But he developed a more fundamental opposition to the Cartesian interpretation of dynamics, in which force was interpreted as measured by quantity of motion (momentum) or its rate of change, with the total quantity of motion in the universe being assumed to be constant. Leibniz came to feel that this association of force with quantity of motion was a distorted one. In essence, in modern times he decided it was more correct to assume that force is something inherent in a body which enables it to do work (although of course he did not use this terminology). In Descartes' view, force was someting that enabled the body to change its momentum. In 1686 Leibniz published in the *Acta Eruditorum* in Leipzig a brief note explaining his point of view in terms of the simple case of a body falling freely under gravity, and insisting that it is the product of mass times velocity squared which is the invariant quantity and the true measure of force. He named this quantity *vis viva* or living force, that is, force connected with motion, as distinct from *vis mortua*, or the dead force of statics. This was the most important paper in the modern theory of energy as a concept in mechanics. It actually gave a new name to a quantity and showed how it is measured. It is true that Christian Huygens in 1673, in his famous treatise *Horologium Oscillatorium*, used effectively the product of mass times the square of the velocity in his treatment of the compound pendulum; but he nowhere singled out such a quantity for special attention because of its general application, and apparently did not think it of enough significance to warrant giving it a name. Leibniz was either prescient enough or lucky enough to grasp the fundamental importance of the new concept and thus gave it a name, which stuck with it well into the nineteenth century, when it was replaced by the term kinetic energy, which indeed is equal to one half the *vis viva*. It is probably folly to stress unduly the question of priority in scientific discovery, although much has been written and will doubtless continue to be written on this subject by enthusiastic historians of science. In this matter of *vis viva* there is certainly enough credit to be shared by two such great natural philosophers as Huygens and Leibniz.

We reproduce here Leibniz's famous 1686 paper in English translation. Leibniz followed this paper with another contribution to the *Acta Eruditorum* in 1695: "A dynamical model for the wonderful laws of nature concerning the forces of bodies and assigning their actions and referring them to their causes." In this rather long article (in two parts), from which we reproduce only a short extract from the beginning, Leibniz discusses in detail the various meanings that have been assigned to the concept of force through the ages. He then lays stress on the ultimate importance of *vis viva*. It is clear he took this idea very seriously. We must recall that in his time, as well as in later times, it was common for many if not most philosophers to think of force as something inherent in the body, not as something that acts from outside on a body. This will be further discussed in connection with the views of d'Alembert, Euler, and Lagrange.

It is fair to say that Leibniz's introduction of the *vis viva* concept inaugurated a controversy which lasted over half a century. It polarized natural philosophers in two camps, those who preferred Descartes' quantity of motion as the really important element connected with force and those who were impressed with the fundamental significance of the *vis viva* of Leibniz. We shall follow this controversy in the succeeding papers by Johann and Daniel Bernoulli, d'Alembert, Euler, and Koenig.

Johann Bernoulli (1667–1748) (also referred to as John or Jean) was one of the elder members of a celebrated Swiss family of pure and applied mathematicians, of whom his son Daniel (1700–1782) is perhaps the best known for his researches in physics. Johann spent most of his professional career as professor in the University of Basel, and for a time after the death of Newton in 1727 was considered the foremost mathematician in Europe. He early became a follower of Leibniz in the latter's reformation of the terminology of the differential and integral calculus and defended him against Newton. Bernoulli also endorsed with enthusiasm Leibniz's decision to adopt *vis viva* as the true measure of force in a moving body. In the period from 1715 to 1730 Bernoulli wrote widely, championing the *vis viva* idea. In his *Discourse on the Laws of the Communication of Motion* he devotes several chapters to this concept and its use. We reproduce here in English translation selections from Chapters V, IX, and X.

Johann Bernoulli made another contribution to mechanics which in the long run was of equal importance to or perhaps even greater than his defense and exploitation of the *vis viva* concept, as far as the evolution of the energy concept is concerned. In a personal letter in 1717 to the French mathematician Pierre Varignon (1654–1722), Bernoulli enunciated in clear-cut fashion the principle of virtual velocities or virtual work, vestiges of which we have already noted in ancient writings. Bernoulli applied the principle to problems in the equilibrium of forces. Varignon was so impressed with Bernoulli's letter that he asked permission to include the contents in his own work, *Nouvelle Mécanique Sur la Statique*, published posthumously in 1725. We reproduce this extract from Varignon's book as our first example of Bernoulli's work bearing on the development of the energy concept, following the selections from the famous *Discourse.*

Bernoulli in his letter to Varignon made an important suggestion about nomenclature. He introduced the term *energie* to refer to the product of a force and the virtual displacement of a particle under the action of the force. This use of the word energy to denote what later came to be called work (*travail* in French) is possibly the first introduction of the term in a sense related to the modern use of the word. Presumably, Bernoulli got the word from Aristotle's use of *εηεργεία*, although there is no direct evidence for this.

It is interesting that in his discussion of *vis viva* in the *Discourse*, Bernoulli makes no use of the word energy, even though he introduces again the principle of virtual displacements for systems in equilibrium. He may well have felt that it was wise to keep problems in statics separate from those of dynamics, a common attitude in his day. It was left to Thomas Young (Paper 18) to tie the word energy to *vis viva* (1803). But the final, firm assignment of the term to phenomena in the sense in which we use it today was delayed to the middle of the nineteenth century.

In assessing the ultimate importance of the contributions of Bernoulli from the selections included here we must bear in mind that he was still thinking of force as something inherent in a moving body; hence the difference in the points of view of the Cartesians and the Leibnizians was very real to him. Bernoulli wanted to make it perfectly clear where his allegiance lay. His work was completed and published, of course, before the appearance of d'Alembert's treatise on dynamics of 1743, in which the latter dismissed the problem as illusory. As we shall see, this did not really end the controversy, nor did it completely clear up the problem of the nature of force.

Jean Le Rond, called d'Alembert (1717–1783), French mathematician, physicist, and philosopher, spent practically the whole of his life in Paris, where he was born. Of illegitimate birth he was brought up by foster parents. Although originally intended for a clerical career, he turned away from religion in favor of law and medicine; he then abandoned his studies in the latter fields in favor of mathematics, in which he was largely self-taught. He became one of the group of celebrated eighteenth-century scientists, which included Euler, the Bernoullis, Clairault, and Lagrange, who developed mathematics to the point where it could successfully solve problems in physical science.

D'Alembert's interest in mechanics developed early in his career. He became particularly concerned with the motions of particles subject to constraints, a subject to which Newton had given little attention. In 1743 d'Alembert published his first important work, *Traité de Dynamique*, in which he set forth his famous principle, designated by him "A general principle for finding the motions of several bodies which react on each other in any fashion." This principle, much misunderstood by many people in nineteenth- and early twentieth-century science, was probably d'Alembert's greatest contribution to mechanics. Lord Kelvin felt that d'Alembert arrived at it from a clear appreciation of Newton's third law of motion, and this may indeed be the case. At any rate, d'Alembert made many successful applications of his principle, and it was successfully employed by Lagrange in his more elaborate analytical treatment of mechanics later in the century.

It was inevitable that d'Alembert in his *Traité* would pay some attention to the concept of *vis viva*, and he did devote a chapter to it. But in a certain sense his most important contribution to the subject was set forth in his preliminary discourse in the treatise. For here he paid his respects to the celebrated controversy that had been raging for so many years between the followers of Descartes, who considered quantity of motion (mass times velocity) as the most important entity in mechanics, and the adherents of Leibniz, who maintained that *vis viva* (mass times the square of velocity) is the only correct quantity to represent force.

D'Alembert expressed the strong opinion that it is folly to think of force as something inherent in a body. If it means anything, force should be considered as an *external* cause of motion. From this point of view the controversy was an illusion, a contention over terminology, and the two points of view were merely different ways of expressing the same thing. He emphasized that change in quantity of motion is connected with the effect produced by a force in *time*, whereas change in *vis viva* is connected with the effect of a force over space. Today we exemplify this point of view by the two theorems: (1) change in momentum of a particle is the *time* integral of the resultant force, and (2) change in one half the *vis viva* is the *space* integral of the resultant force.

We reproduce here in English translation the extract from d'Alembert's *Traité* bearing on the quantity of motion-*vis viva* controversy. Cleary, he felt that he had cleared up the matter and no more needed to be said about it. That this was not actually the case, however, is shown, subsequent to the publication of d'Alembert's *Traité*, by the writings of Daniel Bernoulli, Samuel Koenig, and Leonhard Euler which we reproduce in part later in this volume. These and other similar continuations of the quarrel were evidently overlooked by the nineteenth-century commentators who considered that d'Alembert had closed the controversy.

Even in his youth (he was only 26 when his *Traité* appeared) d'Alembert was an ardent controversialist and became even more so as he grew older. He disagreed strongly with many of his contemporaries on many points in mathematics and physics and was in turn warmly attacked. Actually, his attempt to apply the *vis viva* concept to problems in mechanics was cast in such a form as to make it more difficult to follow than the corresponding work of the Bernoullis and others. Moreover, in his illustrations of what he called the conservation of *vis viva* he somehow failed to grasp, or at any rate to express at all clearly, the necessity for the introduction of what we now call potential energy.

In view of the acknowledged preeminence of Leonhard Euler (1707–1783) in the development of mathematics and mathematical physics, it is appropriate and essential that we devote some attention to his attitude toward the concept of *vis viva*, which had already been introduced by Leibniz (Paper 12a) well before the birth of Euler. This is particularly important in the light of Euler's great influence on the work of other European natural philosophers throughout the eighteenth century, and even beyond. Although a native of Basel, Switzerland, Euler spent practically all his professional career in Berlin and St. Petersburg at the scientific academies in those cities. He was without question the most prolific of all mathematical scientists. His collected works are still in process of publication and to date have reached a total of over 60 volumes.

As a precocious student in Basel, Euler early became interested in mechanics. His critical treatise "De Sono," published in 1727, when the author was a boy of 20, indicated wide reading of the work of the earlier masters, Mersenne and Newton. (See Paper 11 in *Acoustics—Its Historical and Philosophical Development*, R. Bruce Lindsay, ed., in Benchmark Series in Acoustics, Dowden, Hutchinson & Ross, Inc., Stroudsburg, Pa., 1973, p. 103.) That he had not overlooked Galileo, Descartes, and Leibniz is shown in a short Latin treatise "Vera Vires Existimandi Ratio" (The True Way of Measuring Forces), which was found among his papers after his death and was published posthumously. (See Euler's *Opera Omnia*, Ser. 2, Vol. 5, Orell Füssli Verlag, Zurich, 1957, pp. 257–262.) Authorities agree that this paper must have been written in the late 1720s. In it Euler makes reference to the controversy still raging between the adherents of Descartes and those of Leibniz with respect to the proper way to measure the force of a body in motion. This paper was clearly prepared before the appearance of d'Alembert's famous *Traité de Dynamique* (1743), and in it Euler retains the view that force is something inherent in the moving body, an idea discarded by d'Alembert and later by Euler himself. However, in this early paper, after analyzing the respective arguments, Euler concludes that the Leibnizian view is the correct one. It is important to note this, since it demonstrates that Euler was early familiar with the *vis viva* concept. However, in this early paper he nowhere uses the term *vis viva*, nor, for that matter, quantity of motion, contenting himself with referring simply to mass times the square of velocity or mass times velocity.

This lack of reference to the accepted terminology becomes even more interesting when we consider Euler's first serious treatise on mechanics, published in St. Petersburg in 1736: "Mechanica sive Motus Scientia Analytice Exposita" (Mechanics or an Analytical Exposition of the Sciences of Motion). (See *Opera Omnia*, Ser. 2, Vols. 1 and 2, Teubner Verlagsgesellschaft GmbH, Leipzig, 1912.) This was a systematic treatment of the kinematics and dynamics of a point mass in rectilinear motion and in motion on a plane. In a certain sense it was an attempt to develop mechanics in the tradition of Newton's *Principia*, with its theorems, corollaries, scholia, and the like, but with the use of the differential notation introduced by Leibniz. In this treatise Euler made no mention of quantity of motion or of *vis viva*, nor did he pay any attention to the idea of conservation. It seems clear that he felt these ideas were irrelevant to the essential task of mechanics, that is, to trace the motions of a particle as a function of time. When this treatise was composed, he had evidently given up the notion that force is something that resides in a body, and had reached the conclusion that forces are external entities which cause the motions or changes in the motion of particles. This relates him significantly to d'Alembert, although the latter in his famous *Traité* did conclude that *vis viva* and its "conservation" play an important part in mechanics.

Many years later, in 1750 to be exact, Euler wrote another memoir on mechanics, "Recherches sur l'origine des forces," which appeared in *Memoires de l'Académie des Sciences de Berlin* [G] (1750), 1752, pages 419–447. (See *Opera Omnia*, Second Series, Vol. 5. pp. 109–131.) This is a significant treatise from the standpoint of the energy concept; in it Euler uses both the momentum and energy equations, although he again fails to use the terminology of quantity of motion and *vis viva*. We reproduce here some relevant extracts from this memoir, prefaced by a brief editorial introduction.

Daniel Bernoulli (1700–1782) was the son of Johann Bernoulli (1667–1748), some of whose work has been included in this volume (see Paper 13). A man of extraordinary versatility, he made outstanding contributions to medicine, mathematics, and physics. From 1725 to 1733 he worked with Euler at the Saint Petersburg Academy. He taught anatomy, botany, and physiology at the University of Basel and became professor of physics there in 1750, a chair he occupied for the rest of his life. Bernoulli's famous book *Hydrodynamica* (Basel, 1738), in which he laid the foundation for the molecular theory of gases, is mentioned in the later discussion of the nature of heat.

Daniel's father, Johann, became an adherent of the *vis viva* concept introduced by Leibniz, and some of his defense of the idea has been included in Paper 13. It is not surprising that Daniel followed in his father's footsteps in supporting the *vis viva* principle and in particular its conservation in the motions of bodies. We reprint here in translation the larger part of his 1748 memoir to the Berlin Academy. He effectively sets up the energy equation (although not with this name) for the special cases of the falling body and for inverse-square-law gravitational attraction. He also applies *vis viva* to aggregates of particles and solves the two-body problem, as far as the velocities of the bodies are concerned. Considering that this came 40 years before the publication of Lagrange's *Mecanique analytique*, this was work of extraordinary significance for the development of mechanics. It is true that d'Alembert in his *Traitė de Dynamique* (1743) had also treated conservation of *vis viva*, but Bernoulli's analysis is much clearer and easier to follow.

Johann Samuel Koenig (1712–1757) (usually known by his middle name) was a German mathematician and physicist, who actually spent little of his professional life in his native country, traveling extensively and residing in Switzerland, France, and the Netherlands. Although trained under Johann and Daniel Bernoulli in Basel, he found it hard to get an academic job and spent much of his life practicing law. In 1744 he finally secured a professorship in mathematics and philosophy at the University of Franeker in the province of Friesland in the northern part of the Netherlands. This university was founded in 1585 but was closed in 1811 by Napoleon. In 1749 Koenig moved to the Hague to become a privy councillor and librarian, although he evidently maintained his professorship, at least by title. He died in Amerongen in the Netherlands after a rather short but active life.

Koenig was acquainted with most of the scientific notables of his time in Europe. He even hobnobbed with Voltaire and the latter's illustrious and mathematically celebrated mistress, the Marquise du Châtelet. He was acquainted with Euler and Maupertuis, and it was the latter who apparently secured for Koenig his election to membership in the Berlin Academy. Koenig developed strong opinions and managed to quarrel eventually with many of his associates. For example, he made vigorous attacks on Euler and Maupertuis in connection with the principle of least action. He even broke off relations with the Marquise du Châtelet to whom he had acted as instructor in Leibnizian philosophy.

Our principal purpose in including an extract from Koenig's writings is not based on any really high distinction he had as a scientist, but rather because he fought for the viability of the *vis viva* concept at a time when it was ignored or attacked by many continental celebrities. This role ensures him a position of importance in the evolution of the concept of mechanical energy. We reproduce in English translation Koenig's article in the Leipzig *Acta Eruditorum* for 1751. This rather impassioned defence of *vis viva* was composed toward the end of his short life. In it he took Euler and Maupertuis to task for claiming that the *vis viva* concept cannot be satisfactorily applied to the solution of problems in static equilibrium. They were, of course, committed to the use of Maupertuis' principle of least action in this connection. Koenig ingeniously and correctly points out that even the principle of least action can be expressed as well in terms of *vis viva* (in which the action is given as the time integral of the kinetic energy in modern notation) as in terms of quantity of motion (in which the action is given as the space integral of the momentum in modern notation). Koenig also states correctly that the *vis viva* concept can be applied directly to solve problems in static equilibrium. Unfortunately, when he gives a simple illustration, his method is incorrect, as the editorial commentary on his solution indicates. Koenig's memoir produced a decidedly unfavorable reaction from Leonhard Euler. This is discussed in detail in an editorial note at the end of our extract from Koenig's writings. Some modern commentators think that Euler was too harsh in his judgment of Koenig, especially when he leveled charges of unethical behavior at the latter. In this connection, see, for example, the entry on Koenig by E. A. Fellmann in *Dictionary of Scientific Biography*, edited by C. C. Gillispie (Charles Scribner's Sons, New York, 1973) vol. 7, p. 442ff and accompanying notes.

Thomas Young (1773–1829), the versatile English physician who successfully turned his attention to physics and classical philology, was born in Somersetshire but spent his professional life in London.

He was a very successful medical practitioner, but his scientific interests were unusually wide. He is perhaps most famous in physics for his studies of the interference of light, in which he adopted the wave theory. From 1801 to 1803 he served as professor of natural philosophy at the Royal Institution of Great Britain; in his lectures there he developed in detail some of his original ideas, including those on elasticity (Young's modulus) and optics.

Young's lectures were very popular, and a published version was brought out by the author in 1807. This was reprinted in 1845 with editorial notes by the British mathematician Philip Kelland. Young naturally began his lectures with a treatment of motion and relevant mechanical principles. In his eighth lecture, collisions are discussed and both elastic and inelastic cases are treated. The conservation of momentum (this term is now used in place of quantity of motion) is stated for both elastic and inelastic collisions. Young then comments that for perfectly elastic collisions the *vis viva* or living force is conserved. At this point Young proposed that the term *energy* be used to denote the *vis viva*. As we recall from the extract from Johann Bernoulli's writings (Papers 13A and 13B) this was not the first time the word "energy" appeared in connection with physical phenomena in the sense in which it is used today. However, since Bernoulli restricted the usage to the product of force and virtual displacement and did not apply it to *vis viva*, Young should receive credit for his suggestion. Young's assignment ultimately became the kinetic energy of today, whereas Bernoulli's choice led to potential energy. It is, of course, interesting to observe that their suggestions did not catch on and made no impact on actual usage until the middle of the nineteenth century.

In his article on collisions in which he suggests the use of the term *energy*, Young cites numerous earlier authorities, including Galileo, Huygens, Gravesande and Euler, but makes no mention of Johann Bernoulli or of his introduction of the term *energie* in his letter to Varignon. It might have been expected that he would be familiar with Bernoulli's contributions, since his erudition was very great. However, as a scholar well acquainted with the classical languages, Young doubtless obtained the term *energy* from the same source as Bernoulli.

To relate Young's use of the word "energy" as closely as possible to the mechanics of collisions as he understood the subject, we include here almost the whole of Lecture VIII in the 1845 edition of his lectures.

Lazare Nicolas Marguerite Carnot (1753–1823) was a French engineer who became one of the military leaders of the French Revolution and served briefly as Napoleon's minister of war. He was the father of

Sadi Carnot, who achieved greater scientific fame in thermodynamics, and also the grandfather of a president of the second French Republic. In his early days in the army engineers, Carnot became interested in machines and tried to apply scientific principles to their behavior. In an early work entitled "Essai sur les Machines in General" (1786) and in a revised and extended version with the title "Principes fondamentaux de l'équilibre et du mouvement" (1803), he endeavored to apply the concepts of *vis viva* and work to the behavior of machines considered as devices for the transmission of mechanical activity (or what we call energy today) from one body to another. In this work he came close to an enunciation of the principle of conservation of energy in the case of machines, although he did not set up the energy equation in the severely analytical way in which Daniel Bernoulli and Lagrange had deduced it.

We include here the translation of an extract from Carnot's 1803 work just mentioned to show how he, as an engineer, looked at the *vis viva* principle. There was, of course, no scientific novelty in this, as the previous contributions of Leibniz, d'Alembert, the Bernoullis, and Lagrange clearly show. The ultimate significance of the extract is that it made engineers aware of the importance of *vis viva*, and hence helped to pave the way for the use of energy in engineering problems.

It is of interest that Carnot refers to what happens when a man raises a weight to a certain height by the use of the word "work" (French *travail*). He does not actually define the work done in this case as the product of the weight and the height, but he comes rather close to this, and evidently realizes the relation between this product in the case of a falling body and the *vis viva*. In this connection, the French engineer Gaspard Gustave de Coriolis (1792–1843) is usually given the credit for actually defining the product of force and distance traveled under the influence of the force as the "work" done by the force (*travail*, in French). This will be found in his treatise "Du Calcul de l'Effect des Machines" published in Paris in 1829 (p. 18ff). It would appear that Lazare Carnot had come very close to anticipating him.

12A

This article was translated expressly for this Benchmark volume by R. Bruce Lindsay, Brown University, from "Brevis demonstratio erroris mirabilis Cartesii et aliorum circa legem naturalem, secundam quam volunt ad Deo eundam semper quantitatem motis conservandi, qua et omne mechanica abituntir," Acta Eruditorum, *Leipzig, 1686, in* Leibniz Mathematische Schriften, *Vol. 2, C. I. Gerhardt, ed., Halle, Druck und Verlag von H. W. Schmidt, 1860, pp. 117–119*

A Brief Demonstration of the Memorable Error of Descartes and Others Concerning the Natural Law According to Which They Claim That the Same Quantity of Motion Is Always Conserved by God, a Law That They Use Incorrectly in Mechanical Problems.

Gottfried Wilhelm von Leibniz

Most mathematicians, when they see that in the five common machines velocity and mass compensate each other, generally estimate the motive force (*vis motrix*) by the quantity of motion, that is, the product of the mass of a body multiplied by its velocity. Or to speak more mathematically, the forces of two bodies of the same kind, put in motion and acting in the same way with respect to their masses and motion, are, they say, in the compound ratio of the bodies (or of their masses) to their velocities. And so it may be in agreement with reason that the same total motive power (*potentia*) is conserved in nature and is not diminished inasmuch as we never see a force given up by one body without being transferred to another, nor increased, because perpetual mechanical motion never takes place and no machine, not even the world as a whole, is able to maintain its force without an additional external impulse. So it has come about that Descartes, who considered moving force (*vis motrix*) to be equivalent to quantity of motion, has declared that the quantity of motion in the world is conserved by God.

That I may show how great is the difference between these two things, I first assume that a body falling from a certain height acquires a force sufficient to raise it back to that height, if its direction carries it so and there are no external impediments. For example, a pendulum will rise to precisely the same height from which it has fallen unless the resistance of the air or some other impediment absorbs some of its force, which we now leave out of account.

Second, I assume that as much work is required to raise a body A of 1 pound to a height CD of 4 ells (*ulnae*) as it does to raise body B of 4 pounds to a height EF of 1 ell (see figure). These assumptions are accepted as valid by the Cartesians and other philosophers and mathematicians of our time. From this it follows that in falling from height CD body A will acquire just as much force as body B does in falling from height EF. For after its fall from C, body A reaches D. There it has force sufficient to raise it again to C by assumption 1, that is, a force sufficient to raise a mass of 1 pound to a height of 4 ells. Similarly, after its fall from E, body B reaches F, where it has force sufficient to raise it again to E, by assumption 1, that

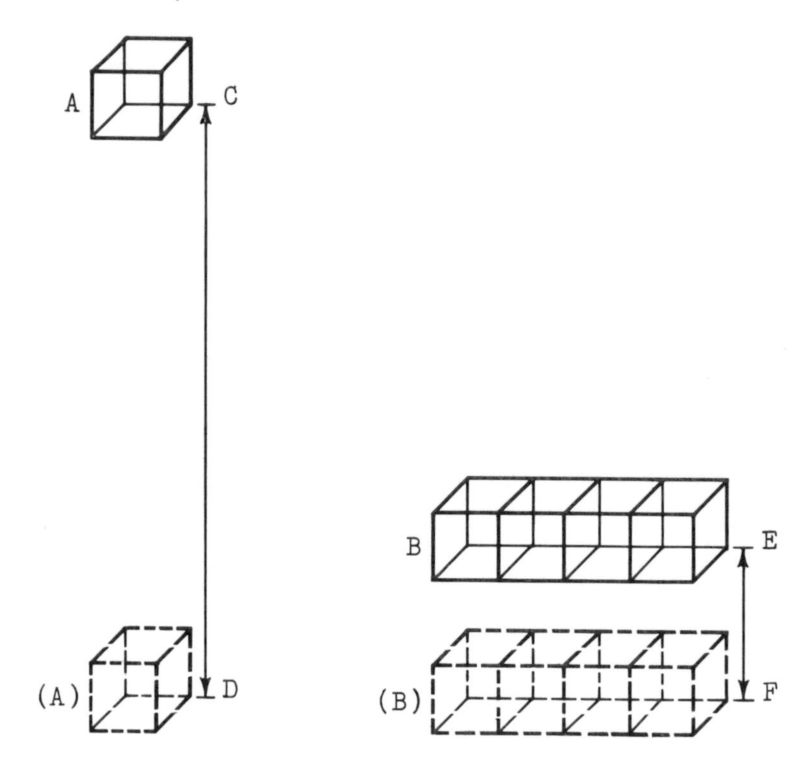

is, a force sufficient to raise a mass of 4 pounds to a height of 1 ell. Hence, by assumption 2, the force of body A when in position D and the force of body B when in position F are equal.

Now let us see whether the quantity of motion is the same for both. Here, contrary to expectation, a very great difference is found to exist. This I make clear as follows. It was demonstrated by Galileo that the velocity acquired in the fall through CD is twice the velocity acquired in the fall through EF. Let us therefore multiply the mass of body A, which is 1, by the velocity acquired by it, which may be taken as 2; the product or the quantity of motion will be 2. Then let us multiply the mass of body B, which is 4, by its acquired velocity, which is 1. The product or the quantity of motion gained by B is therefore 4. Hence the quantity of motion of A in position D is only half that of body B in position F. However, previously the forces of both were found to be equal. And so there is a large discrepancy between the *vis motrix* and the quantity of motion. Hence, one of these quantities cannot be given by the other, a fact we have undertaken to show. From this it appears that however a force may be determined from the magnitude of the effect that it produces, for example, the height to which a heavy body of given size and kind can be raised, a force certainly cannot be determined by the velocity that it imparts to the body. For to give the same body double the velocity requires not double but more than the double the work. No one should wonder why in the case of the common machines, the lever, the wheel and axle, the pulley, the wedge, the screw, and the like, equilibrium holds when the magnitude of one body is compensated by the velocity of the other, as brought about by the arrangement of the machine, or when the sizes are inversely proportional to the velocities (the nature of the bodies being assumed the same), or when the same quantity of motion is produced in any other way. For then the same ultimate quantity of effect for both bodies, or the same height of ascent or descent for both, occurs on whatever side of the system in equilibrium you wish the motion to be produced. And so it happens by accident in this case that the force may be measured by the quantity of motion. But other cases exist, such as the one already described, in which the two are not the same.

For the rest, since nothing seems simpler than our demonstration, it is strange that it never occurred to Descartes or his followers, all most learned men. But too much faith in his own mode of thinking led him astray. For Descartes, by a fault common to great men, eventually became a little too much the presiding genius. And I fear that not a few of the Cartesians have begun to imitate the Peripatetics, whom they laugh at; that is, they have accustomed themselves to consulting the books of the master rather than depending on right reasoning and on the nature of things.

It is therefore proper to say that forces are in compound ratio of the bodies (assumed of the same specific gravity or density) to the height which produces velocity, that is, the height from which, when falling, they are able to acquire such velocities, or more generally (since in the meantime no velocity has as yet been produced) the height that would produce them, but not generally to the velocities themselves, although this at first sight seemed plausible, and was so seen by many. From this many errors were born, which are contained in the writings on mathematical mechanics of Honoratus Faber and Claudius Dechales as well as Joh. Alph. Borelli and other men, otherwise outstanding in these studies. It is owing to this, I believe, that doubt has been cast by not a few learned men on the law of Huygens concerning the center of oscillation of the pendulum, which law is most certainly valid.

12B

This article was translated expressly for this Benchmark volume by R. Bruce Lindsay, Brown University, from "Specimen dynamicum pro admirandis naturae legibus circa corporum vires et mutuas actiones deligendis et suae causas revocandis," Acta Eruditorum, *Leipzig, 1695, in* Leibniz Mathematische Schriften, *Vol. 2, C. I. Gerhardt, ed., Halle, Druck und Verlag von H. W. Schmidt, 1860, pp. 234–235*

A Dynamical Model for the Wonderful Laws of Nature Concerning the Forces of Bodies and Assigning Their Actions and Referring Them to Their Causes

Gottfried Wilhelm von Leibniz

[*Ed. note:* In this memoir Leibniz begins by pointing out that many illustrious men in many places have investigated the problems of the new science of dynamics. He provides an elaborate review of the various views as to the nature of force held by philosophers from ancient times down to his own. After a few pages of this summary, he finally emphasizes his own views as follows.]

Force is dual in character. The elementary variety, which I call dead force (*vis mortua*) because motion does not yet exist in it, but only the instigation toward motion, is like a sphere resting in a tube [*Ed. note:* a shot in a gun that has not yet been fired?], or a stone in a slingshot, held back by a chain. The other variety of force is the ordinary one associated with actual motion. This I call living force (*vis viva*). Examples of dead force are centrifugal force, the force of gravity and centripetal force, and the force with which a stretched spring begins to contract.

[*Ed. note:* These examples indicate how far Leibniz was from the modern point of view in mechanics. However, from his point of view, since in centrifugal and centripetal force there is no actual motion in a straight line away from or toward the center, he is justified in thinking of these forces as not associated with motion in his sense. By the force of gravity here he evidently means only the weight as exhibited by the force with which an object presses against the surface on which it rests.]

But in a collision produced by a heavy body that has been falling for some time or by a bow unbending, or by some other equivalent scheme, the force is living force, produced by an infinite number of applications of dead force. This indeed is what Galileo intended to convey when speaking in an obscure way about collisions. He spoke of infinite force as compared with the simple force of gravity. Although the *impetus* [*Ed. note:* quantity of motion or momentum] is always associated with living force, nevertheless these two are shown in what follows to be quite different.

[*Ed. note:* In what follows Leibniz indulges in a rather elaborate philosophical discussion of the various properties of *vis viva*, scarcely calculated to make his more physically minded contemporaries show greater interest in the concept he had introduced. Nor do these rather abstract considerations have much attraction for us today. Toward the end of his twelve-page article, Leibniz finally introduces some specific examples of the *vis viva* of bodies in motion and emphasizes the fundamental dependence on the square of the velocity.]

13A

This article was translated expressly for this Benchmark volume by R. Bruce Lindsay, Brown University, from Discours sur les lois de la communication du mouvement, *Paris, 1724, Chaps. V, IX, X*

USE OF THE VIS VIVA CONCEPT IN PROBLEMS OF MOTION

Johann Bernoulli

CHAPTER V
On the *Vis Viva* of Bodies in Motion

1. In this chapter I propose to examine one of the most important matters concerning motion. I speak here of that force of bodies which G. W. Leibniz (1646–1716) called *vis viva* (living force), to distinguish it from another force to which he had given the name *vis mortua* (dead force). In an earlier section of this work I have already had occasion to define what I understood by *vis viva* and *vis mortua,* and to determine in passing the true measure of *vis viva.* My aim here is to explain thoroughly the nature and properties of this force. I undertake this the more willingly because a large number of philosophers, who are indeed very enlightened, still confuse these two forces and cannot be convinced of their error.

2. We have seen in Chapter Three that *vis mortua* consists of a simple exertion, and this exertion is such that it continues to exist even if a foreign obstacle at any moment prevents the production of local motion in the body on which the *vis mortua* is exerted. An example of such a force is a weight. A heavy body held up by a horizontal table continually tries to descend. And it would effectively be able to descend were it not that the table provides an obstacle which holds it back. Thus gravity produces a *vis mortua* in the body. The effect of this is only momentary. At every instant gravity exerts an infinitely small amount of energy on the body on which it acts, which is immediately absorbed by the resistance of the obstacle. These small degrees of velocity perish at their birth and are reborn in perishing, and it is in this steady reciprocation, in this cycle of production and destruction, that the force of gravity consists, when it is held back by an immovable object. This force is the one called *vis mortua.* As for the obstacle, when it resists the force of gravity it receives a force equal and opposite to the force of gravity acting on the weight. [*Ed. note:* The language is a bit obscure, but it seems clear that Bernoulli means that the force against the obstacle due to the weight is equal and opposite to the reaction force of the obstacle against the weight.] *Vis mortua* has this peculiar property that it does not produce any effect that endures longer than itself. If the heavy body held up by the table were to lose all its weight at any instant, at that same instant the table would cease to be pressed down upon.

3. The situation is not the same with *vis viva;* its nature is entirely different. It can neither be born nor perish in an instant. It takes more or less time to produce *vis viva* in a body that did not have it to begin with. It also takes time to destroy *vis viva* in a body that has it. A *vis viva* is produced gradually in a body when, the

body being originally at rest, a force is applied and induces by small degrees a local motion. We suppose in this case that no obstacle retards the motion. This motion is acquired in infinitely small degrees and increases to a finite and definite velocity, which continues uniformly until the force producing the motion ceases to act on the body. Thus the *vis viva* produced in a body in a finite time by a force that no obstacle retards is something real. It is equivalent to that part of the cause consumed in producing it, since every efficient cause must equal its effect. [*Ed. note:* This is the famous principle that Leibniz enunciated in 1680 and which over the years has provided the basis of much philosophical discussion.]

4. The body that receives *vis viva,* if not retarded by any obstacle, provides resistance to it only through its inertia, which is always proportional to its mass, so that the small degrees of force exerted on the body are conserved and accumulate so as to produce at last a local motion. A *vis viva* produced by a continual force without any retarding obstacle might be compared to a surface described by the motion of a line or to a solid described by the motion of a surface. There is no more possibility of comparing *vis mortua* with *vis viva* than of comparing a line with a surface or a surface with a solid; they are heterogeneous quantities that do not really admit of comparison.

5. Whatever may be the cause of a force that by the duration of its action finally produces motion, if it is of a definite amount, such as that produced, for example, by a stretched spring, which by its relaxation employs a force to produce an actual velocity in a body that previously had none, I then say, and the matter is self-evident, that to the extent that the body receives new degrees of force, the cause that produced them ought to disappear entirely until all the force of the spring is used up and transferred to the body—until it has been, so to speak, gathered together in the accumulation of all the small degrees that have been successively produced. It is this force entering into the body put into motion by the using up of the tension of the spring that ought properly to be called the *vis viva.* It is in virtue of this that the body moves from one place to another with a certain velocity, more or less great according to the energy of the spring. [*Ed. note:* Although Bernoulli here uses the word *énergie,* which has been translated as energy, we must not take this to mean that he thought of this as corresponding to the modern meaning of the energy of the spring; qualitatively he seems to be close to the modern meaning.]

6. Here again we see the great difference between *vis viva* and *vis mortua.* Pressure or tension alone, or the *vis mortua,* that is received by an immovable obstacle, produced by the effort of a spring trying to unstretch itself, does not decrease the force of the spring to nothing or even decrease it at all. Air, for example, compressed in a vessel, makes a continued effort to expand, without, however, losing any of its force, because the walls of the vessel cannot give way; they can only maintain the pressure without decreasing the elasticity of the air. On the other hand, the force of the spring is used up in giving motion to a body, that is, in producing *vis viva.* The production of the least degree of this force requires the loss or destruction of the force of the spring to an equal degree. The one is the cause, the other the immediate effect. But the cause could not disappear in whole or in part without being refound in the effect in the production of which it has been employed.

7. I Conclude from this that the *vis viva* of a body which has been produced by

the relaxing of such a spring is capable of stretching it again to the same degree of force that it had originally. And if one supposes that this *vis viva* is employed to stretch two, three, or more springs all equal to each other but more weak than the first, I say that the first spring can produce an effect twice, three times, or several times greater than any one of the weak springs. The equality that exists between effect and efficient cause confirms what we have just said.

8. In this equality consists the conservation of forces of bodies that are in motion. It is seen that the smallest part of a positive cause cannot be permanently lost and that it can besides reproduce an effect only to the extent of replacing such loss.

9. Since people have long labored under the opinion that the quantity of motion, that Is, the product of the mass of a body multiplied by its velocity, was the measure of the force of a body, they have erroneously believed that the quantity of motion in the universe must be constant.

10. The origin of this error, as I have already suggested, lies in confusing the nature of *vis mortua* with that of *vis viva.* The fundamental principle of statics demands that in the equilibrium of powers the quantities that enter be made up of the absolute [*Ed. note:* impressed] forces and their virtual velocities. This principle has been extended further than was justified by applying it to bodies with actual velocities.

[*Ed. note:* Bernoulli is here referring to the principle variously called the principle of virtual velocities, virtual displacements or virtual work. It says that if a set of forces $\mathbf{F}_i$ is exerted on a collection of bodies subject to constraints, and if small possible displacements $\delta\mathbf{r}_i$ (consistent with the constraints) are allowed each body, then the collection will be in equilibrium if

$$\Sigma\, \mathbf{F}_i \cdot \delta\mathbf{r}_i = 0$$

where the sum is taken over all the bodies of the system. One can also express it in the form

$$\Sigma\, F_i \cdot \delta\dot{r}_i = 0$$

where $\delta\mathbf{r}_i$ are virtual displacements and $\delta\dot{\mathbf{r}}_i$ corresponding virtual velocities. The meaning of the word "virtual" here is that the displacements are merely imagined as possible displacements consistent with the constraints: they are not *actual* displacements. Bernoulli was presumably the first to give prominence to this principle, stated in precise form, as the basic principle of equilibrium for bodies, although we have seen in earlier articles in this book that the principle was effectively used in connection with machines as early as Aristotle.]

11. It is only within the last 30 or 40 years that some people have perceived that the two forces are of a totally different nature, not having any more relation with each other than a line has to a surface or a surface to a solid. Leibniz was the first who noted that this force was not at all equal to the product of mass and velocity, but that it was measured by the product of mass multiplied by the square of the velocity.

12. The novelty of Leibniz's view attracted opponents. He confirmed his view by showing the perfect agreement that exists between it and Galileo's law of a

body falling under gravity; a law generally accepted. Leibniz could see that a given weight with two units of velocity could rise four times as high as a weight with one unit of velocity; it would rise nine times as high if it had three units of velocity; and so on. Finally, he showed that the heights to which heavy bodies are capable of being raised are always proportional to the squares of their velocities. He maintained that the height to which a weight can rise can be taken as a measure of the force of the body. He concluded that the *vis viva* of a body is proportional to the mass multiplied by the square of the velocity.

13. But the adversaries of Leibniz are not satisfied with his hypothesis concerning the heights that he maintains provide the measure of forces. They bring up cases and insist among other things that one must not neglect the time taken by the weight to reach its height. For example, a weight that with double velocity rises four times as high according to them ought to be considered to have a force only double, since it takes twice as much time to rise. These gentlement believe they are justified in maintaining that in the estimation of forces it is necessary to consider not only height attained but also time spent. They are persuaded that the force of a body is composed of two entities, the direct entity of height and the inverse entity of time. They do not realize that time has nothing to do with the subject of their dispute, since it is easy to have a given heavy body rise to different heights in the same time. In this connection, one needs only to think of a cycloid. We know that all arcs beginning with the lowest point are isochronous, that is, performed in the same time.

14. Leibniz answered the objections of his critics, but made no headway against prejudices in favor of the common and erroneous view that the force of a body in motion is equal to the quantity of motion, that is, the product of mass times velocity. It was in vain that he pointed out to his adversaries that, if the opinion they maintained were to hold good, one could bring about a purely mechanical perpetual motion, which according to Leibniz should be absolutely impossible. But his opponents preferred to admit the possibility of perpetual motion rather than abandon an opinion held for a long time in order to embrace a new one, which they regarded as a kind of heresy in physical science.

15. Shortly before the death of Leibniz, his views were completely rejected in England and indeed treated with contempt. They were attacked in a collection of letters between Clarke and Leibniz issued in two printings with notes. They were attacked, I say, to pour ridicule on the views of this great man concerning *vis viva,* but not without exciting extreme surprise on the part of those who recognized the truth of Leibniz's views.

16. It is true that the number of Leibniz's supporters is still very small in the rest of Europe. I am perhaps the first in about 28 years. It is not that the deductions of Leibniz have themselves appeared strong enough to cause me to embrace his views. For I admit that since his deductions were indirect and not based fundamentally on the matter in hand, they were not able to convince me. But they stimulated my thought, and it was only after a long and serious meditation that I finally found the means of convincing myself by direct deductions beyond every doubt; Leibniz, to whom I communicated these things, wished me well. My efforts have served to attract some followers to him and to reconcile to his views some of those who previously had been engaged in a long dispute with him, at length being fully convinced by his reasoning.

17. As for myself, I embrace with pleasure the occasion of making known my discoveries to the illustrious members of the Royal Academy of Sciences and consider it an honor to submit my work to their judgment. They are judges with minds equally brilliant and penetrating and are incapable of partisanship and prejudice. Equity is the only law in their decision. I flatter myself that they will take the trouble to examine with care what I have the honor to propose as the true way of estimating the quantity of force of a body in motion. This question is an intricate one. It demands all the more attention since philosophers and well-known mathematicians have treated it with contempt. If this discourse has the good fortune to please my judges, I should be willing to add several useful remarks that lack of time has prevented me from communicating here. The matter in question is abundant. It would merit a complete treatise. I have included here the most essential aspects.

CHAPTER IX
General Geometric Demonstration of the Theorem That the Quantity of *Vis Viva* Is Proportional to the Product of Mass by the Square of the Velocity

1. I propose to give here a general demonstration strong enough that, beyond all exception and quite alone, I believe it capable of convincing the most obstinate partisans of the common opinion. This demonstration is also based on the decomposition of motion. I shall prove that when a body has precisely enough velocity needed to bend a spring, against which it strikes perpendicularly, the same body with double the velocity can bend not two but four springs equal to the first; with a velocity three times as great it will be able to bend nine springs; and so on.

2. In order to be convinced of the truth of this, let us take the case of a body C (see figure) that strikes obliquely a spring placed at L with velocity *CL.* Let the angle of obliquity *CLP* be 30°, so that the perpendicular *CP* becomes equal to *CL*/2. Let the velocity *CL* = 2. Let the resistance of the spring at *L* be such that, to bend it, it takes precisely 1 unit of velocity in body *C,* when *C* strikes it perpen-

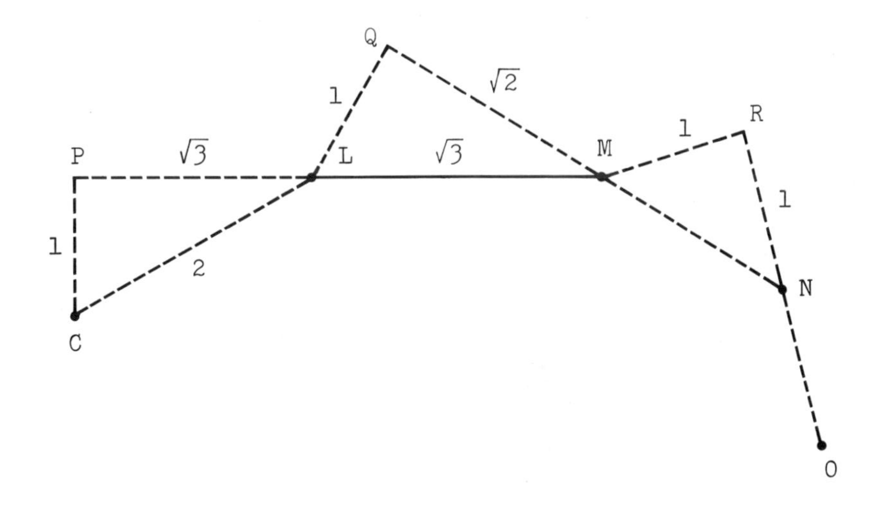

dicularly. We suppose that body C moves on a horizontal plane. This being given, I say that after body C has hit the spring L obliquely with velocity $CL = 2$ units, a velocity which because of the composition of motions is composed of $CP = 1$ unit and $PL = \sqrt{3}$ units, the body C will lose entirely its perpendicular motion along CP and will retain only its motion along PL. Hence the body C, having used up its motion along CP in bending the first spring L, will continue to move in the direction PLM with velocity $LM = PL = \sqrt{3}$. Let us imagine placed at M a second spring like the first, with the angle of obliquity equal to LMQ, such that the perpendicular $LQ = 1$. It is clear that the motion along LM, being compounded of the two motions along LQ and QM with the motion along LQ, will be entirely used up in bending the spring M; the motion along QM will continue in the direction QMN with velocity $MN = QM = \sqrt{2}$. Imagine placed at N a third spring equal to each of the preceding ones and imagine that the body meets it at an angle of 45° (MNR), so that MN (perpendicular to the line along which the spring at N is directed) becomes equal to 1. It is clear that the motion along MN, compounded of motions along MR and RN, will use up the first of these motions along MR in bending the spring at N, and consequently its other motion along RN will continue with a velocity $NO = RN = 1$. The body C will then have retained a velocity of 1 unit in the direction RNO, after having bent the three springs L, M, and N, and it is with this velocity that C will bend the fourth spring O against which it is assumed to strike perpendicularly.

It would appear from all this that the body C with 2 units of velocity has the ability to bend four springs, each of which to be bent requires 1 unit of velocity in C. But the bending of these four springs is the total effect of the force of body C moving with 2 units of velocity. The whole of this velocity of body C is used up in bending these four springs one after the other. The bending of a single spring is the total effect of the force of the same body C moving with 1 unit of velocity. The resistance of each spring is such that it precisely destroys 1 unit of velocity in body C. Since the total effects are combined as the forces that produce these effects, it is necessary that the *vis viva* of body C moving with 2 units of velocity must be four times the *vis viva* of the same body moving with 1 unit of velocity.

3. We can demonstrate in the same way that a triple, quadruple, and quintuple, velocity gives to a body C a force 9, 16, and 25 times greater, respectively, because in this case it will be able to bend before stopping 9, 16, and 25 equal springs. The method of demonstration is similar to the special case just discussed. I draw from all this the general conclusion that the *vis viva* of a body is proportional to the square of the velocity and not to the velocity itself.

CHAPTER X

Concerning the Three Laws That Are Always Observed in the Direct Collision of Two Bodies; One of These Laws Taken at Will Always Has a Necessary Connection with the Other Two

1. Let us now add to what has just been said in the previous chapter some reflections on the three laws that are always obeyed by hard bodies, which I call perfectly rigid, when they collide with each other. The first of these laws states the conserva-

tion of relative velocity before and after collision. One finds the relative velocity by obtaining the difference of the absolute velocities of the bodies, if the bodies are traveling in the same direction and the sum of the bodies are going in the opposite direction.

[*Ed. note:* The author refers to a demonstration of this law in a previous chapter of his discourse. Effectively, he is using Newton's coefficient of restitution by writing the relative-velocity relation in the form

$$x - y = -e(a - b)$$

where a and b represent the velocities of particles with masses A and B, respectively, before collision, and x and y are the corresponding quantities after collision, to use Bernoulli's notation. Newton found from his experiments that for perfectly rigid bodies the coefficient of restitution approaches a limiting value of unity. In this case then

$$x - y = b - a$$

which is the first law cited by Bernoulli. It is of interest that he nowhere makes any mention of Newton in this connection.]

The second law is that of the conservation of the quantity of direction, or conservation of the sum of the masses of the two bodies multiplied by the velocity of their common center of gravity.

[*Ed. note:* This amounts to conservation of momentum or quantity of motion. Thus Bernoulli's formulation would be

$$A + B\frac{Aa + Bb}{A + B} = A + B\frac{Ax + By}{A + B}$$

But this is the same as

$$Aa + Bb = Ax + By$$

or conservation of momentum. It is hard to say why Bernoulli did not use the simpler expression or refer to the law as conservation of momentum.]

The third law is the conservation of *vis viva.* [*Ed. note:* Strictly, this holds only in the limit of the collision of hard, rigid bodies, what are called in modern nomenclature perfectly elastic bodies.] It would only make the law obscure if I were to attempt to derive it here. Everyone regards as an incontrovertible axiom that no efficient cause may be destroyed either in whole or in part, but that it produces an effect equal to its (apparent) loss. The idea we have of *vis viva* [*Ed. note:* Bernoulli always uses the French form, *force vive*], an entity existing in a moving body, is of something absolute and independent that would remain in the body even if the rest of the universe were destroyed. It is then clear that, if the *vis viva* of a body decreases or increases on collision with another body, the *vis viva* of the other body will increase or decrease by the same amount, the increase in the one being the immediate effect of the decrease in the other. This necessarily implies the conservation of the total quantity of *vis viva,* or that the total quantity remains constant during the collision. [*Ed. note:* It must be emphasized that the author is here confining his attention to purely elastic collisions. He takes no account of the much more commonly observed inelastic collisions in which the *vis viva* is certainly not conserved.]

2. But however evident and certain this law is through the very idea involved in *vis viva,* what is not clear up to now is the manner of measuring this force. There has been a general prejudice in favor of the idea that this force is proportional to the product of mass and velocity. It is this prejudice that has led to a false view of the meaning of conservation of quantity of motion. People have abandoned this wrong conception only since some well-known authorities have demonstrated that the quantity of motion can be increased or diminished in the collision of a body, without indeed demonstrating, however, the true method of measuring *vis viva.*

[*Ed. note:* This is a confusing statement. Bernoulli can hardly be denying the validity of conservation of momentum on collision, since he has just previously stated this as one of the laws governing collisions. It must be that he refers to the quantity of motion of a single body undergoing collision. "Prejudice" in favor of quantity of motion refers, of course, to the insistence of the followers of Descartes, even in the eighteenth century, that the only true measure of the force of a body in motion is quantity of motion, and that Leibniz's so-called *vis viva* is not really a force at all, but a mere mathematical device. Bernoulli, like other followers of Leibniz, insists that the latter's version of living force is the genuine force associated with a body in motion. Yet he has to accept the role of quantity of motion in collisions. He was somehow so imbued with the notion that it was physically necessary to have a force inherent in a body in motion that he refused to see the possibility that both quantity of motion and Leibnizian *vis viva* are useful ways of describing the motion of bodies. Here he lacked the broader view of d'Alembert, who, as we have seen earlier, felt that there was really no fundamental conflict between the two notions.]

Leibniz was the first to discover that *vis viva* is measured by the product of mass and the square of the velocity. But, as we have seen, few people agreed with his reasoning. I believe I have established the truth of his reasoning in a manner so evident that henceforth it will be beyond all argument.

3. We must add some reflections on the nature of the threefold law. The three conservation laws involved are (1) conservation of relative velocity, (2) conservation of quantity of direction [*Ed. note:* As we have already seen, this is really conservation of momentum], and (3) conservation of the sum of the product of mass times the square of the velocity. If we grant two of these laws, the third necessarily follows mathematically. I demonstrate this in the following way. Let A and B be two bodies and let their velocities before collision be a and b, respectively. Let their velocities after collision be x and y, respectively. We assume at first that both before and after collision the bodies move in the same direction. The first conservation law then yields the relation

$$a - b = y - x \qquad (1)$$

The second yields

$$Aa + Bb = Ax + By \tag{2}$$

[*Ed. note:* Bernoulli takes the mass of A as A, and similarly for B.]

I then deduce the third conservation law in the following fashion. By transposition of terms in (1), we get

$$a + x = b + y \tag{3}$$

By a similar transposition of terms in (2), we get

$$Aa - Ax = By - Bb \tag{4}$$

Multiplication of the two sides of (3) and (4) yields finally

$$Aa^2 + Bb^2 = Ax^2 + By^2 \tag{5}$$

Formula (5) expresses perfectly what we seek, that is, the conservation of the sum of the products of the mass by the square of the velocity. [*Ed. note:* Bernoulli then goes on to point out that the same result is obtained if the bodies do not move in the same direction before and after collision.]

4. From this calculation it appears that the conservation of the sum of the products of the masses by the squares of the velocities has a necessary connection with the other two conservation principles. Anyone with a little mathematical ability could have drawn this conclusion at once without perhaps grasping its fundamental meaning and usefulness. To such a person the result might appear as merely a sterile bit of mathematics. This is apparently what happened to M. Huygens, the great mathematician and genius of the first order. From this proposition he demonstrated a theorem [*Ed. note:* involving the products of masses by the squares of velocities] (see the long demonstration given in his treatise "De motu corporum ex percussione," Prop. XI), but somehow failed to realize that the conservation of *vis viva* is hidden in it. Huygens was without doubt unacquainted with the fact that the force of a body in motion is proportional to the product of the mass and the square of the velocity, or if not ignorant of it he refused to admit this proposition. Without direct recourse to nature and its first principles the most important theorems degenerate into simple speculation.

[*Ed. note:* Bernoulli is scarcely fair to Huygens here. The latter's work on collisions (originating around 1656) came long before Leibniz called attention to the importances of mv^2 as *vis viva* (1686). Huygens was still thinking of force in the traditional Galilean–Cartesian fashion ans saw no reason to attribute special physical significance to the appearance of mv^2 in his analysis. The same comment applies to Huygens's work on the center of oscillation of a compound pendulum set forth in his *Horologium Oscillatorium* (1673). Here again mv^2 comes out in the analysis, but Huygens evidently saw no fundamental physical significance in this. Bernoulli would seem to be correct in maintaining that Huygens should not be given credit for the discovery of *vis viva* over Leibniz, although certain modern authorities have suggested such priority is justified.]

5. Now that this truth is established beyond all attack, one can only admire the perfect conformity which reigns between the laws of nature and the results of mathematics, a conformity so constantly observed under all circumstances that it almost seems as if nature had consulted the mathematicians in establishing the laws of motion. For it might well have been possible that the forces in bodies in motion were not actualy proportional to the products of the masses multiplied by the square of the velocities, and that nature had been put together in quite another fashion. In that case, mathematical order would have been violated. *Vis viva,* that unique source of the continuation of motion in the universe, would not have been conserved, nor consequently would the equality of efficient cause and its effect have been maintained; in a word, the whole of nature would have fallen into chaos.

[*Ed. note:* In the remaining four chapters of his discourse, Bernoulli attacks the problems presented by the collision of more than two bodies. The details here, however, do not involve any further fundamental points concerning the role of *vis viva.* He also provides a demonstration of the center of oscillation of a compound pendulum, with more emphasis on the role of *vis viva* in it. In an appendix to the discourse he takes up the problem of the physical nature of elasticity.]

13 B

Reprinted from *A Source Book in Physics,* W. F. Magie, ed.,
McGraw-Hill Book Company, New York, 1935, pp. 48–50

THE PRINCIPLE OF VIRTUAL DISPLACEMENTS AND USE OF THE WORD ENERGY (from Letter of Johann Bernoulli to Pierre Varignon, 1717)

Johann Bernoulli

In a letter written from Bâle on January 26, 1717, M. (Jean) Bernoulli, after defining what he means by the word *energy*, in the way that will appear in the following definition, enunciated the principle that in every case of equilibrium of any forces, however they are applied to equilibrate one another, whether immediately or mediately, the sum of the positive energies will be equal to the sum of the negative energies, taken as positive.

This proposition seemed to me so general and so beautiful, that, seeing that I could easily deduce it from the preceding theory, I asked his permission, which he granted, to add it in this place, with the demonstration which my theory furnished of it, and which he did not send me.

. . . .

. . . .

Here is his explanation of what he understands by the word *energy*, in the letter in which he enunciated this beautiful proposition.

Definition XXXII

Conceive (says he) several different forces which are acting along different lines or directions of tendency to maintain in equilibrium a point, a line, a surface, or a body; conceive also that we impress on the whole system of forces a small displacement, either parallel to itself along any direction, or about any fixed point; it is easy to see that by this displacement each of these forces will advance or recede in its direction, unless some one or more of the forces have their directions perpendicular to the direction of the small displacement; in which case the force or the forces will neither advance nor recede: for these advancements or recessions, which I call *virtual velocities*, are nothing other than the amounts by which each line of tendency increases or decreases because of the small displacement; and these increments or decrements are found by drawing a perpendicular from the end of each line of tendency which will cut off from the line of tendency of each force in the neighboring position, to which it has been brought by the small displacement, a small portion which will be the measure of the *virtual velocity* of this force.

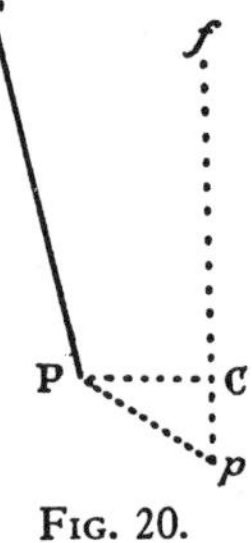

Fig. 20.

For example, let P (Fig. 20) be a point in the system of forces which is in equilibrium; let F be one of these forces, which pushes or pulls the point P in the direction FP or PF, let Pp be a small straight line which the point P describes in a small displacement, by which the line of tendency FP takes the position fp, which will either be exactly parallel to FP, if the displacement of the system is so made that all its points move parallel to a given straight line; or, prolonged, will make with FP an infinitely small angle, if the displacement of the system takes place around a fixed point. Now draw PC perpendicular to fp, and you will have Cp representing the *virtual velocity* of the force F, so that $F \times Cp$ is what I call the *energy*. Notice that Cp is either positive or negative with respect to other similar lines: it is positive, if the point P is pushed by the force F and the angle FPp is obtuse; and negative if the angle FPp is acute; but on the contrary, if the point P is pulled by the force, Cp is negative, if the angle FPp is obtuse, and positive, if it is acute. All this being well understood, I lay down (says M. Bernoulli) the following

General Proposition, Theorem XL

In every case of equilibrium of forces, in whatever way they are applied, and in whatever directions they act on one another, either mediately or immediately, the sum of the positive energies will be equal to the sum of the negative energies, taken as positive.

14

Reprinted from *A Source Book in Physics,* W. F. Magie, ed.,
McGraw-Hill Book Company, New York, 1935, pp. 55–58

THE VIS VIVA CONTROVERSY

Jean Le Rond d'Alembert

Jean le Rond d'Alembert was born on November 17, 1717, and died on October 29, 1783. He was an illegitimate child. His mother, Madame de Tencin, exposed him on the steps of the church of St. Jean le Rond. He was placed in charge of a family from which he took his name. At the age of twelve he entered the Collège Mazarin, where he distinguished himself by ability in his studies. In 1741 he was made a member of the French Academy. He collaborated with Diderot in the preparation of the mathematical part of the *Encyclopédie* and was a prominent leader in the stirring intellectual life of the times.

The extract which follows presents d'Alembert's discussion of the measure of the quantity of motion. This question, first proposed by Leibnitz, had since been discussed by many distinguished mathematicians. D'Alembert shows that the question is of no importance for mechanics and is a mere discussion about words. His statements ended the controversy.

The extract is taken from the *Traité de Dynamique,* published in 1743.

Quantity of Motion

. . . All that we see distinctly in the motion of a body is that the body traverses a certain distance and that it takes a certain time to traverse that distance. It is from this one idea that all the principles of mechanics should be drawn, if we wish to demonstrate them in a clear and accurate way; so no one need be surprised that for this reason I have turned my thought away from causes of motion to consider solely the motions that they produce; and that I have entirely excluded forces inherent in bodies in motion, obscure and metaphysical entities which can only cast shadows on a science that is in itself clear.

It is for this reason that I have thought it best not to take up the consideration of the famous question of *vis viva.* For thirty years mathematicians have been divided in opinion as to whether the force of a body in motion is proportional to the product of the mass by the velocity, or to the product of the mass by the square of the velocity; for example whether a body twice as large as another one, and which has three times as much velocity, has eighteen times as much force or only six times as much. Notwithstanding the disputes to which this question has given rise, its perfect uselessness for mechanics has induced me to make no mention of it in the present work; I believe however that I should not pass over altogether in silence an opinion, which Leibnitz took credit to himself for discovering, which the great Bernoulli has since studied in such a sound and successful way, which MacLaurin has done all he could to overthrow, and about which the writings of a great many illustrious mathematicians have excited public interest. So, without wearying the reader by the detail of all that has been said on this question, it may not be out of place to state briefly the principles by which it can be solved.

When we speak of the force of bodies in motion, either we have no clear idea of what the word means or we can only mean in general the property of moving bodies by which they overcome the obstacles that they encounter, or resist them. It is therefore not by the distance that a body traverses with uniform motion, or by the time that it takes to traverse that distance, or finally by the simple, unique, and abstract consideration of its mass and velocity that we can at once estimate the force; it is solely by the obstacles that a body encounters and by the resistance that these obstacles offer to it. The greater the obstacle that a body can overcome, or that it can resist, the greater may we say is its force, provided that, without meaning to express by this word a hypothetical entity which resides in the body, we use the word only as an abbreviated way of expressing a fact, just as we say that one body has twice as much velocity as another instead of saying that in equal times it traverses twice the distance, without intending to mean by this that the word velocity represents an entity inherent in the body.

This being understood, it is evident that we may oppose to the motion of the body three types of obstacles; impenetrable obstacles, which entirely destroy the motion whatever it may be; or obstacles which have precisely the resistance necessary to destroy the motion of the body and which destroy it instantly, in the case of equilibrium; or finally obstacles which destroy the motion little by little in the case of retarded motion. As the impenetrable obstacles destroy similarly all sorts of motions, they cannot help us to discover the force; it is therefore only in equilibrium or in retarded motion that we should look for the measure of this force. Now everyone agrees that there is equilibrium between two bodies when the products of their masses by their virtual velocities, that is to say, the velocities with which they tend to move, are equal. Therefore in the case of equilibrium the product of the mass by velocity, or what is the same thing, the quantity of motion, may represent the force. Everyone agrees also that in retarded motion the number of obstacles overcome is proportional to the square of the velocity; thus, for example, a body which has compressed a spring with a certain velocity, can with twice that velocity compress either all at once or in succession not two but four springs like the first, nine springs with three times the velocity, etc. From this fact the advocates of *vis viva* conclude that the force of bodies which actually are moving is in general proportional to the product of the mass by the square of the velocity. When we consider the matter, what inconvenience can there be in having the measure of forces different in the two cases of equilibrium and of retarded motion, since, if we think only in terms of clear ideas, the word force should be used to signify only the effect produced by overcoming an obstacle or by resisting it. It must nevertheless be admitted that the opinion of those who consider force as the product of mass by velocity may hold good not only in the case of equilibrium but also in the case of retarded motion, if in the latter case we measure the force, not by the absolute quantity of the obstacles, but by the sum of the resistances of these same obstacles. For there is no doubt that this sum of the resistances is proportional to the quantity of motion, since, as everyone admits, the quantity of motion that a body loses at each instant, is proportional to the product of the resistance by the infinitely short duration of the instant, and that the sum of these products is evidently the total resistance. The whole difficulty is to decide whether we ought to measure force by the absolute quantity of the obstacles or by the sum of their resistances. It seems to be more natural to measure force in this latter way, for an obstacle is only one while it is resisting, and it may properly be said that the sum of the resistances is the obstacle overcome: furthermore, in estimating force in this way we have the advantage of a common measure of force both for equilibrium and for retarded motion; nevertheless as we have no clear and distinct idea of the meaning of the word force, except when we restrict the use of the word to express an effect, I believe that we ought to leave everyone free to make his own choice, and then there will be nothing left in the question except either a futile metaphysical discussion or a dispute about words still more unworthy of the consideration of philosophers.

What we have now said should be enough to make the issue clear to our readers. But another very natural consideration should carry conviction. If a body has a simple tendency to move with a certain velocity, which tendency is arrested by some obstacle; if the body moves really and uniformly with this velocity; or finally if it commences to move with the same velocity, which is gradually consumed and annulled by some cause or other: in these cases the effect produced by the body is different, but the body considered by itself has nothing more in one case than in the other; only the action of the cause which produces the effect is applied differently. In the first case the effect reduces to a simple tendency, which, properly speaking, has no precise measure, since no movement results from it; in the second case, the effect is the distance passed over uniformly in a given time, and this effect is proportional to the velocity; in the third case the effect is the distance passed over before the motion is annulled, and this effect is proportional to the square of the velocity. Now these different effects are evidently produced by one and the same cause; so that those who have said that the force is sometimes proportional to the velocity, and sometimes to its square, can only have been speaking of the effect when they express themselves in that way. These different effects coming all from the same cause may help us, as we may remark in passing, to perceive the lack of precision in the statement so often proposed as an axiom, that causes are proportional to their effects.

Finally even those who are not able to go back to the metaphysical principles of the question of *vis viva* will easily see that it is only a dispute about words if they consider that the two parties to the dispute are in every other way entirely in agreement on the fundamental principles of equilibrium and of motion. If we set the same problem in mechanics before two mathematicians, one of whom is opposed to and the other a partisan of *vis viva*, their solutions of the problem, if they are correct, will always agree; the question of the proper measure of force is therefore entirely useless in mechanics and without any real object. Without doubt it would not have brought to birth so many volumes, if care had been taken to distinguish between the clear and the obscure features of it. If it is taken in that way, no one would have needed more than a few lines to settle the question; but it seems as if most of those who have discussed it have been unwilling to discuss it in a few words.

15

This article was translated expressly for this Benchmark volume by R. Bruce Lindsay, Brown University, from "Recherches sur l'origine des forces," in Mém. Acad. Sciences Berlin, 6, *419–447 (1752)*

THE ORIGIN OF FORCES: CONSERVATION OF THE QUANTITY OF MOTION AND VIS VIVA

Leonhard Euler

[*Ed. Note:* In this paper Euler reaffirms with emphasis his conviction that forces are entities that act from the outside on bodies; they are not inherent in the bodies themselves—a view expressed clearly by d'Alembert in his famous *Traité.* Euler concentrates his attention on problems of the collision of particles and finally convinces himself that forces due to collision are indeed the only forces which exist in nature. This gets him into trouble with interplanetary attractive forces and he has to go to some trouble (invention of an ether of space and the like) to get around this. However, our principal reason for including an extract from this treatise is that in discussing a simple linear collision problem Euler, using the second-order differential-equation equivalent to Newton's second law of motion, deduces the law of conservation of momentum and also the energy equation, although, of course, not with this terminology. In fact, in his derivation he employs neither the words quantity of motion nor *vis viva* and makes no mention of either Descartes or Leibniz. He probably felt that by this time the controversy was over, although this is not borne out by the contemporary writings of Daniel Bernoulli and the slightly later work of Koenig (Papers 16 and 17 in this volume).

Euler realizes that collisions can be elastic or inelastic, depending on whether the colliding bodies are ultimately restored to their original size and shape after collision or are permanently deformed. But he says he can handle both kinds. He then discusses a particular case, and we here reproduce enough of his treatment to show its connection with the concept of mechanical energy.]

35. Whether we consider elastic or inelastic bodies, the calculation will be the same during the time that impact takes place. Figure 1 shows two spherical bodies, A and B, that move in the same direction along the same straight line MN. The velocity of body A is assumed to be greater than the velocity of B, before collision, so A can overtake and collide with B. Suppose that this happens at the instant when the center of the first body is at point A and that of the second body is at point B. . . .

36. Let the velocity of body A before impact be a and that of body B be b, so

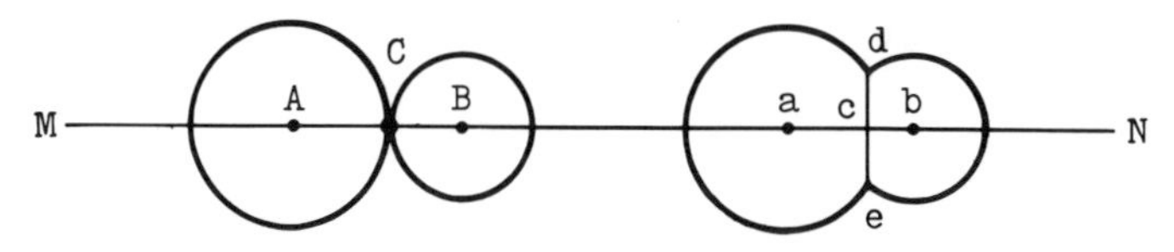

Figure 1

that $a > b$. At the instant of impact the distance between the centers of the two bodies, AB, will be the sum of their semidiameters. If we let the semidiameter of the first body, AC, equal α, and that of the second body, BC, equal β, we shall have $AB = \alpha + \beta$. Let us suppose that at time t after impact, when the bodies are in the situation represented in the figure by ab, the center of A has moved a distance $Aa = x$ and the center of B has moved a distance $Bb = y$. In this state the distance between the centers, ab, becomes

$$AB + Bb - Aa = \alpha + \beta + y - x$$

which will be smaller than $\alpha + \beta$ because of the deformation that the bodies experience during the collision. We shall then have $x > y$; and if we set $x - y = z$, the quantity z will measure the amount of deformation, or the difference between distance ab and AB.

37. We now assume that immediately after impact the velocity of A at a becomes v, and the velocity of B at b becomes u, and that these velocities are different from a and b, respectively. It is also assumed that the velocity v of A after impact is greater than the velocity u of B, so that the bodies are still obliged to act on each other to avoid the penetration of the one by the other. Let the force with which the two bodies act on each other be equal to P. Since the plane of contact of the two bodies (de in Figure 1) is perpendicular to the straight line MN, the body A at a will be acted on by this force P in the direction ca, and the body B at b will be acted on by the same force P in the direction cb.

38. Since the bodies are pressed, one against the other, by the same force P, their deformation will become great in proportion to the distance between the centers $\alpha + \beta - z$. After the lapse of the element of time, dt, it will become smaller; that is, $\alpha + \beta - z - dz$. The force P must be precisely of a magnitude such that in acting on the bodies in time dt it reduces the distance between their centers to $\alpha + \beta - z - dz$. [*Ed. note:* There follows a sentence of qualitative nature, having little to do with the analytical argument.]

39. To find the correct value of the force P, it is only necessary to employ the principles of mechanics. Let the mass of body A be set equal to A and that of body B set equal to B. The velocities u and v will be changed in time dt by du and dv, respectively, so that

$$A\,dv = -P\,dt, \qquad B\,du = P\,dt$$

[*Ed. note:* These are the equations of motions following Newton's second law, which Euler was among the first, if not the first, to write in analytical fashion. (Newton curiously enough never did!).]

Since body A at a is pushed backward by the force, its velocity will be diminished, and its differential will be negative. Similarly, the differential for B will be positive.

40. If we add the two equations, we get

$$A\,dv + B\,du = 0$$

so that on integration we have

$$Av + Bu = \text{constant}$$

This equation holds independent of the value of force P, even indeed if it were indeterminate. At every instant during the collision, the expression $Av + Bu$ keeps the same value and hence is also equal to the value prevailing at the instant of impact: $Aa + Bb$. Hence, one has the equation

$$Av + Bu = Aa + Bb$$

This represents a general property connected with all cases of collision, in accordance with the great principle that the motion of the common center of gravity is not altered by the action suffered by the bodies on impact.

[*Ed. note:* This is, of course, the expression of conservation of momentum, recognized by Descartes long before Euler's time and also effectively by Newton and Huygens and other natural philosophers. It is curious that Euler does not use the term, nor say anything about "quantity of motion." He does not use the word conservation, although he certainly was impressed by the fact that the quantity he was dealing with remains unaffected in its value by the collision.]

41. The equation $Av + Bu = Aa + Bb$ is not sufficient to evaluate the two unknowns u and v. Hence, we turn to the two differential equations in dv and du. Since it is also a question of determining the true value of the force P through the use of the differential dz, we must introduce the space differentials dx and dy such that

$$dt = \frac{dx}{v} = \frac{dy}{u}$$

If we substitute these values for dt into the two differential equations, we get

$$Av\,dv = -P\,dx, \qquad Bu\,du = P\,dy$$

From these we obtain the expressions

$$P = -\frac{Av\,dv}{dx}, \quad P = \frac{Bu\,du}{dy}$$

$$P = -\frac{Au\,dv}{dy}, \quad P = \frac{Bv\,du}{dx}$$

42. But these equations by themselves do not help us to evaluate P. For this we need another integral, which we obtain by adding both sides of

$$Av\,dv = -P\,dx$$

$$Bu\,du = P\,dy$$

to get

$$Av\,dv + Bu\,du = -P\,dx + P\,dy$$

Since $dx - dy = dz$, this becomes

$$Av\,dv + Bu\,du = -P\,dz$$

If we integrate this equation, we get

$$Avv + Buu = \text{constant} - 2 \int P\,dz$$

Since at the instant of impact $z = 0$, $v = a$, and $u = b$, we can write

$$Avv + Buu = Aaa + Bbb - 2 \int P\,dz$$

[*Ed. note:* The integral here is a definite integral with limits $z = 0$, $z = z$. On the left side we have the final *vis viva* of the system of two bodies. The equation says in modern nomenclature that the change in kinetic energy is equal to the work done in the collision. So Euler here effectively glimpsed an important stage in the development of the concept of mechanical energy. But he says nothing about *vis viva* and Leibniz. Possibly this was due to his dislike of Liebniz's assignment of *vis viva* as the true measure of "force" inherent in moving bodies, a view that Euler could no longer accept. He may well have felt it would be confusing to use the name *vis viva* in connection with a collision problem in which the force P was effectively *external* to the bodies and came into play solely because of the impact. The history of *vis viva* is a tangled network in which different authorities insisted on adopting different viewpoints, even when dealing with what we today consider essentially the same analysis. This again emphasizes the slow and tortuous development of physical concepts.]

16

This article was translated expressly for this Benchmark volume by R. Bruce Lindsay, Brown University, from "Remarques sur le principe de la conservation des forces vives pris dans un sens général," in Histoire de l'Académie Royale des Sciences et Belles Lettres de Berlin, *Annee 1748, 1750, pp. 356–364*

THE CONSERVATION OF VIS VIVA

Daniel Bernoulli

1. The usual law of *vis viva* assumes a uniform force of gravity always remaining parallel to itself. As soon as gravity changes because of the change in the situation of the bodies, either with respect to strength or direction, we can no longer with assurance express the *vis viva* by the product of the mass and the distance of fall of the center of gravity, as is usually done. Nevertheless, conservation of *vis viva* still exists if it is correctly interpreted. In a communication sent earlier to the Academy of Sciences in St. Petersburg I have shown how one can explain certain irregularities in the motion of the moon by means of the usual law of conservation of *vis viva*. It will not be superfluous to investigate how this law can be applied to other arrangements than that of uniform gravity in a straight line.

2. If several bodies form an arbitrary system in such a way that no one of them can move independently of the rest, and each is subjected to an arbitrarily changing gravity, the conservation of *vis viva* can be found as follows. Let the system consist of masses m, m', m'', m''', and so on, and let their velocities be v, v', v'', v''', and so on. Consider now a single body as independent of the system, and let the body leave a certain point without being affected by gravity, a point to which the body may return by an arbitrary path. Then it is easy to evaluate the velocity the body must have when free of the system. Let such velocities be called u, u', u'', u''', and so on. The law of the conservation of *vis viva* will then yield, in general form,

$$mv^2 + m'v'^2 + m''v''^2 + \cdots = mu^2 + m'u'^2 + m''u''^2 + \cdots$$

3. I shall not stop here to illustrate and confirm this general law with examples, since my present aim is to demonstrate the wide extent of the law's application, indeed not merely to a system of many bodies, but also in relation to single bodies, when it is a question of relating their *vis viva* to the motions performed by them. This kind of consideration facilitates greatly the determination of the quantities u, u', u'', u''', and so on.

4. I begin with the assumption of a uniform and parallel gravity. The square of the velocity gained in this case is, as is well known, proportional to the displacement, and since this remains independent of the path of the body, there is always a conservation of *vis viva* with respect to the height from which the fall takes place.

Under this assumption, if we have a system of general bodies with masses m, m', m'', and so on, and assume that the acceleration of gravity is equal to unity, and that the vertical fall distances are x, x', x'', and so on, we have

$$u^2 = 2x, \quad u'^2 = 2x', \quad u''^2 = 2x'', \cdots$$

and the general equation in section 2 becomes

$$mv^2 + m'v'^2 + m''v''^2 + \cdots = 2mx + 2m'x' + 2m''x'' + \cdots$$

Thus the total *vis viva* is equal to the product of the total mass of the system with twice the vertical distance the center falls.

5. Let us now assume that the gravity, although still uniform, is directed toward a fixed central point; then in the previous treatment we have to replace the vertical-fall displacements by the displacements toward the gravitational center. The conservation of *vis viva,* both for single bodies and for systems of bodies, can be formulated in the same way with respect to the actual displacements toward the center as in the usual simple assumption of vertical fall. Accordingly, the *vis viva* of a simple body remains the same no matter in what path it moves from one point to another, and for a system of bodies we obtain as before

$$mv^2 + mv'^2 + m''v''^2 + \cdots = 2mx + 2m'x' + 2m''x'' + \cdots$$

in which x, x', x'', and so on, are the displacements of the individual bodies toward the gravitational center.

6. When, however, by some supplementary force the force of gravity is made variable, the situation with respect to the distance to the gravitational center must be treated as follows: a simple body will always have the same *vis viva* if it moves from a given distance from the gravitational center back to this same distance, independent of the path it follows. If the gravitational center is at E in Figure 1, if the body begins its motion at A, and if it reaches point C by the arbitrary path ABC, for the determination of the velocity at C we need only draw the straight line AE, and with E as center draw the arc CD. The velocity at C will be the same as that attained at D by vertical fall toward the center E. It is thus easy to determine the velocity and the *vis viva* of the body, no matter what the form of the law of gravitation. If the body is at a distance x from E and if the acceleration force is ξ, we have

$$u^2 = -2 \int \xi \, dx$$

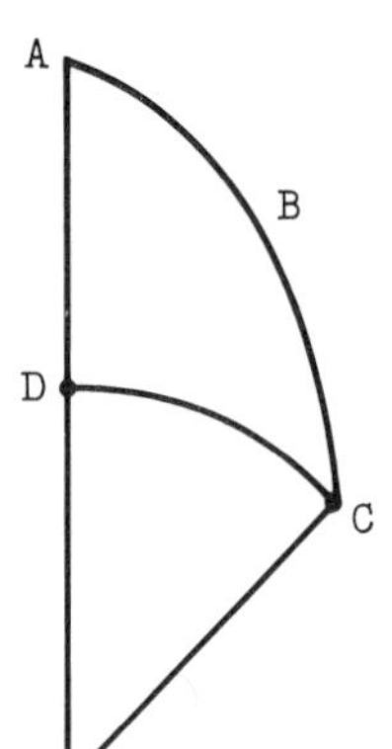

Figure 1

[*Ed. note:* What Bernoulli means by ξ is really the acceleration itself, not the force. Otherwise, the equation could not be dimensionally correct. Actually this error is corrected below.]

If we investigate under the same assumptions a system of arbitrarily moving bodies whose distances from E are x, x', x'', and so on, respectively, and whose accelerations are ξ, ξ', ξ'', and so on, respectively, I maintain that the general law of the conservation of *vis viva* will now be represented by the following equation:

$$mv^2 + m'v'^2 + m''v''^2 + \cdots = -2m \int \xi\, dx - 2m' \int \xi'\, dx' - 2m'' \int \xi''\, dx'' - \cdots$$

In the integration of the various terms one must insert constants of integration according to the initial conditions and also constants for the initial *vis viva* values, in case the motion does not start from rest.

Let the acceleration be inversely proportional to the distance from E and the initial distances be a, a', a''; assume that the system starts from rest, and that the acceleration at distance b is equal to unity. Then

$$\xi = \frac{b^2}{x^2}, \quad \xi' = \frac{b^2}{x'^2}, \quad \xi'' = \frac{b^2}{x''^2}, \cdots$$

The law of the conservation of *vis viva* then yields

$$mv^2 + m'v'^2 + m''v''^2 + \cdots = 2m\left(\frac{b^2}{x} - \frac{b^2}{a}\right) + 2m'\left(\frac{b^2}{x'} - \frac{b^2}{a'}\right) + 2m''\left(\frac{b^2}{x''} - \frac{b^2}{a''}\right) + \cdots$$

7. We now take two force centers and suppose that an arbitrary gravitational law holds for both. Here the same method must be followed to set up the law of the conservation of *vis viva*. We again begin with single bodies. Let points A and B in Figure 2 be the two force centers. Let the body begin at point G and move along any arbitrary path to point H. We draw the infinite straight line ABF, and from B as center draw the arcs GE and HC.

To obtain the *vis viva* at H, we merely have to put together the two expressions for the *vis viva* that the body would obtain if a single gravitational force were impressed, so the whole is provided as in section 6. Accordingly, one calculates the *vis viva* required if the body were attracted to B along path FD, and likewise that required if the body were attracted to A along path EC. The sum of these two will then be the resultant *vis viva* at H.

If we set BH equal to x and AH equal to y and assume that the acceleration toward B is ξ and that toward A is η, and if finally the velocity at H is u, we have

$$u^2 = -2 \int \xi\, dx - 2 \int \eta\, dy$$

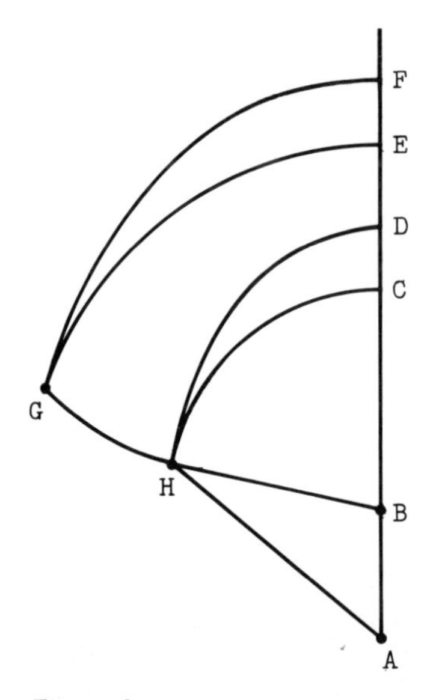

Figure 2

[*Ed. note:* Bernoulli then applies this reasoning to a system of bodies, with the obvious analytical result that need not be repeated here. He then continues.]

We see from the foregoing that our theory can be extended to an arbitrary number of centers of attraction, and we shall always be correct in saying that the total *vis viva* of a body will be found to be the sum of the individual values of *vis viva* with respect to the individual centers, in accordance with section 6. We shall also be able to determine the sum $mu^2 + m'u'^2 + m''u''^2 + \cdots$ as set forth in section 2, that is, the total *vis viva* of a system of bodies.

9. All these results concerning the *vis viva* and its astonishing conservation have stimulated me to investigate what happens if the centers of attraction themselves change their position. This can happen in an infinite variety of ways. However, not to be overzealous in questions that may be answered simply enough, I shall discuss only one case. It will be assumed that the bodies attract each other with a force varying inversely as the square of the distance, a case that I attack with the greatest satisfaction, as it can shed new light on Newton's system.

10. Let there be two bodies. Even though their masses are unequal, their attraction for each other will be equal. Every particle of the one will attract every particle of the other, and the mutual attraction will always be determined by the product of the two masses. Everything else remains as formerly. If one mass increases 10 times and the other 20 times, the mutual attraction will increase 200-fold. For the rest, we assume that every bit of material always exerts the same attraction, an assumption which I do not consider certain for all the material in the universe. Without doubt there is no contradiction in the assumption that the attractive force and inertia of the sun's material are different from the attractive force and inertia of the material of the planets. If we assume that the mass of the planets is much smaller than that of the sun, the latter cannot change its position appreciably for different positions of the planets. Perhaps such an assumption corresponds better

with astronomical observations, which lead to another system than that of Newton. If one wanted to assume that the bodies had different attractions for each other, the mass would still have to be multiplied with that of their attraction. We shall here adhere to the first assumption, since it is the simpler one. To express all quantities by their proper values, we assume that a mass μ at distance ρ produces unit acceleration, that is, the natural acceleration of gravity. Thus μ can denote the mass of the earth and ρ its radius. Let two bodies have masses M and m, respectively, and let their separation be x. They will then attract each other with a force

$$\frac{\rho^2}{x^2} \cdot \frac{Mm}{\mu}$$

The acceleration that M will attain will be

$$\frac{\rho^2}{x^2} \cdot \frac{m}{\mu}$$

and that for m will be

$$\frac{\rho^2}{x^2} \cdot \frac{M}{\mu}$$

11. If the two bodies may move freely so that they move toward each other in a straight line, and if their initial separation is a, then

$$\textit{vis viva of } M = \frac{2Mm}{\mu} \left[\left(\frac{m}{M+m}\right) \cdot \left(\frac{\rho^2}{x} - \frac{\rho^2}{a}\right)\right]$$

Hence, the total *vis viva* of the system of two bodies is

$$\frac{2Mm}{\mu} \quad \frac{p^2}{x} - \frac{p^2}{a}$$

I now maintain that this quantity remains unaltered no matter how the two bodies go from the original separation a to separation x. [*Ed. note:* It is assumed that $x < a$.]

In Figure 3, let ABC be an arbitrary curve along which the body M can move freely. Let the other curve DEF be a similar arbitrary curve for body m. Initially, the bodies are at A and D, respectively. Let them move along arbitrary paths until they have reached B and E, respectively. Let AD equal a and BE equal x. I maintain that the total *vis viva* of the two bodies when they have reached B and E, respectively, is always

$$\frac{2Mm}{\mu} \left(\frac{p^2}{x} - \frac{p^2}{a}\right)$$

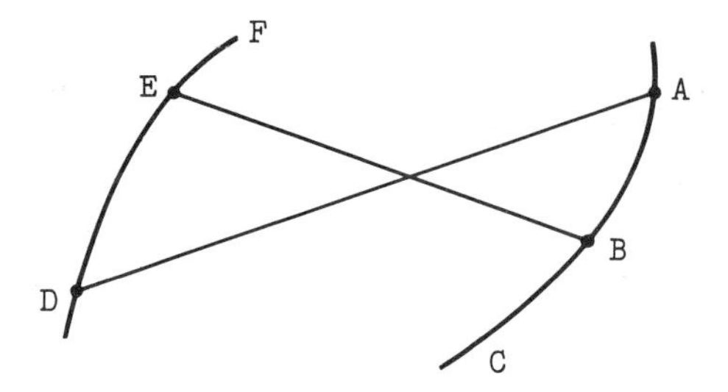

Figure 3

The total *vis viva* depends only on the final and initial separations and not on the path.

[*Ed. note:* Bernoulli then discusses the setting up of the *vis viva* for three bodies in gravitational attraction according to the inverse square law, and generalizes this to *n* bodies. He then continues.]

14. The general law discussed above is valid without exception for every arbitrary law of gravitation [*Ed. note:* He seems to be tacitly assuming *central* forces.] It is only for the sake of producing a simple formula and avoiding integral signs that I have restricted attention to the inverse-square case. Nature never denies the validity of the great law of the conservation of *vis viva,* and that is what I wish to set forth.

15. If the system does not start from rest, what we have calculated as the *vis viva* attained at the final point reached must be considered as the *increase* in *vis viva.* I had hoped to apply the new theory to several problems, but other work has prevented me. Perhaps it will be the subject of the next work I shall have the honor of presenting to the Academy.

17

This article was translated expressly for this Benchmark volume by R. Bruce Lindsay, Brown University, from "De universali principio aequilibrii et motus in vi viva reperto, deque nexa inter vim vivam et actionem, ut riusque minimo," Nova Acta Eruditorum, *Leipzig, 1751, pp. 125–135, 162–176*

VIS VIVA AND THE PRINCIPLE OF LEAST ACTION

Dissertation on a Principle of Universal Equilibrium and Motion Made Plain with the Use of Vis Viva, and Concerning the Connection Between Vis Viva and Action and the Minimum Value of Each

Samuel Koenig
University of Franeker, The Netherlands

When I was recently here at Aquisgranus, where I had come for reasons of health, a friend reported to me (for in our country, Frisia, one hears about such things late or never) that there was much discussion in Germany concerning least action and, indeed, the problem of the most general principle of mechanics. With a great desire to learn what was going on in this matter and happy over such outstanding attempts at the formulation of the most important part of natural philosophy, I learned much from him that I had either neglected or deemed of no importance, owing to the hatred of *vis viva,* openly prevailing in questions of this kind, to the great detriment of our noble science. I now found the business very painful.

I have previously devoted much effort and time to the study of these things and, unless I am mistaken, have been able to lead many to a clearer understanding. I realize that because of the obscurity which prevails, even after such great progress, the skill of even the most acute mathematicians has been held back. I therefore thought I ought not to delay to take advantage of the leisure afforded by my journey to write down a few things strongly fixed in my memory in honor of that celebrated man who made such a great contribution to these matters. It is my desire to speak with greatest force and praise this man [*Ed. note:* Leibniz.] as having come very close to the truth. I maintain that through the light of his genius these things are not only made marvelously clearer, but may be commended with complete assurance to the attention of all learned men.

I heartily applaud those things which have been discovered about *least action* as applied to changes in motion. For within the limits, which I plan to discuss, they are true and involve a fertile and beautiful principle of higher dynamics, whose power to handle difficult questions I have often discussed.

However, when it is affirmed that least action has a place in connection with the equilibrium state, if this assertion is taken in the strict sense, I disagree. For I have determined with certainty that in the state of equilibrium the *vis viva* must be absolutely zero, and the same holds for the action. Hence, if we wish to speak correctly, equilibrium results from the principle that for it both *vis viva* and action vanish, or one may say both are *minimized.*

When, however, in the case of actual motion with a continuing effort to reach a center or centers, *vis viva* and action are given as producers of motion, then, with the system in a state other than equilibrium, the rules of maxima and minima are indeed applicable; but both *vis viva* and action correspond to a maximum rather than a minimum. Thus it develops that this kind of wise parsimony, which mature deliberation seems to suggest, disappears completely in the accepted sense.

But those arguments which I understand are advanced most strongly in support of the universality of the principle of least action, to the detriment of the principle of *vis viva*, in my opinion cannot stand. Rather this view cuts off access to the very sources of the truths of nature. For in our scientific circles, action is so closely connected with *vis viva* that they cannot really be separated. Otherwise, the foundation constructed for these things by the most learned men would utterly collapse. From the native intelligence of the people we are talking about, I doubt that this will happen.

Let us assume in agreement with Leibniz and Wolf, who were the first to write correctly concerning these things, that action is proportional to the product of velocity and space (distance). Hence, since space (distance) is obtained by multiplying velocity by time, it follows that action is proportional to the square of the velocity multiplied by time, that is, proportional to *vis viva* multiplied by time. From the teaching of these same authorities it follows, therefore, that for equal times action is proportional to *vis viva.* This being given, the result is that whenever the action has a maximum or minimum the *vis viva* during the same time must also have a maximum or minimum.

This consideration alone is sufficient to show that anything derived from the principle of least action can also be deduced by the same reasoning from the principle of minimum *vis viva.* This straightway overthrows the laws of motion that have been assigned to rigid bodies by the principle of least action. I plan to discuss later how these laws can be derived from the *vis viva* principle.

The principal argument used to segregate the principle of *vis viva* from the principle of least action is this: the laws of statics or equilibrium can be derived from the principle of least action, whereas it is not possible to derive them from the theory of *vis viva.* I am certain that the facts are quite otherwise. I admit that Leibniz, the author of the doctrine of *vis viva,* and then thereafter John Bernoulli, the teacher and defender of this neglected truth, as well as other promoters and friends, Hermann Bulfinger, Gravesande, and others, did not use *vis viva* in statics, but employed a different kind of force, which they called *vis mortua* (dead force), to be determined by multiplying mass with *initial* velocity. Hence, in the opinion of all, the ability to handle statics was taken away from *vis viva.* But this was a too hasty judgment. For, in ignorance of the universal role of *vis viva* in mechanics, they conceded to its opponents what never should have been conceded. However, what ingenious men declare is so, strong prejudice fixes firmly in the minds of all men, in this case that the laws of statics must be based on the product of initial velocity and mass. This scheme was seen to be in agreement with experiment, and a more probable approach was not apparent. For this reason, I decided to look into new principles, not yet confirmed by the authority of the many, but by no means subject to easy dismissal. Meditating earnestly on the long-term consequences of the perennial fight between the true and the false, I was brought to see that not only is the Cartesian method of defining force as the product of mass and velocity wrong, as was first noted by the genius of Leibniz, but that the reasons, based on *vis mortua,* hitherto given by his opponents as conclusive cannot be further conceded. I insist that the hypothesis that *vis mortua* is determined by the product of mass times velocity is completely devoid of meaning. It had its origin in the foolish and perverse employment of the infinitely small motions exerted by a mere process of thought or devices in equilibrium for the purpose of ascertaining the weight necessary to establish that equilibrium. The perversity of the argument was indeed evident to common sense. Many good mathematicians passed judgment on it to this effect, as they sought other demonstrations of the elements of statics. What there might be of truth in the principle and what of error, no one could say with any certainty.

During a stay in France, through the brilliant studies of *vis viva* of that illustrious woman, the Marquise du Chatelet, I was about to satisfy myself that I ought to examine more carefully the original difficulties with this concept, when I finally came to realize that every state of equilibrium, whether of solids or liquids, is derivable from the theory of *vis viva;* I am convinced that the simplicity of nature and the general principles of dynamics indicate that it cannot logically be done in any other way. I followed up this point of view at length in the treatise that I wrote about these things. While the Marquise was acquainted with me she kept this treatise, but if the statements given out at that time were true, it [later] disappeared. Thereafter, immersed in other business I had no occasion to think of these things until now.

[*Ed. note:* Koenig then goes on to say that he has decided to make evident how the laws of statics can be derived from *vis viva* by discussing in detail some simple examples. He begins with the lever.]

Problem

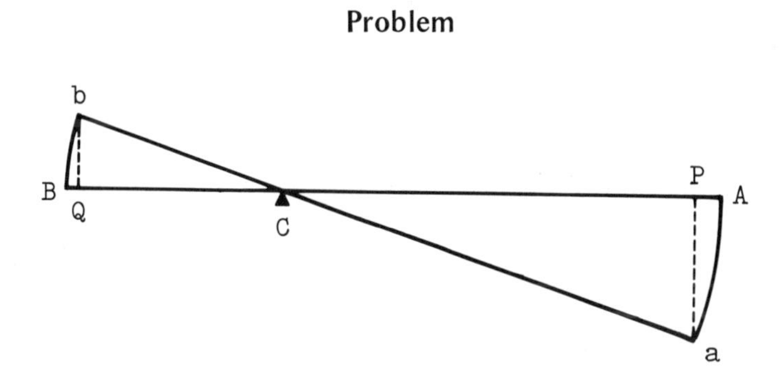

Given the straight line (or rigid weightless rod), as in the figure, with weights *A* and *B* attached at the ends and movable about the fulcrum *C;* determine by the principle of *vis viva* the ratio of the distances of the weights *A* and *B* from the fulcrum necessary to ensure equilibrium.

Solution

Let it be supposed that the line is not in equilibrium, but that A sinks to a, while in the process B moves up to b. Treating A and B as masses, their respective weights will be gA and gB, where g is the acceleration of gravity. Let the distances of A and B from C be CA and CB, respectively, also equal to Ca and Cb, respectively. We shall denote Ca by a and Cb by b for simplicity. Let the descent of the mass A $(= aP)$ be denoted by x. Because of the similarity of the triangles CaP and CbQ, it follows that

$$Qb = (aP) \cdot \frac{bc}{ac}$$

Hence

$$Qb = \frac{b}{a}x$$

Consequently, the *vis viva* of the whole system ACB will be the *vis viva* of the mass A produced by its free descent through aP minus the *vis viva* destroyed by the free ascent of B through Qb. This will be

$$gAx - gB \cdot \frac{b}{a}x = \frac{gx}{a}(Aa - Bb)$$

If it is assumed that this net *vis viva* must vanish for the system to be in equilibrium, we have

$$Aa = Bb$$

This means that the distances of A and B from the fulcrum are inversely proportional to the masses of A and B, respectively. Hence a fundamental theorem of statics has been found by the method of *vis viva.*

[*Ed. note:* Although Koenig gets the right answer, i.e., the law of the lever, one has to be overgenerous in admitting that he has derived it by using the principle of *vis viva.* He is essentially assuming that the *vis viva* gained in a downward free fall of a mass under gravity is proportional to the product of mass times g times the downward displacement (the coefficient of proportionality 2 being usually neglected in eighteenth-century discussions of *vis viva*). This is a valid assumption, following Leibniz, and means that one can safely replace the product of mass and the square of the velocity by a simpler equivalent expression. However, in equating the *vis viva* gained by the one mass (A) to that gained by the other (B), he makes a wholly unwarranted general assumption, holding only in the equilibrium case, which is, of course, what he is trying to establish. His reasoning is therefore at the best circular. It is also true that he talks about the "destruction" of *vis viva* through the ascent of B. This is in complete disagreement with the fact that B actually gains *vis viva* as it rises; certainly it does not involve any loss in *vis viva.* The latter can be "destroyed" only by reducing it to zero. Another obvious defect in Koenig's argument is his failure to say anything about the constraint force involved in the fact that the weights A and B are at the ends of a rigid, weightless rod.

From our standpoint it is clear that what Koenig was actually doing was solving the problem by means of the principle of virtual work or displacements, which does not directly involve the *vis viva* principle. The principle of virtual work says that if a set of forces $\mathbf{F}_i$ is impressed on a system of particles subject to constraints, and if each particle is given a virtual displacement $\delta \mathbf{r}_i$ subject to the constraints, then, for equilibrium to hold,

$$\Sigma \mathbf{F}_i \cdot \delta \mathbf{r}_i = 0$$

In the case of the lever, this reduces to

$$m_A g\, \delta x_A + m_B g\, \delta x_B = 0$$

But from the constraint involved here

$$\delta x_B = -\frac{b}{a}\delta x_A$$

Hence the principle yields

$$m_A g\, \delta x_A - m_B g \cdot \frac{b}{a}\delta x_A = 0$$

Since this must hold for arbitrary δx_A, we have

$$m_A \cdot a - m_B \cdot b = 0$$

which is the law of the lever (with $m_A = A$ and $m_B = B$, in Koenig's notation). The principle of virtual work was well known in Koenig's time, having been first enunciated by John Bernoulli in 1717.

Actually, to treat the problem of the equilibrium of even such a simple system as the lever by direct application of *vis viva* (or kinetic energy, as one half the *vis viva* is now called in modern terminology) is not precisely a trivial exercise, and it is perhaps not surprising that it eluded Koenig. The total mechanical energy of the system of particles A and B is

$$E = \frac{1}{2} m_A \dot{x}_A^2 + \frac{1}{2} m_B \dot{x}_B^2 + m_A g x_A + m_B g x_B$$

This is, however, subject to the constraint

$$x_B = -\frac{b}{a} x_A$$

and hence takes the form

$$E = \frac{1}{2} \dot{x}_A^2 \left(\frac{m_a a^2 + m_B b^2}{a^2} \right) + g x_A (m_A a - m_B b)$$

If equilibrium corresponds to a minimum of E with respect to x_A, we get

$$\frac{\partial E}{\partial x_A} = m_A a - m_B b = 0$$

which again gives the law of the lever. But it is hard to see how Koenig could have reasoned this way. In any case, even this modern method, although it introduces the total energy, does not make direct use of *vis viva.* No wonder there was controversy in Koenig's day over the use of *vis viva* in the establishment of the laws of statics. One can indeed bring the energy more directly into the problem by solving it in terms of Lagrange's equations and showing that the above condition must hold if the total energy is not to involve time (i.e., be a real invariant of the motion). But this is like shooting mosquitoes with heavy artillery!

In the dissertation being discussed here, Koenig went on to consider other problems that he claimed to be able to handle by means of *vis viva* and, indeed, least action as well. In one of these he actually brought in *vis viva* as measured by the product of mass and the square of velocity, that is, a problem of motion. We present this here with critical comments.]

Problem

Given two bodies A and B at the ends of a weightless, rigid rod rotating about the fixed fulcrum C with constant angular velocity; determine the point C for which the *vis viva* and the action become a maximum or minimum.

Solution

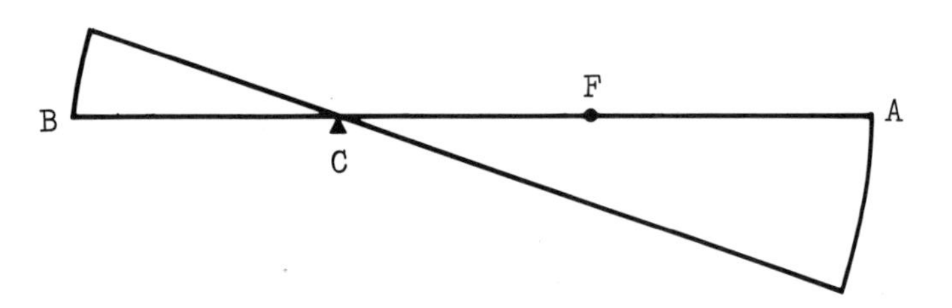

Referring to the figure, let $AB = a$, $AC = z$; thus $BC = a - z$. Let the translational velocity at the arbitrary point F on the rod be denoted by c (where $CF = l$). Since the velocities of points on the rod are proportional to the distances from C, the squares of the velocities will be proportional to the squares of the corresponding distances.

Hence

$$l^2 : z^2 = c^2 : \frac{c^2 z^2}{l^2}$$

where c^2z^2/l^2 is therefore the square of the velocity of body A.

Similarly,

$$l^2 : (a-z)^2 = c^2 : \frac{c^2 (a-z)^2}{l^2}$$

and $\frac{c^2 (a-z)^2}{l^2}$ is the square of the velocity of body B. Hence the *vis viva* of the system will be

$$\frac{c^2}{l^2} [z^2 A + (a-z)^2 B]$$

Because of the equality of the times, this will also be the action. To find the value of z that makes this quantity a maximum or minimum, we differentiate with respect to z and set the result equal to zero. Thus

$$\frac{d}{dz} \left\{ \frac{c^2}{l^2} [z^2 A + (a-z)^2 B] \right\} = \frac{c^2}{l^2} [2zA - 2(a-z)B]$$

When put equal to zero this yields

$$z = \frac{aB}{A+B}$$

This shows that the point C in question is the center of gravity of the system and also that the *vis viva* (as well as the action) for the rotation about point C is a minimum. [*Ed. note:* The fact that the second derivative with respect to z is positive is assurance of the correctness of the last statement.]

The position of C is also that about which A and B are in static equilibrium, as far as rotation is concerned.

[*Ed. note:* The following comments are in order. Leaving out of account the reference to action, we see that the *vis viva* is expressed correctly and the mathematical analysis is correct. The connection of the result with the center of gravity is of interest. But the attempt to tie the result to the state of equilibrium is questionable. The total *vis viva* is certainly not zero under the equilibrium condition, as it ought to be. The statement concerning the equality of *vis viva* and *action* will not hold water. The two quantities are of course of different physical dimensions. Moreover, the action is an integral:

$$\text{action} = \int_{\mathbf{r}_1}^{\mathbf{r}_2} \Sigma\, m_i \dot{\mathbf{r}}_i \cdot d\mathbf{r}_i$$

where the masses of the particles of the system are denoted by m_i and the corresponding position vectors are $\mathbf{r}_i$. This integral can be transformed into a time integral, since

$$\Sigma\, m_i \dot{\mathbf{r}}_i \cdot d\mathbf{r}_i = \Sigma\, m_i\, (\dot{x}_i\, dx_i + \dot{y}_i\, dy_i + z_i\, dz_i)$$

$$= 2T\, dt$$

where T is the total kinetic energy and t is the time. Hence the action becomes

$$\int_{t_1}^{t_2} 2T\, dt = \int_{t_1}^{t_2} (\textit{vis viva})\, dt$$

If the *vis viva* is constant, the action is

$$\text{action} = (\textit{vis viva})\,(t_2 - t_1)$$

But then the action depends directly on the time interval, and it is hard to see what minimization means. Of course, it must be recalled that the use of the term action by Maupertuis and others at the time when Koenig was writing was somewhat lacking in the precision of modern usage. In any case the problem as posed by Koenig is far from realistic. He should have considered the case in which the particles start to move from rest under gravity. But, of course, he did not, and it is perhaps small wonder that his results were not taken too seriously except as a target for attack.

Koenig's dissertation produced an understandable reaction among the defenders of Maupertuis's principle of least action, notably from Euler, who was at that time working at the Berlin Academy of which Maupertuis was then president. In a polemic entitled "Examen Dissertationis Clariss. Professoris Koenig Actis. Erud. Lips. Insertae Pro Mense Martio, 1751," also printed in French as "Examen de la dissertation de M. le Professeur Koenig insérée dans les Actes de Leipzig pour le mois de Mars, 1751" (published in Mémoires de l'Académie des Sciences de Berlin [7] (1751), 1753, pp. 219-245), Euler quite categorically states that in his opinion the arguments set forth by Koenig are worthless. It is clear from his exposition that Euler felt strongly that the *vis viva* idea was of value only for problems involving motion and was of no use in the case of equilibrium. For both motion and equilibrium he believed that the principle of least action of Maupertius was the correct general mode of attack. Euler does not deny the obvious thesis that in static equilibrium, where there is no motion, the total *vis viva* is zero, but considers that in most cases the application of this result to the actual determination of the conditions of equilibrium will be too complicated to be of much use. This is, of course, a correct judgment. It is curious, however, that he seems to accept Koenig's "demonstration" of the law of the lever by the use of *vis viva* as adequate, although he treats it as trivial. It is hard to believe that if he had examined Koenig's argument more carefully he would not have observed the difficulty pointed out in our analysis.

The eighteenth century was a time of strong feelings and intense partisanship with respect to the principles of mechanics. When an able man got an idea that appeared to work successfully in a number of special cases, he tended to cling to it tenaciously and to ignore or attack other ideas that might have application to the same class of problems. Most natural philosophers of the time tended to mingle metaphysical and indeed religious prejudices with their ideas concerning natural phenomena. This made it difficult for most of them to keep what we now call "an open mind." There were exceptions, to be sure. It is well known that d'Alembert stated openly in his *Traité de dynamique* (Paris, 1743) that the controversy over *vis viva* and quantity of motion between the Leibnizians and the Cartesians was merely a useless argument over terminology and had no fundamental significance. Euler must have been familiar with d'Alembert's view, but clearly failed to take the same enlightened attitude. He was evidently not at all impressed by Koenig's emphasis on the fact that action may be expressed not only as quantity of motion times space, but also is *vis viva* times the time. Koenig was obviously unable to utilize this correspondence successfully. This had to wait for people like Lagrange and Hamilton. Modern mechanics tends to place the emphasis on energy in the expressions for and deductions from the principle of least action and its vigorous competitor, Hamilton's principle. This is due to the premier roll energy has come to assume in contemporary science. Euler and his contemporaries could not be expected to foresee this development. In fact, the Cartesians and the followers of Maupertuis could easily and triumphantly point to the fact that in cases of inelastic collisions and explosions, as in gunnery, there is obviously no conservation of *vis viva,* whereas quantity of motion is conserved in all such cases: who could doubt the advantage of momentum as the key idea in motion; before the difficulties confronting *vis viva* could be cleared up, the notion of potential energy had to be developed as well as the puzzle as to what happens when existing *vis viva* appears to disappear completely. Koenig simply did not have sufficient grasp of physics to see through these matters. Unfortunately, he also did not wholly understand what he did see. However, he certainly deserves credit for insisting that attention should continue to be paid to *vis viva* in mechanical problems. Actually, as we see in another article in this book, d'Alembert in his *Traité de dynamique* had already devoted much attention to *vis viva* and its alleged conservation. Koenig makes no reference to this, although he does mention John Bernoulli. Koenig is usually given the credit for having derived the law that the total *vis viva* of a system of particles is equal to the *vis viva* of the center of mass of the system plus the *vis viva* with respect to the center of mass, a result often of use in modern mechanics; so let us not be too harsh on Koenig!]

18

Reprinted from T. Young, *Course of Lectures on Natural Philosophy and the Mechanical Arts*, Vol. 1, London, 1845, pp. 57–61

VIS VIVA AND ENERGY

Thomas Young

LECTURE VIII.

ON COLLISION.

HAVING inquired into the laws and properties of the motions and rest of single bodies under the operation of one or more forces, and into the equilibrium of these forces in different circumstances, we are next to examine some simple cases of the motions of various moveable bodies acting reciprocally on each other. In all problems of this kind, it is of importance to recollect the general principle already laid down respecting the centre of inertia [gravity] that its place is not affected by any reciprocal or mutual action of the bodies constituting the system.

Whenever two bodies act on each other so as to change the direction of their relative motions, by means of any forces which preserve their activity undiminished at equal distances on every side, the relative velocities with which the bodies approach to or recede from each other, will always be equal at equal distances. For example, the velocity of a comet, when it passes near the earth in its descent towards the sun, is the same as its velocity of ascent in its return, although at different distances its velocity has undergone considerable changes. In this case, the force acts continually, and attracts the bodies towards each other; but the force concerned in collision, when a body strikes or impels another, acts only during the time of more or less intimate contact, and tends to separate the bodies from each other. When this force exerts itself as powerfully in causing the bodies to separate as in destroying the velocity with which they meet each other, the bodies are called perfectly elastic: when the bodies meet each other without a re-action of this kind, they are called more or less inelastic. Ivory, metals, and elastic gum, are highly, and almost perfectly elastic: clay, wax mixed with a little oil, and other soft bodies, are almost inelastic: and the effects of inelastic bodies may be imitated by elastic ones, if we cause them to unite or adhere after an impulse, so as to destroy the effect of the repulsive force which tends to separate them.

When two bodies approach to each other, their form is in some degree changed, and the more as the velocity is greater. In general, the repulsive force exerted is exactly proportional to the degree in which a body is compressed; and when a body strikes another, this force continues to be increased until the relative motion has been destroyed, and the bodies are for an instant at rest with respect to each other; the repulsive action then proceeds with an intensity which is gradually diminished, and if the bodies are perfectly elastic they re-assume their primitive form, aud separate with a velocity equal to that with which they before approached each other. Strictly speaking, the repulsion commences a little before the moment of actual contact, but only at a distance which in common cases is imperceptible. The change of form of an elastic substance, during collision, is easily shown by throwing a ball of ivory on a slab of marble or a piece of smooth iron, coloured with black lead or printing ink; or by suffering it to fall from various heights: the degree of compression will then be indicated by the magnitude of the black spot which appears on the ball. It may be shown, from the laws of pendulums, that, on the supposition that the force is proportional to the degree of compression, its greatest exertion is to the weight of a striking body, as the height from which the body must have fallen, in order to acquire its velocity, to half the depth of the impression.

For making experiments on the phenomena of collision, it is most convenient to suspend the bodies employed by threads, in the manner of pendulums; their velocities may then be easily measured by observing the chords of the arcs through which they descend or ascend, since the velocities acquired in descending through circular arcs are always proportional to their chords; and for this purpose, the apparatus is provided with a graduated arc, which is commonly divided into equal parts, although it would be a little more correct to place the divisions at the ends of arcs, of which the chords are expressed by the corresponding numbers. (Plate V. Fig. 72.)

The simplest case of the collision of elastic bodies is when two equal balls descend through equal arcs, so as to meet each other with equal velocities. They recede from each other after collision with the same velocities, and rise to the points from which they before descended, with a small deduction for the resistance of the surrounding bodies.

When a ball at rest is struck by another equal ball, it receives a velocity equal to that of the ball which strikes it, and this ball remains at rest. And if two equal balls meet or overtake each other with any unequal velocities, their motions will be exchanged, each rising to a height equal to that from which the other descended.

The effect of collision takes place so rapidly, that if several equal balls be disposed in a right line in apparent contact with each other, and another ball strike the first of them, they will all receive in succession the whole velocity of the moving ball before they begin to act on the succeeding ones; they will then transmit the whole velocity to the succeeding balls, and remain entirely at rest, so that the last ball only will fly off.

In the same manner, if two or more equal balls, in apparent contact, be in motion, and strike against any number of others placed in a line, the first of the moving balls will first drive off the most remote, and then the second the last but one, of the row of balls which were at rest: so that the same number of balls will fly off together on one side, as descended to strike the row of balls on the other side; the others remaining at rest.

If the line of balls, instead of being loosely in contact, had been firmly united, they would have been impelled with a smaller velocity, and the ball striking them would have been reflected. For when a smaller elastic body strikes a larger, it rebounds with a velocity less than its first velocity, and the larger body proceeds also with a less velocity than that of the body striking it. But if a larger body strikes a smaller, it still proceeds with a smaller velocity, and the smaller body advances with a greater.

The momentum communicated by a smaller elastic body to a larger one is greater than its own, and when the first body is of a magnitude comparatively inconsiderable, it rebounds with a velocity nearly as great as the velocity of its impulse, and the second body acquires a momentum nearly twice as great as that of the first. When a larger body strikes a smaller one, it communicates to it only as much momentum as it loses.

In the communication of motion between inelastic bodies, the want of a repulsive force, capable of separating them with an equal relative velocity, is probably owing to a permanent change of form; such bodies receiving and retaining a depression at the point of contact. When the velocity is too small to produce this change of form, the bodies, however inelastic, may usually be observed to rebound a little.

Bodies which are perfectly inelastic, remain in contact after collision; they must therefore proceed with the same velocity as the centre of inertia [gravity] had before collision. Thus, if two equal balls meet, with equal velocities, they remain at rest; if one is at rest, and the other strikes it, they proceed with half the velocity of the ball which was first in motion. If they are of unequal dimensions, the joint velocity is as much smaller than that of the striking ball, as the weight of this ball is smaller than the sum of the weights of both balls. And in a similar manner the effects of any given velocities in either ball may be determined.

It follows immediately from the properties of the centre of inertia [gravity] that in all cases of collision, whether of elastic or inelastic bodies, the sum of the momenta of all the bodies of the system, that is of their masses or weights multiplied by the numbers expressing their velocities, is the same, when reduced to the same direction, after their mutual collision, as it was before their collision. When the bodies are perfectly elastic, it may also be shown that the sum of their energies or ascending forces, in their respective directions, remains also unaltered.

The term energy may be applied, with great propriety, to the product of the mass or weight of a body, into the square of the number expressing its velocity. Thus, if a weight of one ounce moves with the velocity of a foot in a second, we may call its energy 1; if a second body of two ounces have a velocity of three feet in a second, its energy will be twice the square of three, or 18. This product has been denominated the living or ascending force [the *vis viva*], since the height of the body's vertical ascent is in proportion to it; and some have considered it as the true measure of the quantity of motion; but although this opinion has been very universally rejected, yet the force thus estimated well deserves a distinct denomination. After the considerations and demonstrations which have been premised on the subject of forces, there can be no reasonable doubt with respect to the true measure of motion; nor can there be much hesitation in allowing at once, that since the same force, continued for a double time, is known to produce a double velocity, a double force must also produce a double velocity in the same time. Notwithstanding the simplicity of this view of the subject, Leibnitz,* Smeaton,† and many others have chosen to estimate the force of a moving body by the product of its mass into the square of its velocity; and though we cannot admit that this estimation of force is just, yet it may be allowed that many of the sensible effects of motion, and even the advantage of any mechanical power, however it may be employed, are usually proportional to this product, or to the weight of the moving body, multiplied by the height from which it must have fallen, in order to acquire the given velocity. Thus a bullet, moving with a double velocity, will penetrate to a quadruple depth in clay or tallow: a ball of equal size, but of one fourth of the weight, moving with a double velocity, will penetrate to an equal depth: and, with a smaller quantity of motion, will make an equal excavation in a shorter time. This appears at first sight somewhat paradoxical: but, on the other hand, we are to consider the resistance of the clay or tallow as a uniformly retarding force, and it will be obvious that the motion, which it can destroy in a short time, must be less than that which requires a longer time for its destruction. Thus also when the resistance, opposed by any body to a force tending to break it, is to be overcome, the space through which it may be bent before it breaks being given, as well as the force exerted at every point of that space, the power of any body to break it is proportional to the energy of its motion, or to its weight multiplied by the square of its velocity.

In almost all cases of the forces employed in practical mechanics, the labour expended in producing any motion, is proportional, not to the momentum, but to the energy which is obtained; since these forces are seldom to be considered as uniformly accelerating forces, but generally act at some disadvantage when the velocity is already considerable. For instance, if it be necessary to obtain a certain velocity, by means of the descent of a heavy body from a height to which we carry it by a flight of steps, we must ascend, if we wish to double the velocity, a quadruple number of steps, and this will cost us nearly four times as much labour. In the same manner, if we press with a given force on the shorter end of a lever, in order to move a weight at a greater distance on the other side of the fulcrum, a certain portion of the force is expended in the pressure which is supported by the fulcrum, and we by no means produce the same mo-

* Acta Erudit. Lips. 1686.

† Ph. Tr. 1776, p. 450, and 1782, p. 337. See Desaguliers's Exp. Ph. ii. 92; and Ph. Tr. 1723, xxxii. 269, 285. Eames on the Force of Moving Bodies, Ph. Tr. 1726, xxxiv. 188. Clarke in Ph. Tr. 1728, xxxv. 381. Zendrini, Sulla Inutilità della Questione Intorno alla Misura delle Forze Vivi, 8vo, Venezia, 1804.

mentum as would have been obtained by the immediate action of an equal force on the body to be moved.

An elastic ball of 2 ounces weight, moving with a velocity of 3 feet in a second, possesses an energy, as we have already seen, which may be expressed by 18. If it strike a ball of 1 ounce which is at rest, its velocity will be reduced to 1 foot in a second, and the smaller ball will receive a velocity of 4 feet: the energy of the first ball will then be expressed by 2, and that of the second by 16, making together 18, as before. The momentum of the larger ball after collision is 2, that of the smaller 4, and the sum of these is equal to the original momentum of the first ball.

Supposing the magnitude of an elastic body which is at rest to be infinite, it will receive twice the momentum of a small body that strikes it; but its velocity, and consequently its energy, will be inconsiderable, since the energy is expressed by the product of the momentum into the velocity. And if the larger body be of a finite magnitude, but still much greater than the smaller, its energy will be very small; that of the smaller, which rebounds with a velocity not much less than its original velocity, being but little diminished. It is for this reason that a man, having a heavy anvil placed on his chest, can bear, without much inconvenience, the blow of a large hammer striking on the anvil, while a much slighter blow of the hammer, acting immediately on his body would have fractured his ribs, and destroyed his life. The anvil receives a momentum nearly twice as great as that of the hammer; but its tendency to overcome the strength of the bones and to crush the man, is only proportional to its energy, which is nearly as much less than that of the hammer, as four times the weight of the hammer is less than the weight of the anvil. Thus, if the weight of the hammer were 5 pounds, and that of the anvil 100, the energy of the anvil would be less than [only] one fifth as great as that of the hammer, besides some further diminution, on account of the want of perfect elasticity, and from the effect of the larger surface of the anvil in dividing the pressure occasioned by the blow, so as to enable a greater portion of the chest to cooperate in resisting it.

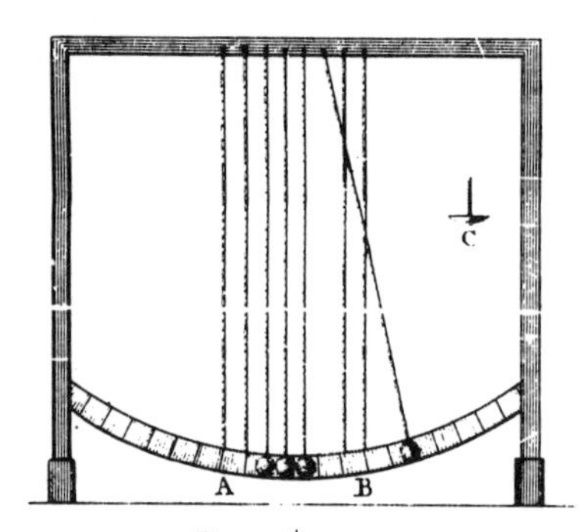

PLATE V. Fig. 72.*

*This figure appears in Thomas Young, *Course of Lectures on Natural Philosophy and the Mechanical Arts,* Vol. 2, London, 1845.

19

This article was translated expressly for this Benchmark volume by R. Bruce Lindsay, Brown University, from Principes fondamentaux de l'équilibre et du mouvement, *Imprimerie De Crapelet, Paris, 1803, pp. 33–37*

VIS VIVA AND MACHINES

Lazare Carnot

CONCERNING MOVING FORCES AND *VIS VIVA*

53. We call forces "moving" (*mouvantes*) or "motive" forces (or forces comparable to weights) in so far as they are applied to machines to overcome resistance or to produce motions of any kind. As we often employ for this purpose men, animals, flowing water, the wind, springs, water turned to vapor by heat, and so on, it is to these agents, or rather to the immediate effects that they produce on the bodies to which they are applied, that we give the name of moving forces. The agent is called the mover, and the force that it exerts or the movement that it produces we call the moving force.

54. Mechanics does not probe into the first causes that produce motion. It does not examine how the will of a man or an animal makes its organs change from rest to motion or spontaneously restores them to rest. Mechanics sees only the resulting phenomenon, and considers only the motion produced. Its only object is to find out how motion, once started, is maintained, propagates, or is modified, leaving aside all new foreign influences. Thus, mechanics does not base its calculations at all on the facultative force of the mover, but only on the effective force that the mover produces, which we have just called *moving force.*

55. We call the *vis viva* (*force vive*) of a body the product of its mass multiplied by the square of its velocity. And here we shall examine this new kind of quantity. Experience shows, as we have just observed, that men, animals, and other such agents can exert forces comparable to weight, either indeed by their own weight or by spontaneous efforts of which they are capable. But two equally natural ways exist of evaluating the action that they effectively exercise. The first consists in seeing what load a man, for example, can carry or what effort, evaluated in terms of weight, he can exert while remaining at rest. Then the force of the man is a stress force equivalent to such and such a weight and is sometimes called *dead force* (*vis mortua*).

56. The second method of evaluating the force of a man, horse, or the like, is to examine the work it is capable of doing in a given time, in a day, for example, of steady labor. From this viewpoint, to reach a precise evaluation as in the previous case, we may still compare the result of the work done with the effect of the weight. For it is natural to evaluate this work either by the weight raised in a given time or by the height to which the weight is raised. This is what is meant when it is said that a horse is equal, as far as force is concerned, to seven men. One should not say that if seven men were to pull in one direction and the horse in the opposite there would be equilibrium. One should rather say that with steady labor the horse by himself will raise as much water from the bottom of a well of given

depth as seven men together in the same time. When one employs laborers, our interest is in knowing what they are able to do in the way of work of a kind analogous to that of which we have just been speaking, rather than knowing how large a load they can carry without moving. This new way of looking at force is then at least as natural and as important as the first. And as it makes sense to say that to raise a 100-kilogram weight to a height of 1,000 meters is the same in this method of evaluating force as to raise a 200-kilogram weight to a height of 500 meters, it follows that forces from this point of view should be considered in direct proportion to the weight to be raised and the heights to which they are raised, or other works comparable to these. It is on this that the notion of *vis viva* is based.

57. Let M be a mass and P its weight. Let g be the acceleration of gravity, dt the element ot time, and H the height to which P has been raised. In consequence of the new way of looking at forces, the force that has to be employed to raise P to height H will be PH. But since H is the space traversed, it can be expressed as the product of a velocity V and a time T. On the other hand, we have

$$P = gM = \frac{g \cdot dt \cdot M}{dt}$$

But $g\ dt$ is a velocity V'; hence

$$PH = \frac{MVV'T}{dt}$$

Since dt and T are two homogeneous quantities, Ph will be the product of a mass multiplied by the product of two velocities or the square of the mean proportional velocity of V and V'. Hence, the force PH is resolved into the product of a mass multiplied by the square of a velocity, which may be written Mu^2, where u is the mean proportional between V and V'. We have here the natural origin of the notion of *vis viva*. There has been a great controversy over the question of whether the force of a body in motion should be represented by the product of the mass multiplied by the velocity or by the product of the mass multiplied by the square of the velocity. As has been seen, this reduces to a dispute about words. Provided we reason logically in accordance with the definition that we have adopted, the conclusion will always be the same, since we always start from the same foundation.

58. By the simple name *force or power*, or *force*, to be exact, we understand quantity of motion (momentum) and motive force (*force motrice*), or, if you wish, force of impact and stress force, since they are subject to the same laws of decomposition. But when we wish to designate a living force (*vis viva*), we must always add the term that characterizes it, "living."

59. We have just seen that living force (*vis viva*) can take either the form Mu^2 (that is, a mass multiplied by the square of a velocity) or the form PH (that is, a motive force multiplied by a distance). In the first case we are dealing with *vis viva* properly speaking. In the second case we should probably use the term *latent vis viva.*

Editor's Comments on Paper 20

The Energy Equation for a General Dynamical System
20 J. L. LAGRANGE

Joseph Louis Lagrange (1736–1813) was born in Turin, Italy, and received his education there. The paternal side of the family was of French origin. Lagrange began his professional career in Italy, where he remained until 1766, when he went to Berlin to take a position in the Berlin Academy. From 1782 until his death in 1813 he lived in Paris and is usually thought of as a French mathematician. He managed to survive the French Revolution and became a favorite of Napoleon, who made him a count of the Empire. His contributions to mathematics, especially to the development of the calculus of variations, are well known. Very early in his career he began to be interested in mechanics. The introduction to his early memoir "Recherches sur la nature et la propagation du son" (Turin, 1759) is reproduced in English translation in *Acoustics, Historical and Philosophical Development*, edited by R. Bruce Lindsay (Benchmark Series on Acoustics, Dowden, Hutchinson & Ross, Inc., Stroudsburg, Pa., 1973), Paper 13, page 125. While still in Turin, Lagrange wrote a brief mathematical memoir "Essai d'une nouvelle méthode pour déterminer les maxima et les minima des formules intégrales indéfinies" (Turin, 1760) (see *Oeuvres*, vol. 1, pp. 336–362). In this essay the method later known as calculus of variations was set forth. Immediately following this, or possibly at the same time, he wrote the memoir "Application de la méthode exposée dans le Mémoire précédent à la solution de différents problèmes de dynamique." In this second memoir (1760) Lagrange used the principle of least action already introduced in a somewhat uncertain fashion by Maupertuis (1698–1759) in 1744. Lagrange put the principle on a firm analytical foundation and used it in a general method for solving problems in the motion of systems of particles. But he needed another principle to complete the solution. For this he used what we now call the energy equation as set up by Daniel Bernoulli (see Paper 16), in which one half the *vis viva* plus an integral that we recognize as the work done by the forces acting on the system is equal to a constant. Although he does not actually use the term *vis viva* (of the equation he merely writes that "all the mathematicians know it"), he was clearly convinced of the usefulness of the *vis viva* concept in mechanics. Curiously enough he does not say he is using least action either, but the equation he writes is clearly an analytical formulation of Maupertuis' principle.

Lagrange summarized his views on mechanics and finally put the subject on a firm analytical basis in his famous treatise *Mécanique analytique* (Paris, 1788). We reproduce here from the second edition (1811), those pertinent extracts in which Lagrange reviews the various fundamental principles of mechanics and his development of what he and his contemporaries called the principle of the conservation of *vis viva* (*force vive* in French). This did not mean the constancy of *vis viva* during the motion of a dynamical system, but rather the invariance of vis viva, plus a function of the position of the particles of the system, what we now call the potential energy. In other words, Lagrange here derived essentially the equation of mechanical energy, whereas in his earlier memoir he had merely assumed it, but he never used the word energy in his treatise. This is a bit strange since, before the second edition of Lagrange's *Mécanique analytique*, Thomas Young had already suggested the employment of this term for *vis viva* (see Paper 18).

20

This article was translated expressly for this Benchmark volume by R. Bruce Lindsay, Brown University, from Mécanique analytique, *2nd ed., Vol. 1, Pt. 2, Sec. 1 & 3, Mme Ve Courcier, Imprimeur-Libraire pour les Mathématiques, Paris, 1811, pp. 221–247, 289–295*

THE ENERGY EQUATION FOR A GENERAL DYNAMICAL SYSTEM

Joseph Louis Lagrange

[*Ed. note:* In Lagrange's *Mécanique analytique,* 2d ed., 1811, we find in Vol. 1, p. 241 (in Sec. 1 of the Second Part, i.e., on dynamics), a discussion of the principle of *vis viva* (*force vive*) as follows. In the previous paragraph Lagrange has cited four principles that have been of importance in mechanics: conservation of *vis viva,* conservation of the motion of the center of gravity, conservation of the moment of rotation (by which he presumably means what in modern terminology is the conservation of moment of momentum or angular momentum), and finally the principle of least action.]

14. The first of these four principles, that of the conservation of *vis viva,* was discovered by Huygens, although in a form somewhat different from that now given to it. We have already made mention of the principle in connection with the problem of the center of oscillation of the compound pendulum. As the principle is employed in the solution of this problem, it consists in the equality between the descent and ascent of the center of gravity of several heavy bodies that fall together and thereafter rise separately, each being thrust up with the velocity it had acquired. But, by the known properties of the center of gravity, the path followed by this center in any direction is expressed by the sum of the products of the mass of each component particle and the distance it moved in the same direction divided by the sum of the masses. On the other hand, by Galileo's theorem, the vertical distance covered by a heavy body in free fall is proportional to the square of the velocity acquired in the fall and with which it could rise to the same height. Thus the principle of Huygens reduces to this: in the motion of heavy bodies the sum of the products of the masses multiplied by the squares of the velocities at any instant is the same whether the bodies move together in any way or whether they move freely from the same vertical height. This is also what Huygens has himself remarked briefly in a small note he wrote about the methods of Jacobus Bernoulli and l'Hôpital concerning the center of oscillations.

Up to that point the theorem had been regarded only as a simple theorem in mechanics. But when John Bernoulli adopted the distinction established by Leibniz between *vis mortua* (dead force), which acts without producing actual movement, and *vis viva,* which accompanies motion, with the measure of the latter the product of the mass times the square of the velocity, he saw in the principle in question only a consequence of the theory of *vis viva* and a general law of nature, according to which the sum of the *viaes vivae* of several bodies remains the same while these bodies act on each other with simple forces, and is always equal to the single *vis viva* resulting from the action of the actual forces moving these bodies. The name conservation of *vis viva* was then given to this principle, and it enabled Bernoulli to solve successfully some previously unsolved problems, which appeared difficult to attack with conventional methods.

Daniel Bernoulli has further extended the principle and deduced from it the laws of motion of liquids in vessels, material that had been treated before his time only in a vague and arbitrary manner. Finally, he generalized the principle considerably in *Memoirs of the Berlin Academy* for 1748 in making clear how one can apply it to the motion of bodies subject to any mutual attractions, or attracted toward fixed centers by forces proportional to arbitrary functions of the distances.

The great advantage of this principle is that it furnishes immediately an equation connecting the velocities of the bodies of the system in question and the variables determining their position in space. When, by the nature of the problem, all these variables reduce to a single one, this equation suffices to solve for this variable. This is the case with the center of oscillation. In general, the conservation of *vis viva* always provides a first integral of the differential equations of motion of each problem, often of great utility.

[*Ed. note:* In subsequent sections of his treatise, Lagrange discusses in detail various properties of motion from the standpoint of general principles. In paragraph 5 of Section III of the Second Part, Lagrange discusses motion properties with respect to *vis viva.* We reproduce the earlier parts of this material next.]

33. In general, no matter in what way the different bodies that make up a system are arranged or joined together, provided that the arrangement is independent of time (that is, the condition equations connecting the coordinates of the different bodies do not contain the variable t), it is clear that in the general formula of dynamics one can always suppose the variations δx, δy, and δz equal to the differentials dx, dy, and dz, which represent the actual effective displacements of the bodies in the time interval dt. It should be understood that the variations of which we are speaking may represent any arbitrary displacements of the bodies of the system in time interval dt, subject to the constraints on the system.

[*Ed. note:* In an earlier paragraph the author has derived the following equation as the "general formula of dynamics":

$$S\left\{\frac{d^2x}{dt^2}\delta x+\frac{d^2y}{dt^2}\delta y+\frac{d^2z}{dt^2}\delta z\right\}m+S\left\{P\,\delta p+Q\,\delta q+R\,\delta r + \text{etc.}\right\}m=0$$

where x, y, and z are the rectangular coordinates of the various bodies (particles) of the system, and δx, δy, and δz are virtual displacements, that is, any arbitrary displacements which the particles may be thought to have, subject to the constraints on the system. P, Q, and R are the rectangular components of the given accelerating forces that act so as to attract each body toward the center to which the forces are directed. The variables p, q, and r are the rectangular coordinates of the bodies with respect to this same center. The quantities δp, δq, and δr are the virtual displacements associated with p, q, and r, respectively. The general expression for the mass of each particle is m, which of course may be different for the various particles. Lagrange uses the symbol S to denote the sum over all particles and all forces. This is equivalent to the more modern symbol Σ.]

This formula can provide only a single equation, but, being independent of the form of the system, it has the advantage of giving a general equation for the motion of any arbitrary system. Substituting dx, dy, and dz for δx, δy, and δz, respectively, and dp, dq, and dr for δp, δq, and δr, respectively, yields the general equation

$$S\left\{\frac{d^2x}{dt^2}dx + \frac{d^2y}{dt^2}dy + \frac{d^2z}{dt^2}dz + P\,dp + Q\,dq + R\,dr + \text{etc.}\right\} m = 0$$

34. If the expression $P\,dp + Q\,dq + R\,dr$ is integrable, we may write

$$P\,dp + Q\,dq + R\,dr + \text{etc.} = d\pi$$

[*Ed. note:* Where π is a function of p, q, and r.] The preceding equation becomes

$$S\left\{\frac{d^2x}{dt^2}dx + \frac{d^2y}{dt^2}dy + \frac{d^2z}{dt^2}dz + d\pi\right\} m = 0$$

On integration, this yields

$$S\left\{\left(\frac{dx}{dt}\right)^2 + \left(\frac{dy}{dt}\right)^2 + \left(\frac{dz}{dz}\right)^2 + 2\pi\right\} m = 2H$$

in which H denotes an arbitrary constant, equal to the value of the left-hand side of the equation at any instant.

This last equation expresses the principle known as the conservation of *vis viva.* For the sum of the final three terms in parentheses is the square of the resultant particle velocity, which when multiplied by m becomes the *vis viva.* The sum of all these terms for all the bodies of the system is the *vis viva* of the whole system. This *vis viva* is equal to the quantity $2H - 2S\pi m$, which depends simply on the accelerating forces acting on the bodies of the system and not at all on their mutual connections (that is, the internal forces). Hence, the *vis viva* of the system is at each instant the same as that which the bodies would have acquired if, being acted on by the same forces, they had moved freely, each on the curve that it actually follows. It is this that has justified giving the name conservation of *vis viva* to this property of motion.

[*Ed. note:* The equation

$$S\left\{\frac{m}{2}\left[\left(\frac{dx}{dt}\right)^2+\left(\frac{dy}{dt}\right)^2+\left(\frac{dz}{dt}\right)^2\right]+m\pi\right\}=H$$

would appear to be the first clear-cut general analytical statement of what came to be called the *equation of mechanical energy* for a system of particles. Here

$$S\left\{\frac{m}{2}\left[\left(\frac{dx}{dt}\right)^2+\left(\frac{dy}{dt}\right)^2+\left(\frac{dz}{dt}\right)\right]^2\right\}$$

is the total kinetic energy of the system, whereas

$$Sm\pi$$

is the potential energy of the system. The sum of the kinetic and potential energy is the constant H, which is the total mechanical energy of the system. It is to be noted, of course, that Lagrange nowhere uses the term energy to refer to this equation. Nor does he call attention to the significance of the constancy of H, save through the term *conservation.*

From the modern viewpoint it is somewhat difficult to understand why the equation repeated at the beginning of this note should have been referred to as an expression of *conservation* of *vis viva.* Strictly speaking, the *vis viva,* of course, does not stay constant during the motion of the system unless π is constant, and constant potential energy is a very special case. Nevertheless, the final statement in Lagrange's Section 34 does reflect a feeling for a definite dynamical situation, so that the equation itself is not reduced to the status of a mere mathematical theorem.

In his next section, Lagrange shows how the principle of *vis viva* can be applied to motion with respect to the center of gravity.

The author then goes on to show how the concept of *vis viva* can be applied to problems of rotational motion. In introducing the principle of least action, he indicates how this can be expressed in terms of *vis viva* by replacing a space integral with a time integral. Finally, he adopts what we now recognize as modern dynamical terminology and introduces the symbol T to denote one half the total *vis viva* and the symbol V in place of $S\pi m$, that is, what we now call the potential energy. To be sure, nothing is said about a real physical meaning for V, but the mathematical groundwork is carefully laid for the subsequent association of the concept of mechanical energy with the quantity $T + V$.

In the latter part of the first volume of *Mécanique analytique,* Lagrange introduces the difference $T - V$, which we now call the Lagrangian function L (he uses the symbol Z for this), and finally derives the famous equations of motion in terms of the dependence of L on generalized coordinates, which now carry his name.

It is clear that in spite of his failure to employ the terminology of energy, Lagrange was essentially aware of the important role that energy plays in the setting up of the fundamental equations governing the motion of dynamical systems.

A review of the historical record indicates that there was little if any progress in the analytical development of the use of the fundamental idea of energy in mechanics after Lagrange until the work of Hamilton (see Paper 30 in this volume). We must not overlook, indeed, the contribution of P. S. Laplace (1749-1827), who showed in 1782 that Lagrange's potential energy function V satisfies the famous equation

$$\frac{\partial^2 V}{\partial x^2} + \frac{\partial^2 V}{\partial y^2} + \frac{\partial^2 V}{\partial z^2} = 0$$

in space free from attracting matter. This was generalized around 1817 by S. D. Poisson (1781-1840) to

$$\frac{\partial^2 V}{\partial x^2} + \frac{\partial^2 V}{\partial y^2} + \frac{\partial^2 V}{\partial z^2} = -4\pi\rho$$

where ρ denotes the density of attracting matter as a function of position. Poisson also applied these equations to the problems of static electricity. Once again the analytical developments that led to the establishment of the concept of energy were proving of importance in the understanding of physical phenomena.

In connection with the immediately preceding remarks it is appropriate to recall the use that Augustin Fresnel (1788–1827) made of the principle of conservation of *vis viva* in his celebrated work on the relative intensities of reflected and refracted polarized light at a plane surface. Fresnel supported the wave theory of light, and during his all too brief professional career wrote numerous memoirs on the behavior of light waves. In 1823 he presented to the French Academy of Sciences a long memoir with the title (in English translation) "The Law of the Modifications Impressed on Polarized Light by Reflection" (Collected Works, Paris, Imprimere Impériale, 1866, Vol. 1, pp. 767–799). In this memoir he deduced the famous Fresnel equations for the reflection and refraction of a plane light wave at a plane interface.

Having defined what he means by the masses associated with the incident reflected and refracted waves, and having decided to call the velocity of the vibrations in the incident light unity and the velocities of the vibrations in the reflected and refracted light by v and u, respectively, Fresnel forms the sum of the living force of the reflected and refracted light waves and equates it to the *vis viva* associated with the incident wave. This is actually one of the first uses of the principle of conservation of energy in a problem involving a physical phenomenon, not strictly mechanical in essence, but treated in accordance with the principles of mechanics. Fresnel's hypothesis may indeed be criticized on the ground that he does not strictly conserve the total energy in the wave but only the kinetic energy, leaving out of account the potential energy. Fortunately, later wave theory (once the concept of energy had been more firmly established) showed that the potential energy for a plane harmonic wave is proportional to the kinetic energy (indeed equal to the latter on the average). Hence, Fresnel was not led astray by his more restricted assumption. Of course, to use his energy-conservation hypothesis, he needed another boundary equation connecting u and v. Here he assumed that $1 + u = v$, or the conservation of vibrational velocity in crossing the boundary. It is interesting that Fresnel was convinced that the intensity of a light wave is associated with the energy it carries.

In his analysis Fresnel used what is commonly called the elastic solid theory of light, which was later replaced by the electromagnetic theory. However, his equations also result (with appropriate changes in the meaning of polarization) from the electromagnetic theory of light.]

Part II

THE NATURE OF HEAT

Editor's Comments on Papers 21 Through 27

The Atomic Theory of Heat
21 P. GASSENDI

The Nature of Heat
22 R. BOYLE

The Nature of Heat
23 J. BLACK

Introduction of the Term "Caloric" for the Substance of Heat
24 A. LAVOISIER

Source of Heat from Friction
25 B. THOMPSON

The Motive Power of Heat
26 S. CARNOT

Evidence Against the Theory of Caloric
27 M. FARADAY and H. DAVY

The historical evolution of the concept of energy is inextricably connected with the problem of the nature of heat. Although this was not realized explicitly in the early days, it is hard to believe that intelligent people did not have their curiosity aroused by the apparent association of increased body heat with physical exertion. In this case, as usual, hindsight is better than foresight.

Of the conflicting views of the nature of heat that persisted for centuries the two most prominent ones were the materialistic and the motional. The materialistic view considered heat to be some form of substance, much more subtle than ordinary matter. It could be either continuous or atomic in character. Its presence made objects hot and its absence made them appear cold. The motional theory assumed that heat was the result of the motion of the constituent parts of matter or, as some held, the motion itself.

A good representative of the materialistic view was Pierre Gassendi (1592–1655), a French priest who wrote widely on mathematics and all aspects of natural phenomena. For example, he participated in experiments on the velocity of sound in air. He was a staunch believer in atomic theory and believed that heat consists of a special kind of atom which can penetrate into the interstices of the atoms of ordinary matter. The selection from his writings included here clearly brings out this point of view.

Conversely, Francis Bacon (1561–1626) was convinced that heat was motion. His views were not clearly expressed and there is little purpose in making a selection from his writings on heat. A more articulate defender of the motional theory of heat was Robert Boyle (1627–1691), who gave definite experimental reasons for believing in this theory, as is evident from the selection from his works presented here.

Boyle was one of the great natural philosophers of the seventeenth century. Interested in all aspects of natural phenomena, he was equally at home in the laboratory, trying to find out how "things really go," and in the study, where he meditated on the causes of things. In physics he is doubtless best known for his experimental study of the behavior of gases, which led to the celebrated Boyle's law for the relation between pressure and volume at constant temperature. But he also studied other phenomena, notably chemical behavior, sound, heat, light, and electricity and magnetism. His views on heat are particularly interesting in light of the later development of the mechanical theory of heat. The last portion of the extract presented here is particularly significant from the standpoint of the development of the idea of energy; in his discussion of the heat developed in driving a nail into wood, Boyle notices that the nail does not get as hot while being driven in as it does once it has been completely driven in and hammering continues. Here he comes close to the differentiation between the change of kinetic energy into work (to overcome the resistance of the wood) and its transformation into heat once the motion of the nail as a whole has ended.

About a century after Boyle, the Swiss scientist Daniel Bernoulli in his famous book *Hydrodynamion* (Basel, 1738) expressed his conviction (Section 10) that heat in a fluid is due to the motion of the constituent particles. This was in connection with his introduction of the molecular theory of gases. He does not here associate the heat, or temperature, with the *vis viva* of the molecules and indeed gives no details. His statement is, however, a definite foreshadowing of the mechanical theory of heat. (For his important contribution to the conservation of *vis viva* in mechanics, see Paper 16.)

Because of its importance in chemical phenomena, it was inevitable that the chemists would originate ideas about heat. The development of calorimetry in the eighteenth century encouraged the view of heat as a substance. This is well brought out in the work of Joseph Black

(1728–1799), who, in his *Lectures on the Elements of Chemistry* given between 1766 and 1797 at the University of Edinburgh, directed much attention in his introduction to the nature of heat. This part of his work is reproduced here in detail since it gives an excellent summary of the status of heat theory at his time. Both the motional and materialistic views of the nature of heat are carefully considered. Black finally concludes that a theory that treats heat as a peculiar substance is the more probable one, but his attitude throughout is more judicious and cautious with respect to jumping to conclusions than has been commonly asserted.

Rather unlike Black in this respect, the French chemist Antoine Laurent Lavoisier (1743–1794) seems to have taken it for granted that the proper view of heat for chemistry is the substantial one. What he felt most keenly was the need for a name to denote heat substance. He and his colleagues introduced the term *caloric* for this purpose. This is well explained in the extract included from Lavoisier's *Traité élementaire de chimie* (1789).

By the end of the eighteenth century the caloric theory, as it soon came to be called, was thoroughly entrenched. However, it did not go unchallenged. In the course of his work on ordnance at the military arsenal in Munich in the last decade of the eighteenth century, Benjamin Thompson, better known as Count Rumford (1753–1814), the American scientist who founded the Royal Institution of Great Britain, was so much impressed by the heat associated with the boring process that he decided to perform carefully designed experiments to see whether he could decide between the caloric theory of heat and the theory which attributes heat to motion. In a paper submitted to the Royal Society of London in 1798, and reproduced here, Rumford set forth the view that his experiments provided convincing evidence of the validity of the motional theory and discredited the caloric theory. Although important in the light of mid-nineteenth-century developments, Rumford's work scarcely proved to be decisive. The upholders of the caloric theory had a ready explanation for his results, as they did for all thermal phenomena. The time was evidently not ripe for the amalgamation of the mechanical idea of energy (still unfortunately denominated "force"), as developed by Leibniz, d'Alembert, the Bernoullis, and Lagrange, with the empirical phenomenon of heat to produce a useful generalization of the concept. This had to wait until the work of Mayer and Joule in the 1840s. Hindsight suggests that this delay is difficult to understand. Perhaps the simplest way to explain it is to assume the plausible hypothesis that all cases of heat produced by friction were looked upon from the standpoint of heat without reference to any associated measurable mechanical effect.

On the other hand, it might have been supposed that the behavior of the steam engine would have suggested an important connection between heat and mechanical effects. In the form of a turbine, the steam engine goes back to antiquity, and in the more modern form of a cylinder engine, to the latter part of the seventeenth century. But here the emphasis was on the production by heat of a vapor that by expanding would push against an obstacle and ultimately move it, or by condensing in a closed vessel would allow the pressure of the atmosphere to achieve the same effect. It must have appeared to early workers with the steam engine that the heat required to produce the steam was only incidentally connected with the operation of the engine. They therefore put little emphasis on it, although they must have realized as a matter of economics that some sort of fuel had to be consumed to make the engine work.

Probably the first really significant attempt to understand the role of heat in the production of motion in an engine was made by the brilliant young French engineer Sadi Carnot (1796–1832) in his memoir *Reflections on the Motive Power of Fire*, published as a small book in Paris in 1824 and scarcely noticed by any one for many years. In this memoir Carnot established the basic result that the work done by a heat engine is proportional to the difference in temperature of the working substance (e.g., steam) as it goes into the engine and as it comes out. He had a clear view of the appropriate measure of what he called the "motive power," a term later replaced as we have seen through the suggestion of Coriolis (1792–1843), by the word "work" (*travail*) in French). The significance of the product of weight times the height to which an object is lifted was well recognized in the seventeenth century, but the term "work" was not applied until the nineteenth century. Carnot took as the measure of the work (or, as he called it, the motive power) involved in the expansion of a gas through the volume dv, at constant pressure p, the quantity $p\, dv$, which is of course the dimensional analog of force times displacement. He naturally made no attempt to associate this with the internal energy of the gas. Daniel Bernoulli, Lagrange, and their successors had established the meaning of the total mechanical energy of a dynamical system (although not employing this name), but its signifiance for those who got work from heat did not begin to become apparent until the middle of the nineteenth century.

Carnot's famous memoir is rightly considered one of the foundation stones of thermodynamics, primarily for its foreshadowing of the second law. With respect to the conservation of energy (first law), the situation is rather dubious. Throughout his memoir, Carnot uses the word "caloric" for heat and insists that there is no loss of caloric when the engine produces "motive power." This was in line with the fundamental hypothesis of conservation of caloric in the materialistic theory of heat. Indeed, Carnot compares the production of motive power by heat with a high temperature at the intake and a low temperature of the exhaust (or condenser) with the production of mechanical action (work) when water at a height falls to a lower point, as in a waterwheel. In the latter case there is no loss in the total amount of water. In the former case there is no loss in caloric. The idea is after all an ingenious one, and much was made of it by Ernst Mach later in the nineteenth century. Unfortunately, it does not provide a lead to the actual historical development of the concept of energy. Much has been written about the "real" meaning of Carnot's use of heat, and some have expressed the opinion that he did not use the word "caloric" in the sense of Lavoisier and other adherents of the caloric theory. The situation is complicated by the fact that in later notes, found at his early death in 1832, Carnot indicated that he had finally adopted the motional interpretation of heat. There is no indication, however, that he ever modified in detail the argument of his 1824 memoir.

In the notes that Carnot left behind, which were not published until 1878, he indicated that he had carried out a calculation of the relation between heat and mechanical work on the basis of the mechanical (anticaloric) theory and had arrived at a result that in modern terminology would correspond to a value of the mechanical equivalent of heat of 3.7 joules per calorie. Unfortunately, he gave no indication of how he had calculated this value. This is, of course, historically of interest, for it has led to claims that Carnot anticipated Mayer in his calculation of the mechanical equivalent. In his notes suggestions are also found for experiments much like those later carried out by Colding and Joule for the measurement of the mechanical equivalent. No one knows when Carnot developed these ideas, nor how seriously he took them. That Carnot began to have serious doubts about the value of the caloric theory is, of course, of great significance. But even if he had expressed these doubts clearly in his published work, it is doubtful that he could have had any essential influence on the development of the mechanical theory of heat and the generalization of the concept of energy, because his work remained essentially unknown for many years. Further attention to Carnot's work will be paid in a subsequent volume on thermodynamics in the Benchmark Series on Energy.

To substantiate the foregoing critique we include the introductory pages of Carnot's memoir.

To return to the problem of heat produced by friction and its relation to a possible explanation of the nature of heat, we call attention to the views of Humphry Davy (1778–1829) and in particular to an experiment attributed to him, the melting produced by rubbing two pieces of ice together in a vacuum. This experiment, performed around 1798 (when Davy was only 20), was never reported in a regular periodical article. However, the results were included in an essay written by Davy, "An Essay on Heat, Light and the Combinations of Light." This essay was printed in 1799 by Thomas Beddoes, who ran a "pnematic" institution in Clifton, England, in a collection entitled *Contributions to Physical and Medical Knowledge, Principally from the West of England*. Davy later repudiated the article, and modern investigations have cast considerable doubt about the claims set forth in the essay. Nevertheless, the essay makes clear Davy's adherence to a motional theory of heat and his disbelief in the caloric theory. In this connection it is worthwhile to call attention to a note which Davy appended to an article, "On Fluid Chlorine," that Michael Faraday (1791–1867) contributed to the *Philosophical Transactions of the Royal Society* in 1823 (**113**, Part II, pp. 160–165). In this paper, Faraday describes how he liquified chlorine gas by *heating* it in a closed vessel. This procedure had been suggested by Davy. His idea was based on his disbelief in the caloric theory and on his confidence that heat is associated with the motion of the molecules of which all substances are composed. Thus, heating increases molecular motion and in a *closed* vessel effectively brings the molecules closer together so that their natural attraction will have a chance to produce a change from gaseous to liquid form. On the other hand, it was part of the caloric theory that, although caloric attracts ordinary matter and tends to surround each molecule with an envelope of caloric, it is self-repulsive and hence tends to keep the molecules apart. Since a heated body contains more caloric than an unheated one, this should tend to keep the molecules farther apart on the average. Indeed this is how the caloric theory was used to explain thermal expansion.

The liquefaction experiment of Faraday, as suggested by Davy, constitutes a valid bit of evidence against the caloric theory and in favor of the motional theory. We therefore include here Davy's note appended to Faraday's article ("Note on the Condensation of Muriatic Acid Gas into the Liquid Form," pp. 164–165 of the article just referred to). It will be noted that nothing is said about caloric or the motional theory of heat as such, but the inference we have drawn appears to be justified. Certainly, no believer in the caloric theory could have reasoned logically as Davy did in his note. Faraday's experimental success was a clincher.

21

This article was translated expressly for this Benchmark volume by R. Bruce Lindsay, Brown University, from "De calore et frigore," in Opera Omnia, *Vol. 1, Corante Nicolao Averanio, Advocato Florentino, Florence, 1727, p. 346*

THE ATOMIC THEORY OF HEAT

Pierre Gassendi

We now take up those qualities which depend on the simultaneous combination of many properties of atoms. To this class belong, first, those commonly called elementary and are indeed closely related to the properties of the Aristotelian elements and considered to be the causes of other properties. Among these are heat, cold, wetness, and dryness. It does not appear possible to explain these except in terms of the size, shape, and mobility of atoms. Thus, we begin with these, since by assuming them we can make other things clearer. Moreover, our sense of touch is particularly affected by these qualities, and without this sense no quality can be perceived by an animal, since the other senses do not operate without contact: every sense has something tactile about it. To begin with heat, it is customary to think of this from the standpoint of its relation to the senses, or, as we may say, how effective it is in producing that very sharp sensation in the skin and other tactile organs in the body when we become hot. Heat, indeed, produces a very special effect on living things. Heating, burning, and the resultant pain are produced in us by heat that manages to get into the pores of the skin and penetrates all parts of the body. By its disintegrating power, heat constitutes a most powerful general effect.

When I say that heat penetrates and disintegrates, I do not refer to some isolated, solitary quality. I mean certain atoms characterized by mass, shape, and motion that penetrate bodies and push their way through the other atoms of which bodies are made, a phenomenon commonly referred to as heating. Heat must be looked upon rather abstractly. Anaximenes and Plutarch thought that neither hotness nor coldness should be thought of as occupying a substance, but rather the atoms themselves, to which must be attributed all motion and hence all action. These atoms do not have heat in themselves or, what amounts to the same thing, are not themselves hot. They are nevertheless able to be sensed. I use the term heat atoms or calorific atoms, because they create heat; that is, they possess the power to penetrate things and to disperse and disintegrate them. The bodies containing such atoms and able to emit them should be considered to be hot, since by this emission they are able to excite the sensation of heat. And when the emitted atoms become free of the body, their rapid motion can produce what is commonly called flame or fire. If the heat atoms are forced into a smaller space by obstacles to their motion, they are said to represent potential heat. Common examples of this are wine, pepper, and similar substances. But even wood, wax, fat, and such like combustible materials are able to discharge heat into other things. Without question, all these substances should be recognized as containing atoms that, as long as they remain fastened together, cannot create heat. As soon as they regain their freedom, they then begin to create the sensation of heat.

What Democritus, Epicurus, and other supporters of the atomic theory taught should not be overlooked: that atoms of heat must be considered very small, round in shape, and rapid in their motion.

[*Ed. note:* Gassendi then gives reasons for the assignment of these properties to heat atoms. They have to be small to penetrate the pores of ordinary matter. They must be spherical to guarantee ease of movement. And they have to be rapid in their movement in order to strike violently against objects in their path, as well as to push such objects out of the way. Here he supports his views by quoting freely from the Greek philosophers. The treatment is qualitative throughout, but Gassendi's emphasis on the motion of heat atoms is at any rate suggestive in the light of the later mechanical theory of heat. Our confidence in his foresight with respect to the nature of heat is indeed somewhat shaken when he goes on to invent atoms of cold!]

22

Reprinted from *Works of the Honorable Robert Boyle*, Vol. 4,
London, 1772, pp. 244–251

THE NATURE OF HEAT

Robert Boyle

SECTION II.

Of the MECHANICAL ORIGIN, *or* PRODUCTION *of* HEAT.

AFTER having difpatched the inftances I had to offer of the production of cold, it remains, that I alfo propofe fome experiments of heat, which quality will appear the more likely to be mechanically producible, if we confider, the nature of it, which feems to confift mainly, if not only, in that mechanical affection of matter we call local motion mechanically modified, which modification, as far as I have obferved, is made up of three conditions.

THE firft of thefe is, that the agitation of the parts be vehement, by which degree of rapidnefs the motion proper to bodies, that are hot, diftinguifhes them from bodies, that are barely fluid. For thefe, as fuch, require not near fo brifk an agitation, as is wont to be neceffary to make bodies deferve the name of hot. Thus we fee, that the particles

particles of water, in its natural (or usual) state, move so calmly, that we do not feel it at all warm, though it could not be a liquor, unless they were in a restless motion; but when water comes to be actually hot, the motion does manifestly and proportionably appear more vehement, since it does not only briskly strike our organs of feeling, but ordinarily produces store of very small bubbles, and will melt butter or coagulated oil cast upon it, and will afford vapours, that, by the agitation they suffer, will be made to ascend into the air. And if the degree of heat be such, as to make the water boil, then the agitation becomes much more manifest by the confused motions, and waves, and noise, and bubbles, that are excited, and by other obvious effects, and phænomena of the vehement and tumultous motion, which is able to throw up visibly into the air great store of corpuscles, in the form of vapours or smoke. Thus, in a heated iron, the vehement agitation of the parts may be easily inferred, from the motion and hissing noise it imparts to drops of water, or spittle, that fall upon it. For it makes them hiss and boil, and quickly forces their particles to quit the form of a liquor, and fly into the air in the form of steams. And, lastly, fire, which is the hottest body we know, consists of parts so vehemently agitated, that they perpetually and swiftly fly abroad in swarms, and dissipate or shatter all the combustible bodies they meet with in their way; fire making so fierce a dissolution, and great a dispersion of its own fuel, that we may see whole piles of solid wood (weighing perhaps many hundred pounds) so dissipated, in very few hours, into flame and smoke, that, oftentimes, there will not be one pound of ashes remaining. And this is the first condition required to heat.

THE second is this, that the determinations be very various, some particles moving towards the right, some to the left hand, some directly upwards, some downwards, and some obliquely, &c. This variety of determinations appears to be in hot bodies, both by some of the instances newly mentioned, and especially that of flame, which is a body; and by the diffusion, that metals acquire, when they are melted, and by the operations of heat, that are exercised by hot bodies upon others, in what posture or situation soever the body to be heated be applied to them. As a thoroughly ignited coal will appear every way red, and will melt wax, and kindle brimstone, whether the body be applied to the upper or to the lower, or to any other part of the burning coal. And congruously to this notion, though air and water be moved never so vehemently, as in high winds and cataracts; yet we are not to expect, that they should be manifestly hot, because the vehemency belongs to the progressive motion, of the whole body; notwithstanding which, the parts it consists of may not be near so much quickened in their motions, made according to other determinations, as to become sensibly hot. And this consideration may keep it from seeming strange, that, in some cases, where the whole body, though rapidly moved, tends but one way, it is not by that swift motion perceived to be made hot.

NAY, though the agitation be very various, as well as vehement, there is yet a third condition required to make it calorific; namely, that the agitated particles, or at least the greatest number of them, be so minute, as to be singly insensible. For though a heap of sand, or dust itself, were vehemently and confusedly agitated by a whirl-wind, the bulk of the grains or corpuscles, would keep their agitation from being properly heat, though, by their numerous strokes upon a man's face, and the brisk commotion of the spirits and other small particles, that may thence ensue, they may perchance occasion the production of that quality.

IF some attention be employed, in considering the formerly proposed notion of the nature of heat, it may not be difficult to discern, that the mechanical production of it may be divers ways effected. For, excepting in some few anomalous cases, (wherein the regular course of things happens to be over-ruled,) by whatever ways the insensible parts

parts of a body are put into a very confufed and vehement agitation, by the fame ways heat may be introduced into that body: agreeably to which doctrine, as there are feveral agents and operations, by which this calorific motion (if I may fo call it) may be excited, fo there may be feveral ways of mechanically producing heat, and many experiments may be reduced to almoft each of them, chance itfelf having, in the laboratories of chemifts, afforded divers phænomena, referable to any one or other of thofe heads. Many of the more familiar inftances, applicable to our prefent purpofe, have been long fince collected by our juftly-famous *Verulam*, in his fhort, but excellent paper *de forma calidi*, wherein (though I do not acquiefce in every thing I meet with there) he feems to have been, at leaft among the moderns, the perfon, that has firft handled the doctrine of heat like an experimental philofopher. I fhall therefore decline accumulating a multitude of inftances of the production of heat, and I fhall alfo forbear to infift on fuch known things, as the incalefcence, obfervable upon the pouring either of oil of vitriol upon falt of tartar, (in the making of tartarum vitriolatum) or of aqua fortis upon filver or quickfilver, (in the diffolution of thefe metals,) but fhall rather chufe to mention fome few inftances not fo notorious as the former, but not fo unfit, by their variety, to exemplify feveral of the differing ways of exciting heat.

AND yet I fhall not decline the mention of the moft obvious and familiar inftance of all, namely, the heat obferved in quick-lime, upon the affufion of cold water, becaufe, among learned men, and efpecially Peripateticks, I find caufes to be affigned, that are either juftly queftionable, or manifeftly erroneous. For, as to what is inculcated by the fchools, about the incalefcence of a mixture of quick-lime and water, by virtue of a fuppofed Antiperiftafis, or invigoration of the internal heat of the lime, by its being invironed by cold water, I have elfewhere fhewn, that this is but an imaginary caufe, by delivering, upon experiment, (which any man may eafily make,) that if, inftead of cold water, the liquor be poured on very hot, the ebullition of the lime will not be the lefs, but rather the greater: and oil of turpentine, which is a lighter, and is looked upon as a fubtiler liquor than water, though it be poured quite cold on quick-lime will not, that I have obferved, grow fo much as fenfibly hot with it.

AND now I have mentioned the incalefcence of lime, which, though an obvious phænomenon, has exercifed the wits of divers philofophers and chemifts, I will add two or three obfervations, in order to an enquiry, that may be fome other time made into the genuine caufes of it; which are not fo eafy to be found, as many learned men may, at firft fight, imagine. The acute *Helmont* indeed, and his followers, have ingenioufly enough attempted to derive the heat under confideration from the conflict of fome alcalizate and acid falts, that are to be found in quick-lime, and are diffolved, and fo fet at liberty, to fight with one another by the water that flakes the lime. But, though we have fome manifeft marks of an alcalizate falt in lime, yet, that it contains alfo an acid falt, has not, that I remember, been proved; and if the emerging of heat be a fufficient reafon to prove a latent acid falt in lime, I know not, why I may not infer, that the like falt lies concealed in other bodies, which the chemifts take to be of the pureft or mereft fort of alcalies.

EXPERIMENT I.

FOR I have purpofely tried, that by putting a pretty quantity of dry falt of tartar in the palm of my hand, and wetting it well in cold water, there has been a very fenfible heat produced in the mixture; and when I have made the trial with a more confiderable quantity of falt and water in a vial, the heat proved troublefomely intenfe, and continued to be at leaft fenfible a good while after.

THIS

THIS experiment ſeems to favour the opinion, that the heat produced in lime, whilſt it is quenching, proceeds from the empyreuma, as the chemiſts call it, or impreſſion left by the violent fire, that was employed to reduce the ſtone to lime. But if by empyreuma be meant a bare impreſſion made by the fire, it will be more requiſite than eaſy, to declare intelligibly, in what that impreſſion conſiſts, and how it operates to produce ſuch conſiderable effects. And if the effect be aſcribed to ſwarms of atoms of fire, that remain adherent to the ſubſtance of the lime, and are ſet at liberty to fly away by the liquor, which ſeems to be argued by the ſlaking of lime without water, if it be for ſome time left in the air, whereby the atoms of fire get opportunity to fly away by little and little: if this, I ſay, be alledged, I will not deny, but there may be a ſenſe, which I cannot explicate in few words, wherein the co-operation of a ſubſtantial effluvium, (for ſo I call it,) of the fire, may be admitted in giving an account of our phænomenon. But the cauſe formerly aſſigned, as it is crudely propoſed, leaves in my mind ſome ſcruples. For it is not ſo eaſy to apprehend, that ſuch light and minute bodies, as thoſe of fire, are ſuppoſed, ſhould be ſo long detained, as by this hypotheſis they muſt be allowed to be, in quick-lime, kept in well-ſtopped veſſels, from getting out of ſo lax and porous a body as lime, eſpecially ſince we ſee not a great incaleſcence or ebullition enſue upon the pouring of water upon minium, or *crocus Martis per ſe*, though they have been calcined by violent and laſting fires, whoſe effluviums or emanations appear to adhere to them by the increaſe of weight, that lead, if not alſo *Mars*, does manifeſtly receive from the operation of the fire. To which I ſhall add, that, whereas one would think, that the igneous atoms ſhould either fly away, or be extinguiſhed by the ſupervening of water, I know, and elſewhere give account, of an

EXPERIMENT II.

IN which two liquors, whereof one was furniſhed me by nature, did by being ſeveral times ſeparated and reconjoined without additament, at each congreſs produce a ſenſible heat.

EXPERIMENT III.

AND an inſtance of this kind, though not ſo odd, I purpoſely ſought and found in ſalt of tartar, from which, after it had been once heated by the affuſion of water, we abſtracted or evaporated the liquor, without violence of fire, till the ſalt was again dry; and then putting on water a ſecond time, the ſame ſalt grew hot again in the vial, and, if I miſremember not, it produced this incaleſcence the third time, if not the fourth; and might probably have done it oftner, if I had had occaſion to proſecute the experiment. Which ſeems at leaſt to argue, that the great violence of fire is not neceſſary to impreſs what paſſes for an empyreum upon all calcined bodies, that will heat with water.

AND on this occaſion I ſhall venture to add, that I have ſometimes doubted, whether the incaleſcence may not much depend upon the particular diſpoſition of the calcined body, which being deprived of its former moiſture, and made more porous by the fire, doth by the help of thoſe igneous effluviums, for the moſt part of a ſaline nature, that are diſperſed through it, and adhere to it, acquire ſuch a texture, that the water impelled by its own weight, and the preſſure of the atmoſphere, is able to get into a multitude of its pores at once, and ſuddenly diſſolve the igneous and alcalizate ſalt it every where meets with there, and briſkly diſjoin the earthy and ſolid particles, that were blended with them; which being exceeding numerous, though each of them perhaps be very minute, and moves but a very little way, yet their multitude makes the confuſed

confused agitation of the whole aggregate of them, and of the particles of the water and salt vehement enough to produce a sensible heat; especially if we admit, that there is such a change made in the pores, as occasions a great increase of this agitation, by the ingress and action of some subtile ethereal matter, from which alone Monsieur *Des Cartes* ingeniously attempts to derive the incalescence of lime and water, as well as that of metals dissolved in corrosive liquors; though as to the phænomena we have been considering, there seems at least to concur a peculiar disposition of body, wherein heat is to be produced to do one or both of these two things, namely, to retain good store of the igneous effluvia, and to be, by their adhesion or some other operation of the fire, reduced to such a texture of its component particles, as to be fit to have them easily penetrated, and briskly, as well as copiously, dissipated, by invading water. And this conjecture (for I propose it as no other) seems favoured by divers phænomena, some whereof I shall now annex. For here it may be observed, that both the dissolved salt of tartar lately mentioned, and the artificial liquor, that grows hot with the natural, re-acquires that disposition to incalescence upon a bare constipation, or closer texture of the parts from the superfluous moisture they were drowned in before; the heat, that brought them to this texture, having been so gentle, that it is no way likely, that the igneous exhalations could themselves produce such a heat, or at least, that they should adhere in such numbers, as must be requisite to such an effect, unless the texture of the salt of tartar, or other body, did peculiarly dispose it to detain them; since

EXPERIMENT IV.

I have found by trial, that sal armoniac dissolved in water, though boiled up with a brisker fire to a dry salt, would, upon its being again dissolved in water, not produce any heat, but a very considerable degree of cold. I shall add, that though one would expect a great cognation between the particles of fire adhering to quick-lime, and those of high rectified spirit of wine, which is of so igneous a nature, as to be totally inflammable; yet I have not found, that the affusion of alkaol of wine upon quick-lime would produce any sensible incalescence, or any visible dissolution or dissipation of the lime, as common water would have done, though it seemed to be greedily enough soaked in by the lumps of lime. And I further tried, that, if on this lime so drenched I poured cold water, there ensued no manifest heat, nor did I so much as find the lump swelled, and thereby broken, till some hours after; which seems to argue, that the texture of the lime was such, as to admit the particles of the spirit of wine into some of its pores, which were either larger or more congruous, without admitting it into the most numerous ones, whereinto the liquor must be received, to be able suddenly to dissipate the corpuscles of lime into their minuter particles, into which (corpuscles) it seems, that the change, that the aqueous particles received by associating with the spirituous ones, made them far less fit to penetrate and move briskly there, than if they had entered alone.

I made also an experiment, that seems to favour our conjecture, by shewing, how much the disposition of lime to incalescence may depend upon an idoneous texture, and the experiment, as I find it, registered in one of my memorials, is this.

EXPERIMENT V.

[UPON quick-lime we put in a retort as much moderately strong spirit of wine, as would drench it, and swim a pretty way above it; and then distilling with a gentle fire, we drew off some spirit of wine much stronger than that, which had been put on, and then the phlegm following it, the fire was encreased, which brought over a good deal of phlegmatick strengthless liquor; by which one would have thought, that the quick-lime

lime had been ſlacked; but when the remaining matter had been taken out of the retort, and ſuffered to cool, it appeared to have a fiery diſpoſition, that it had not before. For, if any lump of it, as big as a nutmeg, or an almond, was caſt into the water, it would hiſs as if a coal of fire had been plunged into the liquor, which was ſoon thereby ſenſibly heated. Nay, having kept divers lumps of this prepared calx well covered from the air for divers weeks, to try, whether it would retain this property, I found, as I expected, that the calx operated after the ſame manner, if not more powerfully. For ſometimes, eſpecially when it was reduced to ſmall pieces, it would upon its coming into the water make ſuch a briſk noiſe, as might almoſt paſs for a kind of exploſion.]

THESE phænomena ſeem to argue, that the diſpoſition, that lime has to grow hot with water, depends much on ſome peculiar texture, ſince the aqueous parts, that one would think capable of quenching all, or moſt of the atoms of fire, that are ſuppoſed to adhere to quick-lime, did not near ſo much weaken the diſpoſition of it to incaleſcence, as the acceſſion of the ſpirituous corpuſcles and their contexture, with thoſe of the lime, encreaſed that igneous diſpoſition. And that there might intervene ſuch an aſſociation, ſeems to me the more probable, not only becauſe much of the diſtilled liquor was as phlegmatick, as if it had been robbed of its more active parts, but becauſe I have ſometimes had ſpirit of wine come over with quick-lime not in unobſerved ſteams, but white fumes. To which I ſhall add, that beſides, that the taſte, and perhaps odour of the ſpirit of wine, is often manifeſtly changed by a well-made diſtillation from quick-lime; I have ſometimes found that liquor to give the lime a kind of alcalizate penetrancy, not to ſay fierineſs of taſte, that was very briſk and remarkable. But I will not undertake, that every experimenter, nor I myſelf, ſhall always make trials of this kind with the ſame ſucceſs, that I had in thoſe above recited, in regard, that I have found quick-limes to differ much, not only according to the degree of their calcination, and to their recentneſs, but alſo, and that eſpecially, according to the differing natures of the ſtones and other bodies calcined. Which obſervation engages me the more to propoſe what hath been hitherto delivered about quick lime, as only narratives and a conjecture; which I now perceive has detained us ſo long, that I am obliged to haſten to the remaining experiments, and to be the more ſuccinct in delivering them.

EXPERIMENT VI.

AND it will be convenient to begin with an inſtance or two of the production of heat, wherein there appears not to intervene any thing in the part of the agent or patient, but local motion, and the natural effects of it. And as to this ſort of experiments, a little attention and reflection may make ſome familiar phænomenon appoſite to our preſent purpoſe. When, for example, a ſmith does haſtily hammer a nail, or ſuch like piece of iron, the hammered metal will grow exceeding hot, and yet there appears not any thing to make it ſo, ſave the forcible motion of the hammer, which impreſſes a vehement, and variouſly determined agitation of the ſmall parts of the iron; which being a cold body before, by that ſuperinduced commotion of its ſmall parts, becomes in divers ſenſes hot; firſt, in a more lax acceptation of the word in reference to ſome other bodies, in reſpect of whom it was cold before, and then ſenſibly hot; becauſe this newly gained agitation, ſurpaſſes that of the parts of our fingers. And in this inſtance, it is not to be overlooked, that oftentimes neither the hammer, by which, nor the anvil, on which a cold piece of iron is forged, (for all iron does not require precedent ignition to make it obey the hammer) continue cold, after the operation is ended; which ſhews, that the heat acquired by the forged piece of iron was not communicated

municated by the hammer or anvil as heat, but produced in it by motion, which was great enough to put ſo ſmall a body, as the piece of iron, into a ſtrong and confuſed motion of its parts, without being able to have the like operation upon ſo much greater maſſes of metal, as the hammer and the anvil; though, if the percuſſions were often and nimbly renewed, and the hammer were but ſmall, this alſo might be heated, (though not ſo ſoon, nor ſo much, as the iron;) by which one may alſo take notice, that it is not neceſſary, a body ſhould be itſelf hot, to be calorifick. And now I ſpeak of ſtriking an iron with a hammer, I am put in mind of an obſervation, that ſeems to contradict, but does indeed confirm our theory: namely, that if a ſomewhat large nail be driven by a hammer into a plank, or piece of wood, it will receive divers ſtrokes on the head before it grow hot; but when it is driven to the head, ſo that it can go no further, a few ſtrokes will ſuffice to give it a conſiderable heat; for whilſt, at every blow of the hammer, the nail enters further and further into the wood, the motion, that is produced, is chiefly progreſſive, and is of the whole nail tending one way; whereas, when that motion is ſtopped, then the impulſe given by the ſtroke, being unable either to drive the nail further on, or deſtroy its intireneſs, muſt be ſpent in making a various vehement and inteſtine commotion of the parts among themſelves, and in ſuch an one we formerly obſerved the nature of heat to conſiſt.

EXPERIMENT VII.

IN the foregoing experiment, the briſk agitation of the parts of a heated iron was made ſenſible to the touch. I ſhall now add one of the attempts, that I remember I made, to render it diſcoverable to the eye itſelf. In order to this, and that I might alſo ſhew, that not only a ſenſible, but an intenſe degree of heat, may be produced in a piece of cold iron by local motion, I cauſed a bar of that metal to be nimbly hammered by two or three luſty men, accuſtomed to manage that inſtrument; and theſe ſtriking with as much force, and as little intermiſſion, as they could, upon the iron, ſoon brought it to that degree of heat, that not only it was a great deal too hot to be ſafely touched, but probably would, according to my deſign, have kindled gun-powder, if that, which I was fain to make uſe of, had been of the beſt ſort: for, to the wonder of the by-ſtanders, the iron kindled the ſulphur of many of the grains of the corns of powder, and made them turn blue, though I do not well remember, that it made any of them go off.

EXPERIMENT VIII.

BESIDES the effects of manifeſt and violent percuſſions, ſuch as thoſe we have been taking notice of to be made with a hammer, there are among phænomena obvious enough, ſome, that ſhew the producibleneſs of heat, even in cold iron, by cauſing an inteſtine commotion of its parts: for we find, that, if a piece of iron, of a convenient ſhape and bulk, be nimbly filed with a large rough file, a conſiderable degree of heat will be quickly excited in thoſe parts of the iron where the file paſſes to and fro, the many prominent parts of the inſtrument giving a multitude of ſtrokes or puſhes to the parts of the iron, that happen to ſtand in their way, and thereby making them put the neighbouring parts into a briſk and confuſed motion, and ſo into a ſtate of heat. Nor can it be well objected, that, upon this account, the file itſelf ought to grow as hot as the iron, which yet it will not do; ſince, to omit other anſwers, the whole body of the file being moved to and fro, the ſame parts, that touch the iron this moment, paſs off the next; and, beſides, have leiſure to cool themſelves, by communicating their newly received agitation to the air, before they are brought to grate again upon the iron, which, being ſuppoſed to be held immoveable, receives almoſt perpetual ſhakes in the ſame place.

WE

WE find alſo, that attrition, if it be any thing vehement, is wont to produce heat in the ſolideſt bodies; as when the blade of a knife, being nimbly whetted, grows preſently hot. And if, having taken a braſs nail, and driven it as far as you can to the end of the ſtick, to keep it faſt, and gain a handle, you then ſtrongly rub the head to and fro againſt the floor, or a plank of wood, you may quickly find it to have acquired a heat intenſe enough to offend, if not burn one's fingers. And I remember, that going once, in exceeding hot weather, in a coach, which, for certain reaſons, we cauſed to be driven very faſt; the attrition of the nave of the wheel, againſt the axeltree, was ſo vehement, as obliged us to light out of the coach, to ſeek for water to cool the overchafed parts, and ſtop the growing miſchief the exceſſive heat had begun to do.

THE vulgar experiment, of ſtriking fire with a flint and ſteel, ſufficiently declares, what a heat, in a trice, may be produced in cold bodies by percuſſion, or colliſion; the latter of which ſeems but mutual percuſſion.

BUT inſtances of the ſame ſort, with the reſt mentioned in the VI. experiment, being obvious enough, I ſhall forbear to multiply and inſiſt on them.

EXPERIMENT IX.

FOR the ſake of thoſe, that think the attrition of contiguous air is neceſſary to the production of manifeſt heat, I thought, among other things, of the following experiment, and made trial of it.

WE took ſome hard black pitch, and having, in a baſon, porringer, or ſome ſuch veſſel, placed it a convenient diſtance under water, we caſt on it, with a good burning-glaſs, the ſun-beams, in ſuch a manner, that, notwithſtanding the refraction, that they ſuffered in the paſſage through the interpoſed water, the focus fell upon the pitch; wherein it would produce ſometimes bubbles, ſometimes ſmoke, and quickly communicated a degree of heat capable to make pitch melt, if not alſo to boil.

EXPERIMENT X.

THOUGH the firſt and ſecond experiments of Section I. ſhew, that a conſiderable degree of cold is produced by the diſſolution of ſal armoniac in common water; yet, by an additament, though but ſingle, the texture of it may be ſo altered, that, inſtead of cold, a notable degree of heat will be produced, if it be diſſolved in that liquor. For the manifeſtation of which, we deviſed the following experiment.

WE took quick-lime, and ſlaked it in common cold water, that all the igneous, or other particles, to which its power of heating that liquor is aſcribed, might be extracted and imbibed, and ſo the calx freed from them; then, on the remaining powder, freſh water was often poured, that all adhering reliques of ſalt might be waſhed off. After this, the thus dulcified calx, being again well dried, was mingled with an equal weight of powdered ſal armoniac, and having, with a ſtrong fire, melted the maſs, the mixture was poured out; and, being afterwards beaten to powder, having given it a competent time to grow cold, we put two or three ounces of it into a wide-mouthed glaſs; and pouring water upon it, within about a minute of an hour, the mixture grew warm, and quickly attained ſo intenſe a heat, that I could not hold the glaſs in my hand. And though this heat did not long laſt at the ſame height, it continued to be very ſenſible for a conſiderable time after.

23

Reprinted from J. Black, *Lectures on the Elements of Chemistry, Given at the University of Edinburgh, 1766–1797,* Matthew Carey, Philadelphia, 1807, pp. 21–34

THE NATURE OF HEAT

Joseph Black

INTRODUCTION.

OF HEAT IN GENERAL.

THAT this extensive subject may be treated in a profitable manner, I propose

1*st.* To ascertain what I mean by the word HEAT in these lectures.

2*dly.* To explain the meaning of the term cold, and ascertain the real difference between cold and heat.

3*dly.* To mention some of the attempts which have been made to discover the nature of heat, or to form an idea of what may be the immediate cause of it.

4*th*, and lastly, I shall begin to describe the sensible effects produced by heat on the bodies to which it is communicated.

Any person who reflects on the ideas which we annex to the word heat will perceive that this word is used for two meanings, or to express two different things. It either means a sensation excited in our organs, or a certain quality, affection, or condition, of the bodies around us, by which they excite in us that sensation. The word is used in the first sense when we say, we feel heat; in the second when we say, there is heat in the fire, or in a hot stone. There cannot be a sensation of heat in the fire, or in the hot stone, but the matter of the fire, or of the stone, is in a state or condition by which it excites in us the sensation of heat.

Now, in beginning to treat of heat and its effects, I propose to use the word in this second sense only, or as expressing that state, condition, or quality of matter, by which it excites in us the sensation of heat. This idea of heat will be modified a little, and extended as we proceed, but the meaning of the word will continue at bottom the same, and the reason of the modification will be easily perceived.

All the experience we have relating to this quality or affection of matter shews, that it is the most communicable from one body to another of any quality that we know. Hot bodies cannot be placed in the contact or neighbourhood of colder ones, without communicating to these a part of their heat.

When a lump of hot iron is taken out of the fire, how can we prevent it from communicating its heat to the surrounding matter? Lay it on the ground, or on a stone, it very quickly communicates to them a part of its heat; lay it on wood, or any other vegetable or animal matter, it heats them in a very short time to such a degree as to set them on fire; let it be suspended in the air by a wire, a little attention will soon convince us that it communicates heat very fast to the air in contact with it.

Thus heat is perpetually communicated from hotter bodies to the colder around them, and, while it passes from the one to the other, it penetrates all kinds of matter without exception: density and compactness are no obstacle to its progress: it appears to pass even faster into dense bodies, in most cases, than into rare ones; but the rare and the dense are all affected by it, and transmit it to others: Even the vacuum formed by the air-pump is pervaded by it. Sir Isaac Newton first discovered this by an experiment. He suspended an instrument for measuring heat in a large glass vessel, and exhausted the air, and suspending at the same time another similar instrument in another glass vessel, equal to the former, but not exhausted, he perceived that the one was affected by the variations of heat as well as the other. (Newton's Optics, Query 18th.)

Much more lately some experiments on the same subject were made by the celebrated Dr. Franklin and some of his friends at Paris. They suspended a hot body under the exhausted receiver of an air-pump, and another similar body, equally hot, in the air of the room near the air-pump, and these bodies being such as to shew exactly the variations of heat that happened in them, it was perceived that both of them gradually lost a part of their heat, until they were reduced to the temperature of the room in which the experiment was made, but that the one which hung in the air lost its heat faster than the one which was suspended *in vacuo*.

The thermometers fell from 60° (Reamur).

	IN VACUO.	IN THE AIR.
to 50°	in 17 min.	in 7 min.
37	54	22
30	85	29
20	167	63

The times of cooling are nearly in the proportion of 5 to 2. This is further confirmed by a set of similar experiments, made by Sir Benjamin Thompson. (Phil. Trans. for the year 1786.)

Sir Isaac Newton thought that such experiments gave a proof that the vacuum of an air-pump is not perfect, but that there is in it some subtile matter by which the heat is transmitted. This opinion probably was founded on a very general association in our minds, between the ideas of heat and matter; for, when we think of heat, we always conceive it as residing in some kind of matter; or possibly this notion of Sir Isaac might be founded on some opinion which he had formed concerning the nature of heat.

There is great reason, however, independently of this experiment, for believing that the vacuum of an air-pump is not a perfect vacuum, and for thinking that there is always some subtile matter, or vapor, present in it; but I can easily imagine, and we shall afterwards see abundant reason to believe, that heat may be communicated, or pass through a vacuum, or a space empty of all other matter.

In this manner, therefore, and upon all occasions without exception, is heat communicated from hotter bodies to colder ones, when they are in contact, or in the vicinity of one another; and the communication goes on until the bodies are reduced to an equal temperature, indicating an equilibrium of heat with one another.

When we consider this communication of heat from hot bodies to colder ones, the first question which may naturally occur to our mind, is, In what manner have these two bodies acted, the one on the other, on this occasion? Has one of them lost something, which the other has gained? And which of them has lost, or which has received?

The vulgar opinion is, that the hot body has lost something which has been added to the other. And those who have attempted to reason more profoundly on the nature of heat, have agreed with the multitude on this point; and have supposed that heat is a positive quality, and depends, either upon an exceedingly subtile and active matter, introduced into the pores of bodies, or upon a tremor or vibration excited among their particles, or perhaps among the particles of a peculiar substance present in all bodies; which subtile matter, or tremulous motion, they have supposed to be communicated from the hot body to the colder, agreeably to our general experience of the communication of matter or of motion.

But although many philosophers have thus agreed with the indistinct notion of the vulgar concerning heat, that it is a positive quality, or an active power residing in the hot body, and by which it acts on the cold one; some of them have not been altogether consistent in this opinion. They have not adhered to it, with respect to all the various cases in which bodies of different temperatures act one on the other. They have supposed that, in some cases, the colder body is the *active* mass, or contains the *active* matter; and that the warmer body is the passive subject which is acted upon, or into which something is introduced. When a mass of ice, for example, or a lump of very cold iron, is laid on the warm hand, instead of heat being communicated from the warm hand to the ice, or cold iron, they have supposed that there is in the ice, or cold iron, a multitude of minute particles, which they call particles of frost, or frigorific particles, and which have a tendency to pass from the very cold bodies into any others that are less cold; and that many of the effects, or consequences of cold, particularly the freezing of fluids, depend on the action of these frigorific particles. They call them Spiculæ, or little darts, imagining that this form will explain the acutely painful sensation, and some other effects of intense cold.

This, however, is the groundless work of imagination.

To form a well-grounded judgment on this subject, we must begin by laying aside all prejudices and suppositions concerning the nature of heat and cold, and then propose to ourselves this simple question. From whence do these two seemingly distinct qualities of bodies originally proceed; Where are the sources of heat and cold? It will immediately occur, that heat has a manifest source, or cause, in the sun and in fires. The sun is evidently the principal, and perhaps ultimately, the only source of the heat diffused through this globe. When the sun shines, we feel that it warms us, and we cannot miss to observe that every thing else is warmed around us. It is also plain that those seasons are the hottest, during which it shines the most, as well as those climates which are the most directly exposed to its light. When the sun disappears, the heat abates, and abates the more the longer his influence is intercepted. We must therefore acknowledge the sun as a manifest cause, acting on all the matter around us, and introducing something into it, or bringing it into a condition which is not its most spontaneous state. We cannot therefore avoid considering this new condition or heat, thus induced in the matter around us, as a positive quality, or real affection, of which the sun is the primary cause, and which is afterwards communicated from those bodies, thus first affected to others.

But, after having formed this conclusion with regard to heat, where shall we find any primary cause or fountain of cold? I am ignorant of any general occasion or cause of cold, except the absence or diminished action of the sun, or winds blowing from those regions on which his light has the weakest power. I therefore see no reason for considering cold as any thing but a diminution of heat. The frigorific atoms, and particles of frost, which have been supposed to be brought by the cold winds, are altogether imaginary. We have not the smallest evidence of their existence, and none of the phenomena, on account of which they have been supposed to exist, require such a fiction in order to their being explained.

Some persons, however, may perhaps still find it difficult to divest themselves entirely of the prejudice, that in certain cases, cold acts in a positive manner. Such persons may perhaps appeal to our feelings, which give us a striking proof of the reality of cold as well as of heat. When we touch a lump of ice, we feel distinctly that it has a quality of coldness, as well as that hot iron has the quality of heat.

But let us examine what we mean by this quality of coldness. We mean a quality, or condition by which the ice produces a disagreeable sensation in the hand which touches it; to which sensation we give the name of cold, and consider it as contrary to heat, and to be as much a reality. So far we are right. The sensation of cold in our organs is no doubt as real a feeling as the sensation of heat. But if we thence conclude that it must be produced by an active or positive cause, an emanation from the ice into our organs, or in any other way than by a diminution of heat, we form a hasty judgment. Of this we may be convinced by several experiments. We can, for instance, take a quantity of water, and reduce it to such a state that it will appear warm to one person, and cold to another, and neither warm nor cold to a third; the first person must be prepared for the experiment by bathing his hand in cold water immediately before; the second, by bathing his hand in hot water, or by a feverish heat in his blood; and the hand of the third person must be in its ordinary natural state, while the water with which these experiments are made is of lukewarm temperature. Even to the same person, such water might be made to appear warm, when felt with one hand, and cold, when felt with the other. We are therefore under the necessity of concluding from these facts, that our sensations of heat and cold do not depend on two different active causes, or positive qualities in those bodies which excite these sensations, but upon certain differences of heat between those bodies and our organs. And, in general, everybody appears hot or warm on being touched, which is more heated than the hand, and communicates heat to it; and every body which is less heated than the hand, and which draws heat from the hand which touches it, appears cold, or is said to be cold. The sensation is in some cases agreeable, and in others disagreeable, according to its intensity, and the state of our organs; but it proceeds always from the same cause, the communication of heat from other bodies to our organs, or from our organs to them. What can we more reasonably expect than that the sensation produced by the intro-susception of the cause of heat, whatever that may be, will be different from the sensation that accompanies its emission from our bodies? The sensations of hunger and repletion are equally distinct.

Besides the uneasiness produced by the touch of very cold bodies, the freezing of water has induced many to believe the existence of frigorific particles. Water, they imagined to be naturally, or essentially fluid, and to have its fluidity in consequence of the round figure and fine polish of its particles; and they thought that to give it solidity, some powerful agent must be employed, which can pervert it from its natural state. They have therefore supposed the existence of frigorific atoms, of angular, pointed, and wedge-like forms, which, being introduced among those of the water, entangle, and fix them one with another.

But the whole of this too is imagination and fiction. We have not the least proof that the particles of water are round, or any good reason for imagining that they have that form. An assemblage of small round bodies, however smooth or polished, would not have the properties which are well known in water; and the supposition, that fluidity is a natural or essential quality of water, is a great mistake, occasioned by our seeing it in these parts of the world much more frequently fluid than solid. In some other parts of the world, its most common or natural state is a state of solidity; there are parts of the globe in which it rarely or never is seen fluid; and the one or the other state of the substance, as of all other bodies, depends on the degree of heat to which it is exposed. Pure ice never melts but when we attempt to heat it above a certain degree; and if we cool pure water to the same degree, or below it, we are sure to see it sooner or later completely congealed.

On these two facts alone, however, the sensation we have of cold, and the freezing of water, has been commonly founded the belief of the existence of frigorific atoms, among the greater number of those who have thought proper to adopt such an opinion.

But some of them have been influenced also by the effect of salts upon ice or snow. Many experiments have shewn, that certain salts, or strong saline liquors, if they be added to ice or snow, occasion these last to melt very quickly, and, at the same time, to become much colder; in consequence of which, this mixture of ice and salts is employed occasionally for freezing many liquids which cannot be frozen by ordinary colds. The liquid which is to be frozen is put into a vessel, and this vessel is plunged into the mixture of ice and salt.

These, and a few other facts which we shall afterwards consider, are enumerated by Professor Muschenbroek, among the reasons which he gives for his belief of the existence of frigorific or congealing particles; but they are not a good foundation for such an opinion; and we shall in the sequel have an opportunity to explain these facts, without having recourse to such a supposition.

We have, therefore, reason to conclude, that when bodies unequally heated are approached to one another, it is always the warmer or less cold body which acts on the other, and communicates to it a real something, which we call heat. Coldness is only the absence or deficiency of heat. It is the state the most proper to common matter; the state which it would assume were it left to itself, and were it not affected by any external cause. Heat is plainly something extraneous to it: It is either something superadded to common matter, or some alteration of it from its most spontaneous state.

Having arrived at this conclusion, it may perhaps be required of me, in the next place, to express more distinctly this something; to give a full description or definition of what I mean by the word, heat in matter.

This, however, is a demand which I cannot satisfy entirely. Yet I shall mention by and by, the supposition relating to this subject, which appears to me the most probable. But our knowledge of heat is not brought to that state of perfection that might enable us to propose with confidence a theory of heat, or assign an immediate cause for it. Some ingenious attempts have been made in this part of our subject, but none of them have been sufficient to explain the whole of it. This however should not give us much uneasiness. It is not the immediate manner of acting, dependent on the ultimate nature of this peculiar substance, or the particular condition of common matter, that we are most interested in; we are far removed as yet from that extent of chemical knowledge, which makes this a necessary step of farther improvement. We have still before us an abundant field of research in the various general facts or laws of action, which constitute the real objects of pure chemical science, namely, the distinctive characters of bodies, as affected by heat and mixture. And, I apprehend, that it is only when we have nearly completed this catalogue, that we shall have a sufficient number of resembling facts to lead us to a clear knowledge of the manner of acting peculiar to this substance, or this modification of matter; and, when we have at last attained it, I presume that the discovery will not be chemical but mechanical. It would, however, be unpardonable, to pass without notice, some of the most ingenious attempts which have had a certain currency among the philosophical chemists.

The first attempt I think was made by Lord Verulam; next after him, Mr. Boyle gave several dissertations on heat; and Dr. Boerhaave, in his lectures on chemistry, endeavored to prosecute the subject still farther, and to improve on the two former authors.

Lord Verulam's attempt may be seen in his treatise *De forma Calidi*, which he offers to the public as a model of the proper manner of prosecuting investigations in natural philosophy. In this treatise he enumerates all the principal facts then known relating to heat, or to the production of heat, and endeavors, after a cautious and mature consideration of these, to form some well founded opinion of its cause.

The only conclusion, however, that he is able to draw from the whole of his facts, is a very general one, viz. that heat is motion.

This conclusion is founded chiefly on the consideration of several means by which heat is produced, or made to appear, in bodies; as the percussion of iron, the friction of solid bodies, the collision of flint and steel.

The first of these examples is a practice to which blacksmiths have sometimes recourse for kindling a fire; they take a rod of soft iron, half an inch or less in thickness, and laying the end of it upon their anvil, they turn and strike that end very quickly on its different sides, with smart blows of a hammer. It very soon becomes red hot, and can be employed to kindle shavings of wood, or other very combustible matter.

The heat producible by the strong friction of solid bodies, occurs often in some parts of heavy machinery, when proper care is not taken to diminish that friction as much as possible, by the interposition of lubricating substances; as in the axles of wheels that are heavy themselves, or heavily loaded. Thick forests are said to have taken fire sometimes by the friction of branches against one another in stormy weather. And savages, in different parts of the world, have recourse to the friction of pieces of wood for kindling their fires. A proper opportunity will afterwards occur for considering this manner of producing heat, with some attention.

The third example above adduced in the collision of flint and steel, is universally known.

In all these examples, heat is produced or made to appear suddenly, in bodies which have not received it in the usual way of communication from others, and the only cause of its production is a mechanical force or impulse, or mechanical violence.

It was, therefore, very natural for Lord Verulam to form his conclusion, as the most usual : nay, perhaps the sole effect of mechanical force or impulse, applied to a body, is to produce some sort of motion of that body. This eminent philosopher has had a great number of followers on this subject.

But his opinion has been adopted with two different modifications.

The greater number of the English philosophers supposed this motion to be in the small particles of the heated bodies, and imagine that it is a rapid tremor, or vibration of these particles among one another. Mr. Macquer also, and Mons. Fourcroy, both incline, or did incline, to this opinion. I acknowledge that I cannot form to myself a conception of this internal tremor, that has any tendency to explain, even the more simple effects of heat, or those phenomena which indicate its presence in a body ; and I think that Lord Verulam and his followers have been contented with very slight resemblances indeed, between those most simple effects of heat, and the legitimate consequences of a tremulous motion. I also see many cases, in which intense heat is produced in this way, where I am certain that the internal tremor is incomparably less than in other cases of percussion, similar in all other respects. Thus the blows, which make a piece of soft iron intensely hot, produce no heat in a similar piece of very elastic steel.

But the greater number of French and German philosophers, and Dr. Boerhaave, have supposed that the motion in which heat consists is not a tremor, or vibration of the particles of the hot body itself, but of the particles of a subtile, highly elastic, and penetrating fluid matter, which is contained in the pores of hot bodies, or interposed among their particles : a matter, which they imagine to be diffused through the whole universe, pervading with ease the densest bodies ; a matter, which some suppose, when modified in different ways, produces light, and the phenomena of electricity.

But neither of these suppositions were fully and accurately considered by their authors, or applied to explain the whole of the facts and phenomena relating to heat. They did not, therefore, supply us with a proper *theory* or *explication* of the nature of heat.

A more ingenious attempt has lately been made, the first outlines of which, so far as I know, were given by the late Dr. Cleghorn, in his inaugural dissertation, published here on the subject of heat. He supposes, that heat depends on the abundance of that subtile fluid elastic matter, which had been imagined before by other philosophers to be present in every part of the universe, and to be the cause of heat. But these other philosophers had assumed, or supposed one property only belonging to this subtile matter, viz. its great elasticity, or the strong repellency of its particles for one another; whereas, Dr. Cleghorn supposed it possessed another property also, that is, a strong attraction for the particles of the other kinds of matter in nature, which have in general more or less attraction for one another. He supposes, that the common grosser kinds of matter consist of attracting particles, or particles which have a strong attraction for one another, and for the matter of heat; while the subtile elastic matter of heat is self-repelling matter, the particles of which have a strong repulsion for one another, while they are attracted by the other kinds of matter, and that with different degrees of force.

This opinion, or supposition, can be applied to explain many of the remarkable facts relating to heat; and it is conformable to those experiments of Dr. Franklin, and of Sir Benjamin Thompson, quoted above. For, wherever there is but a very small quantity of common matter, as in the vacuum of an air-pump, there we may expect to find the matter of heat excessively rarefied, in consequence of its own very great elasticity and self-repellency, which, in this case, is little counteracted by the attraction of other matter.

A cold body, therefore, placed in such a vacuum, is supplied more slowly with heat, or with the matter of heat, than when placed in contact with common matter in a denser state, which, by its attraction for the matter of heat, condenses a much greater quantity of it into the same space. And a hot body, placed in such a vacuum, will retain its heat longer than in ordinary circumstances, in consequence of the scarcity of common matter in contact with it, by the attraction of which, its heat would be drawn off more quickly than if there were no other matter present but the matter of heat.

Such an idea of the nature of heat is, therefore, the most probable of any that I know; and an ingenious attempt to make use of it has been published by Dr. Higgins, in his book on vegetable acid, and other subjects. It is, however, altogether a suppostion; and I cannot at present make you understand the application of this theory, or the manner in which it has been formed; the greater number of you not being yet acquainted with the effects of heat, and the different phenomena which this theory is meant to explain, nor with some discoveries which preceded this theory, and gave occasion to it.

Our first business must, therefore, necessarily be, to *study the facts* belonging to our subject, and to attend to the manner in which heat enters various bodies, or is communicated from one to another, together with the consequences of its entrance, that is, the effects that it produces on the bodies.

These particulars, when considered with attention, will lead us to some more adequate knowledge and information upon the subject....which again will enable you to examine and understand the attempts that have been made to explain it, and put you in the way to form a judgment of their validity.

When we attend to the effects produced by heat in the bodies to which it is communicated, we see that they are very various in the different kinds of matter. But there are some effects which are produced in all kinds, or in a great variety of bodies, in a similar manner or with such inconsiderable variations, that the similarity of its action is sufficiently evident. This is true, especially with regard to the simpler kinds of matter, such as water, salts, stones, metals, air, and many others. These similar effects, produced by heat upon such bodies of the more simple kind, may therefore be considered as the *general* effects of heat; and thus distinguished from many which it produces on certain particular bodies only.

These general effects of heat are, EXPANSION, FLUIDITY, VAPOR, IGNITION, or INCANDESCENCE, and INFLAMMATION, or COMBUSTION.

24

This article was translated expressly for this Benchmark volume by R. Bruce Lindsay, Brown University, from Traité élémentaire de chimie, *Pt. 1, Chap. 1, Chez Chuchet, Libraire, Paris, 1789, pp. 4–6*

INTRODUCTION OF THE TERM "CALORIC" FOR THE SUBSTANCE OF HEAT

Antoine Laurent Lavoisier

[*Ed. note:* After discussing the numerous changes of state brought about by the agency of heat, the author discusses the nature of heat itself, as follows.]

It is difficult to conceive these phenomena (that is, changes of state) without admitting that they are the effect of a real, material substance, a very subtle fluid that can penetrate between the molecules of all bodies and scatter them. Even admitting that the existence of this fluid is a hypothesis, we shall see in what follows that it explains the phenomena of nature in a very satisfactory fashion.

This substance, whatever it may be, being the cause of heat, or in other words the senation that we call heat being the effect of the accumulation of this substance, we should not in a rigorously logical language designate it by the name "heat." For the same designation cannot express both cause and effect. This led me in the memoir that I published in 1777 (*Recueil de l'Academie,* p. 420) to designate it by the name fiery fluid or matter of heat. Since then, in the joint work that M. de Morveau, M. Berthollet, and M. de Fourcroy have carried out to reform chemical terminology, we have felt we ought to avoid this circumlocution, which lengthens description and makes it less precise and clear. We have therefore designated the cause of heat, the elastic fluid that produces heat, by the name *caloric.* Independently of the fact that this expression fulfills our object in the system we have adopted, it has still another advantage in that it can adapt itself to all sorts of opinions. Since, strictly speaking, we are not obliged to suppose that caloric is real material, it suffices, as will be grasped best by the reading of that which follows, that it should be some repulsive cause which pushes the molecules of matter apart, and one can thus envisage the effects in an abstract and mathematical manner.

Is light a modification of caloric or is caloric a modification of light? It is impossible to settle this definitely in the present state of our knowledge. What is certain is that, in a system in which one agrees to admit facts only and where one refrains from supposing anything not flowing directly from such facts, one really ought to designate by different names things that produce different effects. We therefore shall distinguish light from caloric, but we nonetheless admit that light and caloric have some common qualities and that under certain circumstances they are combined in the same manner and produce in part the same effects.

What we have just said should be sufficient to fix the idea that we should attach to the word caloric. But a more difficult task remains, to provide accurate ideas as to how caloric acts on bodies. Since this subtle material penetrates into the pores of all the substances we know about, and since no vessels exist from which it is unable to escape, and that are therefore unable to hold it without loss, one can learn its properties only through its effects, which for the most part are fleeting and difficult to grasp. In the case of things that one can neither see nor feel, one must be on one's guard against the errors of the imagination, which always tends to leap ahead beyond what is true and finds it painful to stay within the narrow circle to which the facts restrict it.

We have just seen that the same body becomes solid, liquid, or aeriform fluid according to the quantity of fluid that has penetrated it, or, to speak more rigorously, depending on whether the repulsive force of the molecules is equal to the attraction of the molecules, or stronger than this attraction, or weaker than this attraction.

25

Reprinted from S. Brown, *Men of Physics: Benjamin Thompson, Count Rumford,* Pergamon Press, Oxford, 1967, pp. 52–73

SOURCE OF HEAT FROM FRICTION

Benjamin Thompson

AMONG the most famous experiments conducted for the purpose of disproving the materiality of heat were those of Count Rumford, described by him in the following pages. We have already seen in Chapter 1 the place this experiment enjoys in the historical perspective of the development of our modern theory of heat.

Rumford's name is often associated with that of his protégé, Humphrey Davy, in connection with experiments of heat produced by friction. A study of Davy's experiments, which were carried out by Davy in 1798 when he was a lad of 19, show that the value of his contribution has been grossly overrated.† They were published in *Contributions to Physical and Medical Knowledge Principally from the West of England* which was edited by Davy's employer, Dr. Thomas Beddoes. Even though these experiments cannot be taken seriously by any physicist who cares to look into the actual experiments, they did have the very important result of calling Rumford's attention to Davy and starting him on his road to fame.

ESSAY IX

An Experimental Inquiry concerning the Source of the Heat which is Excited by Friction

Read before the Royal Society, January 25, 1798

Philosophical Transactions LXXXVIII, 80–102 (1798)
Bibliotheque Britannique VIII, 3–34 (1798)
Nicholson's quarto Journal I, 459–468, 515–518 (1798)
Gilbert's Annalen der Physik IV, 257–281, 377–399 (1798)
Scherer's Journal der Chemie I, 9–31 (1798)
Voigt's Magazin I, 94–106 (1798)

† E. N. daC. Andrade, *Nature*, March 9, 1935, p. 359.

It frequently happens that in the ordinary affairs and occupations of life, opportunities present themselves of contemplating some of the most curious operations of Nature; and very interesting philosophical experiments might often be made, almost without trouble or expense, by means of machinery contrived for the mere mechanical purposes of the arts and manufactures.

I have frequently had occasion to make this observation; and am persuaded that a habit of keeping the eyes open to everything that is going on in the ordinary course of the business of life has oftener led, as it were by accident, or in the playful excursions of the imagination, put into action by contemplating the most common appearances, to useful doubts and sensible schemes for investigation and improvement, than all the more intense meditations of philosophers in the hours expressly set apart for study.

It was by accident that I was led to make the experiments of which I am about to give an account; and, though they are not perhaps of sufficient importance to merit so formal an introduction, I cannot help flattering myself that they will be thought curious in several respects, and worthy of the honour of being made known to the Royal Society.

Being engaged lately in superintending the boring of cannon in the workshops of the military arsenal at Munich, I was struck with the very considerable degree of Heat which a brass gun acquires in a short time in being bored, and with the still more intense Heat (much greater than that of boiling water, as I found by experiment) of the metallic chips separated from it by the borer.

The more I meditated on these phænomena, the more they appeared to me to be curious and interesting. A thorough investigation of them seemed even to bid fair to give a farther insight into the hidden nature of Heat; and to enable us to form some reasonable conjectures respecting the existence, or non-existence, of an *igneous fluid*,—a subject on which the opinions of philosophers have in all ages been much divided.

In order that the Society may have clear and distinct ideas of the speculations and reasonings to which these appearances gave rise in my mind, and also of the specific objects of philosophical investigation they suggested to me, I must beg leave to state them at some length, and in such manner as I shall think best suited to answer this purpose.

From *whence comes* the Heat actually produced in the mechanical operation above mentioned?

Is it furnished by the metallic chips which are separated by the borer from the solid mass of metal?

If this were the case, then, according to the modern doctrines of latent Heat, and of caloric, the *capacity for Heat* of the parts of the metal, so reduced to chips, ought not only to be changed, but the change undergone by them should be sufficiently great to account for *all* the Heat produced.

But no such change had taken place; for I found, upon taking equal quantities, by weight, of these chips, and of thin slips of the same block of metal separated by means of a fine saw, and putting them at the same temperature (that of boiling water) into equal quantities of cold water (that is to say, at the temperature of 59½°F.), the portion of water into which the chips were put was not, to all appearance, heated either less or more than the other portion in which the slips of metal were put.

This experiment being repeated several times, the results were always so nearly the same that I could not determine whether any, or what change had been produced in the metal, *in regard to its capacity for Heat*, by being reduced to chips by the borer.*

* As these experiments are important, it may perhaps be agreeable to the Society to be made acquainted with them in their details.

One of them was as follows:

To 4590 grains of water, at the temperature of 59½F. (an allowance as compensation, reckoned in water, for the capacity for Heat of the containing cylindrical tin vessel being included), were added 1016⅛ grains of gun-metal in thin slips, separated from the gun by means of a fine saw, being at the temperature of 210°F. When they had remained together 1 minute, and had been well stirred about, by means of a small rod of light wood, the Heat of the mixture was found to be = 63°.

From this experiment the *specific Heat* of the metal, calculated according to the rule given by Dr. Crawford, turns out to be = 0.1100, that of water being = 1.0000.

An experiment was afterwards made with the metallic chips as follows:

From hence it is evident that the Heat produced could not possibly have been furnished at the expence of the latent Heat of the metallic chips. But, not being willing to rest satisfied with these trials, however conclusive they appeared to me to be, I had recourse to the following still more decisive experiment.

Taking a cannon (a brass six-pounder), cast solid, and rough as it came from the foundry (see Fig. 1 [Plate 5]), and fixing it (horizontally) in the machine used for boring, and at the same time finishing the outside of the cannon by turning (see Fig. 2 [Plate 5]), I caused its extremity to be cut off, and, by turning down the metal in that part, a solid cylinder was formed, $7\frac{3}{4}$ inches in diameter, and $9\frac{8}{10}$ inches long, which, when finished, remained joined to the rest of the metal (that which, properly speaking, constituted the cannon) by a small cylindrical neck, only $2\frac{1}{5}$ inches in diameter, and $3\frac{8}{10}$ inches long.

This short cylinder, which was supported in its horizontal position and turned round its axis by means of the neck by which it remained united to the cannon, was now bored with the horizontal borer used in boring cannon; but its bore, which was 3.7 inches in diameter, instead of being continued through its whole length (9.8 inches) was only 7.2 inches in length; so that a solid bottom was left to this hollow cylinder, which bottom was 2.6 inches in thickness.

This cavity is represented by dotted lines in Fig. 2 [Plate 5]; as also in Fig. 3 [Plate 5], where the cylinder is represented on an enlarged scale.

This cylinder being designed for the express purpose of generating Heat *by friction*, by having a blunt borer forced against its solid bottom at the same time that it should be turned round its

---

To the same quantity of water as was used in the experiment above mentioned, at the same temperature (*viz.* $59\frac{1}{2}$°), and in the same cylindrical tin vessel, were now put $1016\frac{1}{8}$ grains of metallic chips of gun-metal bored out of the same gun from which the slips used in the foregoing experiment were taken, and at the same temperature (210°). The Heat of the mixture at the end of 1 minute was just 63°, as before; consequently the specific Heat of these metallic chips was = 0.1100. Each of the above experiments was repeated three times, and always with nearly the same results.

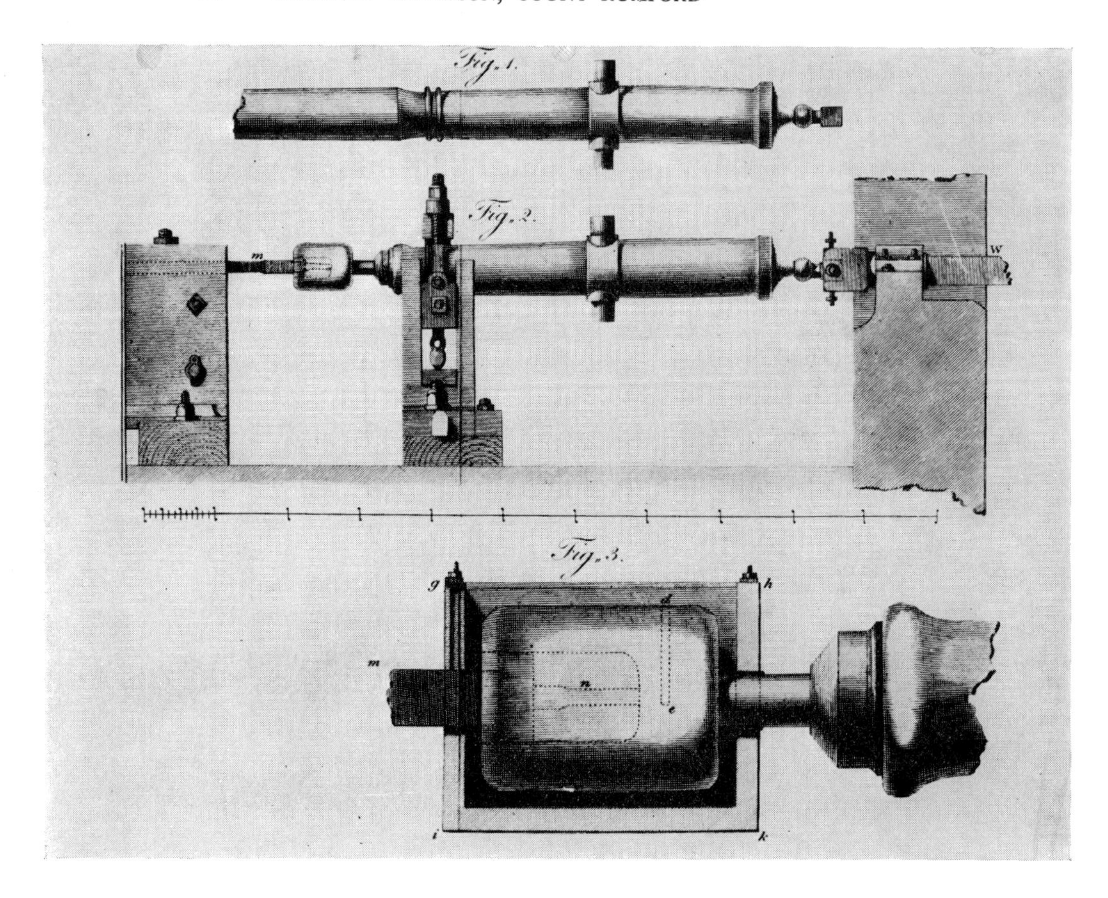
Fig. 1.
Fig. 2.
m
W
Fig. 3.
g
h
d
m
n
e
i
k

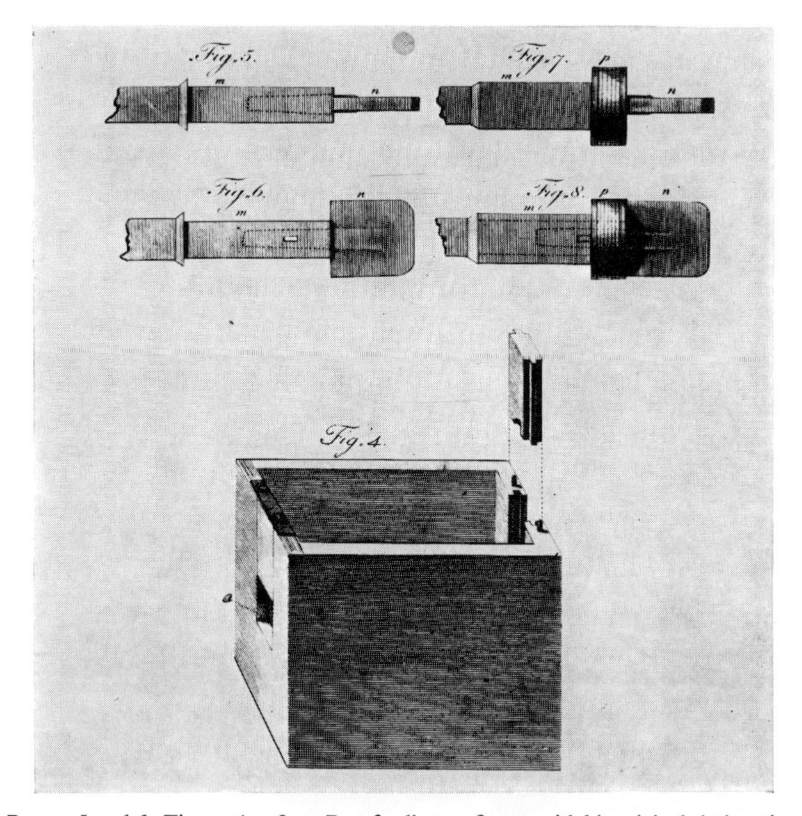

PLATES 5 and 6. Figures 1 to 8 are Rumford's own figures with his original designations.

axis by the force of horses, in order that the Heat accumulated in the cylinder might from time to time be measured, a small round hole (see *d*, *e*, Fig. 3 [Plate 5]), 0.37 of an inch only in diameter, and 4.2 inches in depth, for the purpose of introducing a small cylindrical mercurial thermometer, was made in it, on one side, in a direction perpendicular to the axis of the cylinder, and ending in the middle of the solid part of the metal which formed the bottom of its bore.

The solid contents of this hollow cylinder, exclusive of the cylindrical neck by which it remained united to the cannon, were 385¾ cubic inches, English measure, and it weighed 113.13 lb., avoirdupois; as I found on weighing it at the end of the course of experiments made with it, and after it had been separated from the cannon with which, during the experiments, it remained connected.*

Experiment No. 1

This experiment was made in order to ascertain how much Heat was actually generated by friction, when a blunt steel borer being so forcibly shoved (by means of a strong screw) against the bottom of the bore of the cylinder, that the pressure against it was equal to the weight of about 10,000 lb., avoirdupois, the cylinder was turned round on its axis (by the force of horses) at the rate of about 32 times in a minute.

This machinery, as it was put together for the experiment, is

* For fear I should be suspected of prodigality in the prosecution of my philosophical researches, I think it necessary to inform the Society that the cannon I made use of in this experiment was not sacrificed to it. The short hollow cylinder which was formed at the end of it was turned out of a cylindrical base of metal, about 2 feet in length, projecting beyond the muzzle of the gun, called in the German language the *verlorner kopf* (the head of the cannon to be thrown away), and which is represented in Fig. 1 [Plate 5].

This original projection, which is cut off before the gun is bored, is always cast with it, in order that, by means of the pressure of its weight on the metal in the lower part of the mould during the time it is cooling, the gun may be the more compact in the neighbourhood of the muzzle, where, without this precaution, the metal would be apt to be porous, or full of honeycombs.

represented by Fig. 2 [Plate 5]. W is a strong horizontal iron bar, connected with proper machinery carried round by horses, by means of which the cannon was made to turn round its axis.

To prevent, as far as possible, the loss of any part of the Heat that was generated in the experiment, the cylinder was well covered up with a fit coating of thick and warm flannel, which was carefully wrapped round it, and defended it on every side from the cold air of the atmosphere. This covering is not represented in the drawing of the apparatus, Fig. 2 [Plate 5].

I ought to mention that the borer was a flat piece of hardened steel, 0.63 of an inch thick, 4 inches long, and nearly as wide as the cavity of the bore of the cylinder, namely, $3\frac{1}{2}$ inches. Its corners were rounded off at its end, so as to make it fit the hollow bottom of the bore; and it was firmly fastened to the iron bar (*m*) which kept it in its place. The area of the surface by which its end was in contact with the bottom of the bore of the cylinder was nearly $2\frac{1}{3}$ inches. This borer, which is distinguished by the letter *n*, is represented in most of the figures.

At the beginning of the experiment, the temperature of the air in the shade, as also that of the cylinder, was just 60°F.

At the end of 30 minutes, when the cylinder had made 960 revolutions about its axis, the horses being stopped, a cylindrical mercurial thermometer, whose bulb was $\frac{32}{100}$ of an inch in diameter, and $3\frac{1}{4}$ inches in length, was introduced into the hole made to receive it, in the side of the cylinder, when the mercury rose almost instantly to 130°.

Though the Heat could not be supposed to be quite equally distributed in every part of the cylinder, yet, as the length of the bulb of the thermometer was such that it extended from the axis of the cylinder to near its surface, the Heat indicated by it could not be very different from that of the *mean temperature* of the cylinder; and it was on this account that a thermometer of that particular form was chosen for this experiment.

To see how fast the Heat escaped out of the cylinder (in order to be able to make a probable conjecture respecting the quantity given off by it during the time the Heat generated by the friction was accumulating), the machinery standing still, I suffered the thermometer to remain in its place near three quarters of an hour, observing and noting down, at small intervals of time, the height of the temperature indicated by it.

		The Heat, as shown by the thermometer, was
Thus at the end of	4 minutes	126°
after	5 minutes, always reckoning from the first observation	125
at the end of	7 minutes	123
	12 „	120
	14 „	119
	16 „	118
	20 „	116
	24 „	115
	28 „	114
	31 „	113
	34 „	112
	37½ „	111
and when	41 minutes had elapsed	110

Having taken away the borer, I now removed the metallic dust, or, rather, scaly matter, which had been detached from the bottom of the cylinder by the blunt steel borer, in this experiment; and, having carefully weighed it, I found its weight to be 837 grains, Troy.

Is it possible that the very considerable quantity of Heat that was produced in this experiment (a quantity which actually raised the temperature of above 113 lb. of gun-metal at least 70 degrees of Fahrenheit's thermometer, and which, of course, would have been capable of melting 6½ lb. of ice, or of causing near 5 lb. of ice-cold water to boil) could have been furnished by so inconsiderable a quantity of metallic dust? and this merely in consequence of *a change* of its capacity for Heat?

As the weight of this dust (837 grains, Troy) amounted to no more than $\frac{1}{948}$ part of that of the cylinder, it must have lost no less than 948 degrees of Heat, to have been able to have raised the temperature of the cylinder 1 degree; and consequently it must have given off 66,360 degrees of Heat to have produced the effects which were actually found to have been produced in the experiment!

But without insisting on the improbability of this supposition, we have only to recollect, that from the results of actual and decisive experiments, made for the express purpose of ascertaining that fact, the capacity for Heat of the metal of which great guns are cast *is not sensibly changed* by being reduced to the form of metallic chips in the operation of boring cannon; and there does not seem to be any reason to think that it can be much changed, if it be changed at all, in being reduced to much smaller pieces by means of a borer that is less sharp.

If the Heat, or any considerable part of it, were produced in consequence of a change in the capacity for Heat of a part of the metal of the cylinder, as such change could only be *superficial*, the cylinder would by degrees be *exhausted*; or the quantities of Heat produced in any given short space of time would be found to diminish gradually in successive experiments. To find out if this really happened or not, I repeated the last-mentioned experiment several times with the utmost care; but I did not discover the smallest sign of exhaustion in the metal, notwithstanding the large quantities of Heat actually given off.

Finding so much reason to conclude that the Heat generated in these experiments, or *excited*, as I would rather choose to express it, was not furnished *at the expense of the latent Heat* or *combined caloric* of the metal, I pushed my inquiries a step farther, and endeavoured to find out whether the air did, or did not, contribute anything in the generation of it.

Experiment No. 2

As the bore of the cylinder was cylindrical, and as the iron bar (*m*), to the end of which the blunt steel borer was fixed, was square, the air had free access to the inside of the bore, and even to the bottom of it, where the friction took place by which the Heat was excited.

As neither the metallic chips produced in the ordinary course of the operation of boring brass cannon, nor the finer scaly particles produced in the last-mentioned experiments by the friction of the blunt borer, showed any signs of calcination, I did not see how the air could possibly have been the cause of the heat that was produced; but, in an investigation of this kind, I thought that no pains should be spared to clear away the rubbish, and leave the subject as naked and open to inspection as possible.

In order, by one decisive experiment, to determine whether the air of the atmosphere had any part, or not, in the generation of the Heat, I contrived to repeat the experiment under circumstances in which *it was evidently impossible for it to produce any effect whatever*. By means of a piston exactly fitted to the mouth of the bore of the cylinder, through the middle of which piston the square iron bar, to the end of which the blunt steel borer was fixed, passed in a square hole made perfectly air-tight, the access of the external air to the inside of the bore of the cylinder was effectually prevented. (In Fig. 3 [Plate 5], this piston (*p*) is seen in its place; it is likewise shown in Fig. 7 and 8 [Plate 6].)

I did not find, however, by this experiment, that the exclusion of the air diminished, in the smallest degree, the quantity of Heat excited by the friction.

There still remained one doubt, which, though it appeared to me to be so slight as hardly to deserve any attention, I was however desirous to remove. The piston which closed the mouth of the bore of the cylinder, in order that it might be air-tight, was fitted into it with so much nicety, by means of its collars of leather, and pressed against it with so much force, that, notwithstanding its being oiled, it occasioned a considerable degree of friction when the hollow cylinder was turned round its axis. Was not the Heat produced, or at least some part of it, occasioned by this friction of the piston? and, as the external air had free access to the extremity of the bore, where it came in contact with the piston, is it not possible that this air may have had some share in the generation of the Heat produced?

Experiment No. 3

A quadrangular oblong deal box (see Fig. 4 [Plate 6]), water-tight, $11\frac{1}{2}$ English inches long, $9\frac{4}{10}$ inches wide, and $9\frac{6}{10}$ inches deep (measured in the clear), being provided with holes or slits in the middle of each of its ends, just large enough to receive, the one the square iron rod to the end of which the blunt steel borer was fastened, the other the small cylindrical neck which joined the hollow cylinder to the cannon; when this box (which was occasionally closed above by a wooden cover or lid moving on hinges) was put into its place, that is to say, when, by means of the two vertical openings or slits in its two ends (the upper parts of which openings were occasionally closed by means of narrow pieces of wood sliding in vertical grooves), the box (*g*, *h*, *i*, *k*, Fig. 3 [Plate 5]) was fixed to the machinery in such a manner that its bottom (*i*, *k*) being in the plane of the horizon, its axis coincided with the axis of the hollow metallic cylinder; it is evident, from the description, that the hollow metallic cylinder would occupy the middle of the box, without touching it on either side (as it is represented in Fig. 3 [Plate 5]); and that, on pouring water into the box, and filling it to the brim, the cylinder would be completely covered and surrounded on every side by that fluid. And farther, as the box was held fast by the strong square iron rod (*m*) which passed in a *square hole* in the center of one of its ends (*a*, Fig. 4 [Plate 6]), while the round or cylindrical neck, which joined the hollow cylinder to the end of the cannon, could turn round freely on its axis in the *round hole* in the center of the other end of it, it is evident that the machinery could be put in motion without the least danger of forcing the box out of its place, throwing the water out of it, or deranging any part of the apparatus.

Everything being ready, I proceeded to make the experiment I had projected in the following manner.

The hollow cylinder having been previously cleaned out, and the inside of its bore wiped with a clean towel till it was quite dry, the square iron bar, with the blunt steel borer fixed to the end of it, was put into its place; the mouth of the bore of the cylinder being closed at the same time by means of the circular piston, through the center of which the iron bar passed.

This being done, the box was put in its place, and the joinings of the iron rod and of the neck of the cylinder with the two ends of the box having been made watertight by means of collars of oiled leather, the box was filled with cold water (*viz.* at the temperature of 60°), and the machine was put in motion.

The result of this beautiful experiment was very striking, and the pleasure it afforded me amply repaid me for all the trouble I had had in contriving and arranging the complicated machinery used in making it.

The cylinder, revolving at the rate of about 32 times in a minute, had been in motion but a short time, when I perceived, by putting my hand into the water and touching the outside of the cylinder, that Heat was generated; and it was not long before the water which surrounded the cylinder began to be sensibly warm.

At the end of 1 hour I found, by plunging a thermometer into the water in the box (the quantity of which fluid amounted to 18.77 lb., avoirdupois, or $2\frac{1}{4}$ wine gallons), that its temperature had been raised no less than 47 degrees; being now 107° of Fahrenheit's scale.

When 30 minutes more had elapsed, or 1 hour and 30 minutes after the machinery had been put in motion, the Heat of the water in the box was 142°.

At the end of 2 hours, reckoning from the beginning of the experiment, the temperature of the water was found to be raised to 178°.

At 2 hours 20 minutes it was at 200°; and at 2 hours 30 minutes it ACTUALLY BOILED!

It would be difficult to describe the surprise and astonishment expressed in the countenances of the bystanders, on seeing so large a quantity of cold water heated, and actually made to boil, without any fire.

Though there was, in fact, nothing that could justly be considered as surprising in this event, yet I acknowledge fairly that it afforded me a degree of childish pleasure, which, were I ambitious of the reputation of a *grave philosopher*, I ought most certainly rather to hide than to discover.

The quantity of Heat excited and accumulated in this experiment was very considerable; for, not only the water in the box, but also the box itself (which weighed 15¼ lb.), and the hollow metallic cylinder, and that part of the iron bar which, being situated within the cavity of the box, was immersed in the water, were heated 150 degrees of Fahrenheit's scale; *viz.* from 60° (which was the temperature of the water and of the machinery at the beginning of the experiment) to 210°, the Heat of boiling water at Munich.

The total quantity of Heat generated may be estimated with some considerable degree of precision as follows:—

	Quantity of ice-cold water which, with the given quantity of Heat, might have been heated 180 degrees, or made to boil. In avoirdupois weight.
Of the Heat excited there appears to have been actually accumulated—	
In the water contained in the wooden box, 18¾ lb., avoirdupois, heated 150 degrees, namely, from 60° to 210°F. .	lb. 15.2
In 113.13 lb. of gun-metal (the hollow cylinder), heated 150 degrees; and, as the capacity for Heat of this metal is to that of water as 0.1100 to 1.0000, this quantity of Heat would have heated 12½ lb. of water the same number of degrees .	10.37
In 36.75 cubic inches of iron (being that part of the iron bar to which the borer was fixed which entered the box), heated 150 degrees; which may be reckoned equal in capacity for Heat to 1.21 lb. of water	1.01
N.B. No estimate is here made of the Heat accumulated in the wooden box, nor of that dispersed during the experiment.	
Total quantity of ice-cold water which, with the Heat actually generated by friction, and accumulated in 2 hours and 30 minutes, might have been heated 180 degrees, or made to boil .	26.58

From the knowledge of the *quantity* of Heat actually produced in the foregoing experiment, and of the *time* in which it was generated, we are enabled to ascertain *the velocity of its production*, and to determine how large a fire must have been, or how much fuel must have been consumed, in order that, in burning equably, it should have produced by combustion the same quantity of Heat in the same time.

In one of Dr. Crawford's experiments (see his Treatise on Heat, p. 321), 37 lb. 7 oz., Troy, = 181,920 grains of water, were heated $2\frac{1}{10}$ degrees of Fahrenheit's thermometer with the Heat generated in the combustion of 26 grains of wax. This gives 382,032 grains of water heated 1 degree with 26 grains of wax, or $14{,}693\frac{14}{26}$ grains of water heated 1 degree, or $\frac{14693}{180}$ = 81.631 grains heated 180 degrees, with the Heat generated in the combustion of 1 grain of wax.

The quantity of ice-cold water which might have been heated 180 degrees with the Heat generated by friction in the before-mentioned experiment was found to be 26.58 lb., avoirdupois, = 188,060 grains; and, as 81,631 grains of ice-cold water require the Heat generated in the combustion of 1 grain of wax to heat it 180 degrees, the former quantity of ice-cold water, namely 188,060 grains, would require the combustion of no less than 2303.8 grains (= $4\frac{8}{10}$ oz., Troy) of wax to heat it 180 degrees.

As the experiment (No. 3) in which the given quantity of Heat was generated by friction lasted 2 hours and 30 minutes, = 150 minutes, it is necessary, for the purpose of ascertaining how many wax candles of any given size must burn together, in order that in the combustion of them the given quantity of Heat may be generated in the given time, and consequently *with the same celerity* as that with which the Heat was generated by friction in the experiment, that the size of the candles should be determined, and the quantity of wax consumed in a given time by each candle in burning equably should be known.

Now I found, by an experiment made on purpose to finish these computations, that when a good wax candle, of a moderate size, $\frac{3}{4}$ of an inch in diameter, burns with a clear flame, just 49 grains of wax are consumed in 30 minutes. Hence it appears that 245 grains of wax would be consumed by such a candle in 150 minutes;

and that, to burn the quantity of wax (=2303.8 grains) necessary to produce the quantity of Heat actually obtained by friction in the experiment in question, and in the given time (150 minutes), *nine candles*, burning at once, would not be sufficient; for 9 multiplied into 245 (the number of grains consumed by each candle in 150 minutes) amounts to no more than 2205 grains; whereas the quantity of wax necessary to be burnt, in order to produce the given quantity of Heat, was found to be 2303.8 grains.

From the result of these computations it appears, that the quantity of Heat produced equably, or in a continual stream (if I may use that expression), by the friction of the blunt steel borer against the bottom of the hollow metallic cylinder, in the experiment under consideration, was *greater* than that produced equably in the combustion of *nine wax candles*, each $\frac{3}{4}$ of an inch in diameter, all burning together, or at the same time, with clear bright flames.

As the machinery used in this experiment could easily be carried round by the force of one horse (though, to render the work lighter, two horses were actually employed in doing it), these computations shew further how large a quantity of Heat might be produced, by proper mechanical contrivance, merely by the strength of a horse, without either fire, light, combustion, or chemical decomposition; and, in a case of necessity, the Heat thus produced might be used in cooking victuals.

But no circumstances can be imagined in which this method of procuring Heat would not be disadvantageous; for more Heat might be obtained by using the fodder necessary for the support of a horse as fuel.

As soon as the last-mentioned experiment (No. 3) was finished, the water in the wooden box was let off, and the box removed; and the borer being taken out of the cylinder, the scaly metallic powder which had been produced by the friction of the borer against the bottom of the cylinder was collected, and, being carefully weighed, was found to weigh 4145 grains, or about $8\frac{2}{3}$ oz., Troy.

As this quantity was produced in $2\frac{1}{2}$ hours, this gives 824 grains for the quantity produced *in half an hour.*

In the first experiment, which lasted only *half an hour*, the quantity produced was 837 grains.

In the experiment No. 1, the quantity of Heat generated in *half an hour* was found to be equal to that which would be required to heat 5 lb., avoirdupois, of ice-cold water 180 degrees, or cause it to boil.

According to the result of the experiment No. 3, the Heat generated in *half an hour* would have caused 5.31 lb. of ice-cold water to boil. But, in this last-mentioned experiment, the Heat generated being more effectually confined, less of it was lost; which accounts for the difference of the results of the two experiments.

It remains for me to give an account of one experiment more, which was made with this apparatus. I found, by the experiment No. 1, how much Heat was generated when the air had free access to the metallic surfaces which were rubbed together. By the experiment No. 2, I found that the quantity of Heat generated was not sensibly diminished when the free access of the air was prevented; and by the result of No. 3, it appeared that the generation of the Heat was not prevented or retarded by keeping the apparatus immersed in water. But as, in this last-mentioned experiment, the water, though it surrounded the hollow metallic cylinder on every side, externally, was not suffered to enter the cavity of its bore (being prevented by the piston), and consequently did not come into contact with the metallic surfaces where the Heat was generated; to see what effects would be produced by giving the water free access to these surfaces, I now made the

Experiment No. 4

The piston which closed the end of the bore of the cylinder being removed, the blunt borer and the cylinder were once more put together; and the box being fixed in its place, and filled with water, the machinery was again put in motion.

There was nothing in the result of this experiment that renders it necessary for me to be very particular in my account of it. Heat was generated as in the former experiments, and, to all appearance, quite as rapidly; and I have no doubt but the water in the box would have been brought to boil, had the experiment been continued as long as the last. The only circumstance that surprised me was, to find how little difference was occasioned in the noise made by the borer in rubbing against the bottom of the bore of the cylinder, by filling the bore with water. This noise, which was very grating to the ear, and sometimes almost insupportable, was, as nearly as I could judge of it, quite as loud and as disagreeable when the surfaces rubbed together were wet with water as when they were in contact with air.

By meditating on the results of all these experiments, we are naturally brought to that great question which has so often been the subject of speculation among philosophers; namely,—

What is Heat? Is there any such thing as an *igneous fluid*? Is there anything that can with propriety be called *caloric*?

We have seen that a very considerable quantity of Heat may be excited in the friction of two metallic surfaces, and given off in a constant stream or flux *in all directions* without interruption or intermission, and without any signs of diminution or exhaustion.

From whence came the Heat which was continually given off in this manner in the foregoing experiments? Was it furnished by the small particles of metal, detached from the larger solid masses, on their being rubbed together? This, as we have already seen, could not possibly have been the case.

Was it furnished by the air? This could not have been the case; for, in three of the experiments, the machinery being kept immersed in water, the access of the air of the atmosphere was completely prevented.

Was it furnished by the water which surrounded the machinery? That this could not have been the case is evident: *first*, because this water was continually *receiving Heat* from the machinery, and could not at the same time be *giving to*, and *receiving Heat from*, the same body; and, *secondly*, because there was no chemical decomposition of any part of this water. Had any such decomposition taken place (which, indeed, could not reasonably have been expected), one of its component elastic fluids (most probably inflammable air) must at the same time have been set at liberty, and, in making its escape into the atmosphere, would have been detected; but though I frequently examined the water to see if any air-bubbles rose up through it, and had even made preparations for catching them, in order to examine them, if any should appear, I could perceive none; nor was there any sign of decomposition of any kind whatever, or other chemical process, going on in the water.

Is it possible that the Heat could have been supplied by means of the iron bar to the end of which the blunt steel borer was fixed? or by the small neck of gun-metal by which the hollow cylinder was united to the cannon? These suppositions appear more improbable even than either of those before mentioned; for Heat was continually going off, or *out of the machinery*, by both these passages, during the whole time the experiment lasted.

And, in reasoning on this subject, we must not forget to consider that most remarkable circumstance, that the source of the Heat generated by friction, in these experiments, appeared evidently to be *inexhaustible*.

It is hardly necessary to add, that anything which any *insulated* body, or system of bodies, can continue to furnish *without limitation*, cannot possibly be *a material substance*; and it appears to me to be extremely difficult, if not quite impossible, to form any distinct idea of anything capable of being excited and communicated in the manner the Heat was excited and communicated in these experiments, except it be MOTION.

I am very far from pretending to know how, or by what means or mechanical contrivance, that particular kind of motion in bodies which has been supposed to constitute Heat is excited, continued, and propagated; and I shall not presume to trouble the Society with mere conjectures, particularly on a subject which, during so many thousand years, the most enlightened philosophers have endeavoured, but in vain, to comprehend.

But, although the mechanism of Heat should, in fact, be one of those mysteries of nature which are beyond the reach of human intelligence, this ought by no means to discourage us or even lessen our ardour, in our attempts to investigate the laws of its operations. How far can we advance in any of the paths which science has opened to us before we find ourselves enveloped in those thick mists which on every side bound the horizon of the human intellect? But how ample and how interesting is the field that is given us to explore!

Nobody, surely, in his sober senses, has even pretended to understand the mechanism of gravitation; and yet what sublime discoveries was our immortal Newton enabled to make, merely by the investigation of the laws of its action!

The effects produced in the world by the agency of Heat are probably *just as extensive*, and quite as important, as those which are owing to the tendency of the particles of matter towards each other; and there is no doubt but its operations are, in all cases, determined by laws equally immutable.

Before I finish this Essay, I would beg leave to observe, that although, in treating the subject I have endeavoured to investigate, I have made no mention of the names of those who have gone over the same ground before me, nor of the success of their labours, this omission has not been owing to any want of respect for my predecessors, but was merely to avoid prolixity, and to be more at liberty to pursue, without interruption, the natural train of my own ideas.

Description of the Figures

[Plate 5, Figs. 1–3; Plate 6, Figs. 4–8]

Fig. 1 shews the cannon used in the foregoing experiments in the state it was in when it came from the foundry.

Fig. 2 shews the machinery used in the experiments No. 1 and No. 2. The cannon is seen fixed in the machine used for boring cannon. W is a strong iron bar (which, to save room in the drawing, is represented as broken off), which bar, being united with machinery (not expressed in the figure) that is carried round by horses, causes the cannon to turn round its axis.

m is a strong iron bar, to the end of which the blunt borer is fixed; which, by being forced against the bottom of the bore of the short hollow cylinder that remains connected by a small cylindrical neck to the end of the cannon, is used in generating Heat by friction.

Fig. 3 shews, on an enlarged scale, the same hollow cylinder that is represented on a smaller scale in the foregoing figure. It is here seen connected with the wooden box (*g*, *h*, *i*, *k*) used in the experiments No. 3 and No. 4, when this hollow cylinder was immersed in water.

p, which is marked by dotted lines, is the piston which closed the end of the bore of the cylinder.

n is the blunt borer seen sidewise.

d, *e*, is the small hole by which the thermometer was introduced that was used for ascertaining the Heat of the cylinder. To save room in the drawing, the cannon is represented broken off near its muzzle; and the iron bar to which the blunt borer is fixed is represented broken off at *m*.

Fig. 4 is a perspective view of the wooden box, a section of which is seen in the foregoing figure. (See *g*, *h*, *i*, *k*, Fig. 3.)

Figs. 5 and 6 represent the blunt borer *n*, joined to the iron bar *m*, to which it was fastened.

Fig. 7 and 8 represent the same borer, with its iron bar, together with the piston which, in the experiments No. 2 and No. 3, was used to close the mouth of the hollow cylinder.

Count Rumford realized that his experiments would not go unchallenged. In fact he raised a number of questions in the paper just given in order to answer them himself. A number of writers in the field disagreed with Rumford's conclusions, and to give the flavor of their objections let us quote from a paper by Emmett published in the *Annals of Philosophy* **16,** 137 (1820).

First, in answer to Rumford's question "Is the heat furnished by the metallic chips which are separated by the borer from the solid mass of metal?" Emmett comments:

> "In the commencement of this reasoning, an assumption is made, which is particularly unfortunate: namely, that if heat being an elastic fluid be evolved by the compression of solid matter, the capacity of that solid for heat must be diminished in proportion to the quantity which has been separated. The whole quantity of heat contained in the solid is doubtless diminished, but why is the capacity to be changed? . . . That the quantity of heat evolved in this experiment was great cannot be disputed, yet it was

by no means sufficient to warrant the conclusions that have been drawn. . . . In these experiments, a very large mass of metal was submitted to an excessive pressure and of the mass, fresh strata was continually exposed to the compression by the wearing off of the brass: hence a definite quantity of heat was separated from each stratum in succession. Now if we admit the existence of caloric in a state of great density in the metals, this cause would be quite adequate to the production of the observed effect. The greatest error appears to be the assumption that the source of the heat thus generated is inexhaustible; the quantity that can be thus excited is finite"

but will not cease, according to this pictures, until all the brass is worn away.

Let us return to more of Rumford's questions. "From whence came the Heat which was continually given off in this manner in the foregoing experiments? Was it furnished by the air, was it furnished by the water which surrounded the machinery? Is it possible that the heat could have been supplied by means of the iron bar to the end of which the blunt steel borer was fixed? Or the small neck of gun-metal by which the hollow cylinder was united to the cannon? These suppositions appear most improbable." The answer to these questions seemed to lie at the very foundation of the caloric theory, namely that caloric pervaded all matter, and therefore could be furnished by all those sources which the Count appeared to discredit. His whole apparatus was bathed in an atmosphere of caloric. A physical picture of the manner in which the caloric was resupplied to the brass cup during the experiment was described as follows by a correspondent to the *Philosophical Magazine* for 1816.†

"If heat be a material fluid, the effect of force on a body containing it would be similar to the force on a body containing any other fluid diffused through its pores in a similar manner. Water being a fluid which in many instances produces effects similar to those produced by heat, it appears best adapted to illustrate the generation of heat by friction.

"I procured a piece of light and porous wood . . . and having immersed it in water until it was saturated, I fixed it firmly over a vessel filled with water, the lower end being . . . below the surface of the water, and then moved a piece of hard wood backwards and forwards on the upper end, with a considerable degree of pressure. I thus found that water could be raised through the pores of wood by friction. The process is easily understood: the piece of hard wood, as it is moved along, presses the water out of the pores, and closes them, driving out the water which is pressed out before it; but when the hard wood has pressed over these pores, the water from below rushes into them to restore the equilibrium.

"The action of the blunt borer in Count Rumford's experiments appears to have produced a similar kind of effect, the heat having been forced out of the pores of the metal by the borer, its place would be supplied by heat from the adjacent parts. Gun-metal being a good conductor, the neck which connected the cylinder with the cannon would be capable of giving passage to all the heat which was accumulated from the cannon, and the other.conductors with which it was connected."

† *Phil. Mag.* **48,** 29 (1816).

26

Reprinted by permission of the publisher from S. Carnot, *Reflections on the Motive Power of Fire and Other Papers,* E. Mendoza, ed., Dover Publications, Inc., New York, 1960, pp. 3–20

THE MOTIVE POWER OF HEAT

Sadi Carnot

Reflections on the Motive Power of Fire, and on Machines Fitted to Develop that Power

EVERY one knows that heat can produce motion. That it possesses vast motive-power no one can doubt, in these days when the steam-engine is everywhere so well known.

To heat also are due the vast movements which take place on the earth. It causes the agitations of the atmosphere, the ascension of clouds, the fall of rain and of meteors, the currents of water which channel the surface of the globe, and of which man has thus far employed but a small portion. Even earthquakes and volcanic eruptions are the result of heat.

From this immense reservoir we may draw the moving force necessary for our purposes. Nature, in providing us with combustibles on all sides, has given us the power to produce, at all times and in all places, heat and the impelling power which is the result of it. To develop this power, to appropriate it to our uses, is the object of heat-engines.

The study of these engines is of the greatest interest, their importance is enormous, their use is continually increasing, and they seem destined to produce a great revolution in the civilized world.

Already the steam-engine works our mines, impels our ships, excavates our ports and our rivers, forges iron, fashions wood, grinds grains, spins and weaves our cloths, transports the heaviest burdens, etc. It appears that it must some day serve as a universal motor, and be substituted for animal power, waterfalls, and air currents.

Over the first of these motors it has the advantage of economy, over the two others the inestimable advantage that it can be used at all times and places without interruption.

If, some day, the steam-engine shall be so perfected that it can be set up and supplied with fuel at small cost, it will combine all desirable qualities, and will afford to the industrial arts a range the extent of which can scarcely be predicted. It is not merely that a powerful and convenient motor that can be procured and carried anywhere is substituted for the motors already in use, but that it causes rapid extension in the arts in which it is applied, and can even create entirely new arts.

The most signal service that the steam-engine has rendered to England is undoubtedly the revival of the working of the coal mines, which had declined, and threatened to cease entirely, in consequence of the continually increasing difficulty of drainage, and of raising the coal.* We should rank second the benefit to iron manufacture, both by the abundant supply of coal substituted for wood just when the latter had begun to grow scarce, and by the powerful machines of all kinds, the use of which the introduction of the steam-engine has permitted or facilitated.

Iron and heat are, as we know, the supporters, the bases, of the mechanic arts. It is doubtful if there be in England a single industrial establishment of which the existence does not depend on the use of these agents, and which does not freely employ them. To take away today from England her steam-engines would be to take away at the same time her coal and iron. It would be to dry up all her sources of wealth, to ruin all on which her prosperity depends, in short, to annihilate that colossal power. The destruction of her navy, which she considers her strongest defence, would perhaps be less fatal.

The safe and rapid navigation by steamships may be regarded as an entirely new art due to the steam-engine. Already this art has permitted the establishment of prompt and regular communications across the arms of the sea, and on the great rivers of the old and new continents. It has made it possible to traverse savage regions where before we could scarcely penetrate. It has enabled us to carry the fruits of civilization over portions of the globe where they would else have been wanting for years. Steam navigation brings nearer together the most distant nations. It tends to unite the nations of the earth as inhabitants of one country. In fact, to lessen the time,

* It may be said that coal-mining has increased tenfold in England since the invention of the steam-engine. It is almost equally true in regard to the mining of copper, tin, and iron. The results produced in a half-century by the steam-engine in the mines of England are today paralleled in the gold and silver mines of the New World—mines of which the working declined from day to day, principally on account of the insufficiency of the motors employed in the draining and the extraction of the minerals.

the fatigues, the uncertainties, and the dangers of travel—is not this the same as greatly to shorten distances?*

The discovery of the steam-engine owed its birth, like most human inventions, to rude attempts which have been attributed to different persons, while the real author is not certainly known. It is, however, less in the first attempts that the principal discovery consists, than in the successive improvements which have brought steam-engines to the conditions in which we find them today. There is almost as great a distance between the first apparatus in which the expansive force of steam was displayed and the existing machine, as between the first raft that man ever made and the modern vessel.

If the honor of a discovery belongs to the nation in which it has acquired its growth and all its developments, this honor cannot be here refused to England. Savery, Newcomen, Smeaton, the famous Watt, Woolf, Trevithick, and some other English engineers, are the veritable creators of the steam-engine. It has acquired at their hands all its successive degrees of improvement. Finally, it is natural that an invention should have its birth and especially be developed, be perfected, in that place where its want is most strongly felt.

Notwithstanding the work of all kinds done by steam-engines, notwithstanding the satisfactory condition to which they have been brought today, their theory is very little understood, and the attempts to improve them are still directed almost by chance.

The question has often been raised whether the motive power of heat† is unbounded, whether the possible improvements in steam-engines have an assignable limit—a limit which the nature of things will not allow to be passed by any means whatever; or whether, on the contrary, these improvements may be carried on indefinitely. We have long sought, and are seeking today, to ascertain whether there are in existence agents preferable to the vapor of water for developing the motive power of heat; whether atmospheric air, for

---

* We say, to lessen the dangers of journeys. In fact, although the use of the steam-engine on ships is attended by some danger which has been greatly exaggerated, this is more than compensated by the power of following always an appointed and well-known route, of resisting the force of the winds which would drive the ship towards the shore, the shoals, or the rocks.

† We use here the expression motive power to express the useful effect that a motor is capable of producing. This effect can always be likened to the elevation of a weight to a certain height. It has, as we know, as a measure, the product of the weight multiplied by the height to which it is raised.

example, would not present in this respect great advantages. We propose now to submit these questions to a deliberate examination.

The phenomenon of the production of motion by heat has not been considered from a sufficiently general point of view. We have considered it only in machines the nature and mode of action of which have not allowed us to take in the whole extent of application of which it is susceptible. In such machines the phenomenon is, in a way, incomplete. It becomes difficult to recognize its principles and study its laws.

In order to consider in the most general way the principle of the production of motion by heat, it must be considered independently of any mechanism or any particular agent. It is necessary to establish principles applicable not only to steam-engines* but to all imaginable heat-engines, whatever the working substance and whatever the method by which it is operated.

Machines which do not receive their motion from heat, those which have for a motor the force of men or of animals, a waterfall, an air current, etc., can be studied even to their smallest details by the mechanical theory. All cases are foreseen, all imaginable movements are referred to these general principles, firmly established, and applicable under all circumstances. This is the character of a complete theory. A similar theory is evidently needed for heat-engines. We shall have it only when the laws of physics shall be extended enough, generalized enough, to make known beforehand all the effects of heat acting in a determined manner on any body.

We will suppose in what follows at least a superficial knowledge of the different parts which compose an ordinary steam-engine; and we consider it unnecessary to explain what are the furnace, boiler, steam-cylinder, piston, condenser, etc.

The production of motion in steam-engines is always accompanied by a circumstance on which we should fix our attention. This circumstance is the re-establishing of equilibrium in the caloric; that is, its passage from a body in which the temperature is more or less elevated, to another in which it is lower. What happens in fact in a steam-engine actually in motion? The caloric developed in the furnace by the effect of the combustion traverses the walls of the boiler, produces steam, and in some way incorporates itself with it.

---

* We distinguish here the steam-engine from the heat-engine in general. The latter may make use of any agent whatever, of the vapor of water or of any other, to develop the motive power of heat.

The latter carrying it away, takes it first into the cylinder, where it performs some function, and from thence into the condenser, where it is liquefied by contact with the cold water which it encounters there. Then, as a final result, the cold water of the condenser takes possession of the caloric developed by the combustion. It is heated by the intervention of the steam as if it had been placed directly over the furnace. The steam is here only a means of transporting the caloric. It fills the same office as in the heating of baths by steam, except that in this case its motion is rendered useful.

We easily recognize in the operations that we have just described the re-establishment of equilibrium in the caloric, its passage from a more or less heated body to a cooler one. The first of these bodies, in this case, is the heated air of the furnace; the second is the condensing water. The re-establishment of equilibrium of the caloric takes place between them, if not completely, at least partially, for on the one hand the heated air, after having performed its function, having passed round the boiler, goes out through the chimney with a temperature much below that which it had acquired as the effect of combustion; and on the other hand, the water of the condenser, after having liquefied the steam, leaves the machine with a temperature higher than that with which it entered.

The production of motive power is then due in steam-engines not to an actual consumption of caloric, but *to its transportation from a warm body to a cold body*, that is, to its re-establishment of equilibrium—an equilibrium considered as destroyed by any cause whatever, by chemical action such as combustion, or by any other. We shall see shortly that this principle is applicable to any machine set in motion by heat.

According to this principle, the production of heat alone is not sufficient to give birth to the impelling power: it is necessary that there should also be cold; without it, the heat would be useless. And in fact, if we should find about us only bodies as hot as our furnaces, how can we condense steam? What should we do with it if once produced? We should not presume that we might discharge it into the atmosphere, as is done in some engines;* the atmosphere would not receive it. It does receive it under the actual condition of things, only because it fulfils the office of a vast

* Certain engines at high pressure throw the steam out into the atmosphere instead of the condenser. They are used specially in places where it would be difficult to procure a stream of cold water sufficient to produce condensation.

condenser, because it is at a lower temperature; otherwise it would soon become fully charged, or rather would be already saturated.*

Wherever there exists a difference of temperature, wherever it has been possible for the equilibrium of the caloric to be re-established, it is possible to have also the production of impelling power. Steam is a means of realizing this power, but it is not the only one. All substances in nature can be employed for this purpose, all are susceptible of changes of volume, of successive contradictions and dilatations, through the alternation of heat and cold. All are capable of overcoming in their changes of volume certain resistances, and of thus developing the impelling power. A solid body—a metallic bar for example—alternately heated and cooled increases and diminishes in length, and can move bodies fastened to its ends. A liquid alternately heated and cooled increases and diminishes in volume, and can overcome obstacles of greater or less size, opposed to its dilatation. An aeriform fluid is susceptible of considerable change of volume by variations of temperature. If it is enclosed in an expansible space, such as a cylinder provided with a piston, it will produce movements of great extent. Vapors of all substances capable of passing into a gaseous condition, as of alcohol, of mercury, of sulphur, etc., may fulfil the same office as vapor of water. The latter, alternately heated and cooled, would produce motive power in the shape of permanent gases, that is, without ever returning to a liquid state. Most of these substances have been proposed, many even have been tried, although up to this time perhaps without remarkable success.

We have shown that in steam-engines the motive-power is due to a re-establishment of equilibrium in the caloric; this takes place not only for steam-engines, but also for every heat-engine—that is,

* The existence of water in the liquid state here necessarily assumed, since without it the steam-engine could not be fed, supposes the existence of a pressure capable of preventing this water from vaporizing, consequently of a pressure equal or superior to the tension of vapor at that temperature. If such a pressure were not exerted by the atmospheric air, there would be instantly produced a quantity of steam sufficient to give rise to that tension, and it would be necessary always to overcome this pressure in order to throw out the steam from the engines into the new atmosphere. Now this is evidently equivalent to overcoming the tension which the steam retains after its condensation, as effected by ordinary means.

If a very high temperature existed at the surface of our globe, as it seems certain that it exists in its interior, all the waters of the ocean would be in a state of vapor in the atmosphere, and no portion of it would be found in a liquid state.

for every machine of which caloric is the motor. Heat can evidently be a cause of motion only by virtue of the changes of volume or of form which it produces in bodies.

These changes are not caused by uniform temperature, but rather by alternations of heat and cold. Now to heat any substance whatever requires a body warmer than the one to be heated; to cool it requires a cooler body. We supply caloric to the first of these bodies that we may transmit it to the second by means of the intermediary substance. This is to re-establish, or at least to endeavor to re-establish, the equilibrium of the caloric.

It is natural to ask here this curious and important question: Is the motive power of heat invariable in quantity, or does it vary with the agent employed to realize it as the intermediary substance, selected as the subject of action of the heat?

It is clear that this question can be asked only in regard to a given quantity of caloric,* the difference of the temperatures also being given. We take, for example, one body A kept at a temperature of 100° and another body B kept at a temperature of 0°, and ask what quantity of motive power can be produced by the passage of a given portion of caloric (for example, as much as is necessary to melt a kilogram of ice) from the first of these bodies to the second. We inquire whether this quantity of motive power is necessarily limited, whether it varies with the substance employed to realize it, whether the vapor of water offers in this respect more or less advantage than the vapor of alcohol, of mercury, a permanent gas, or any other substance. We will try to answer these questions, availing ourselves of ideas already established.

We have already remarked upon this self-evident fact, or fact which at least appears evident as soon as we reflect on the changes of volume occasioned by heat: *wherever there exists a difference of temperature, motive power can be produced.* Reciprocally, wherever we can consume this power, it is possible to produce a difference of temperature, it is possible to occasion destruction of equilibrium in the caloric. Are not percussion and the friction of bodies actually means of raising their temperature, of making it reach spontaneously a

* It is considered unnecessary to explain here what is quantity of caloric or quantity of heat (for we employ these two expressions indifferently), or to describe how we measure these quantities by the calorimeter. Nor will we explain what is meant by latent heat, degree of temperature, specific heat, etc. The reader should be familiarized with these terms through the study of the elementary treatises of physics or of chemistry.

higher degree than that of the surrounding bodies, and consequently of producing a destruction of equilibrium in the caloric, where equilibrium previously existed? It is a fact proved by experience, that the temperature of gaseous fluids is raised by compression and lowered by rarefaction. This is a sure method of changing the temperature of bodies, and destroying the equilibrium of the caloric as many times as may be desired with the same substance. The vapor of water employed in an inverse manner to that in which it is used in steam-engines can also be regarded as a means of destroying the equilibrium of the caloric. To be convinced of this we need to observe closely the manner in which motive power is developed by the action of heat on vapor of water. Imagine two bodies *A* and *B*, kept each at a constant temperature, that of *A* being higher than that of *B*. These two bodies, to which we can give or from which we can remove the heat without causing their temperatures to vary, exercise the functions of two unlimited reservoirs of caloric. We will call the first the furnace and the second the refrigerator.

If we wish to produce motive power by carrying a certain quantity of heat from the body *A* to the body *B* we shall proceed as follows:*

(1) To borrow caloric from the body *A* to make steam with it—that is, to make this body fulfil the function of a furnace, or rather of the metal composing the boiler in ordinary engines—we here assume that the steam is produced at the same temperature as the body *A*.

(2) The steam having been received in a space capable of expansion, such as a cylinder furnished with a piston, to increase the volume of this space, and consequently also that of the steam. Thus rarefied, the temperature will fall spontaneously, as occurs with all elastic fluids; admit that the rarefaction may be continued to the point where the temperature becomes precisely that of the body *B*.

(3) To condense the steam by putting it in contact with the body *B*, and at the same time exerting on it a constant pressure until it is entirely liquefied. The body *B* fills here the place of the injection-water in ordinary engines, with this difference, that it condenses the vapor without mingling with it, and without changing its own temperature.†

* [This is only a sketch and Carnot accidentally leaves the cycle incomplete. E. M.]

† We may perhaps wonder here that the body *B* being at the same temperature as the steam is able to condense it. Doubtless this is not strictly possible, but the slightest difference of temperature will determine the condensation, which suffices

The operations which we have just described might have been performed in an inverse direction and order. There is nothing to prevent forming vapor with the caloric of the body *B*, and at the temperature of that body, compressing it in such a way as to make it acquire the temperature of the body *A*, finally condensing it by contact with this latter body, and continuing the compression to complete liquefaction.

By our first operations there would have been at the same time production of motive power and transfer of caloric from the body *A* to the body *B*. By the inverse operations there is at the same time expenditure of motive power and return of caloric from the body *B* to the body *A*. But if we have acted in each case on the same quantity of vapor, if there is produced no loss either of motive power or caloric, the quantity of motive power produced in the first place will be equal to that which would have been expended in the second, and the quantity of caloric passed in the first case from the body *A* to the body *B* would be equal to the quantity which passes back again in the second from the body *B* to the body *A*; so that an indefinite number of alternative operations of this sort could be carried on without in the end having either produced motive power or transferred caloric from one body to the other.

Now if there existed any means of using heat preferable to those which we have employed, that is, if it were possible by any method whatever to make the caloric produce a quantity of motive power greater than we have made it produce by our first series of operations, it would suffice to divert a portion of this power in order by the method just indicated to make the caloric of the body *B* return

---

to establish the justice of our reasoning. It is thus that, in the differential calculus, it is sufficient that we can conceive the neglected quantities indefinitely reducible in proportion to the quantities retained in the equations, to make certain of the exact result.

The body *B* condenses the steam without changing its own temperature—this results from our supposition. We have admitted that this body may be maintained at a constant temperature. We take away the caloric as the steam furnishes it. This is the condition in which the metal of the condenser is found when the liquefaction of the steam is accomplished by applying cold water externally, as was formerly done in several engines. Similarly, the water of a reservoir can be maintained at a constant level if the liquid flows out at one side as it flows in at the other.

One could even conceive the bodies *A* and *B* maintaining the same temperature, although they might lose or gain certain quantities of heat. If, for example, the body *A* were a mass of steam ready to become liquid, and the body *B* a mass of ice ready to melt, these bodies might, as we know, furnish or receive caloric without thermometric change.

to the body A from the refrigerator to the furnace, to restore the initial conditions, and thus to be ready to commence again an operation precisely similar to the former, and so on: this would be not only perpetual motion, but an unlimited creation of motive power without consumption either of caloric or of any other agent whatever. Such a creation is entirely contrary to ideas now accepted, to the laws of mechanics and of sound physics. It is inadmissible.* We should then conclude that *the maximum of motive power resulting from the employment of steam is also the maximum of motive power realizable by any means whatever.* We will soon give a second more vigorous demonstration of this theory. This should be considered only as an approximation. (See page 15.)

We have a right to ask, in regard to the proposition just enunciated, the following questions: What is the sense of the word *maximum* here? By what sign can it be known that this maximum is attained? By what sign can it be known whether the steam is employed to greatest possible advantage in the production of motive power?

Since every re-establishment of equilibrium in the caloric may be the cause of the production of motive power, every re-establish-

* The objection may perhaps be raised here, that perpetual motion, demonstrated to be impossible by mechanical action alone, may possibly not be so if the power either of heat or electricity be exerted; but is it possible to conceive the phenomena of heat and electricity as due to anything else than some kind of motion of the body, and as such should they not be subjected to the general laws of mechanics? Do we not know besides, *à posteriori*, that all the attempts made to produce perpetual motion by any means whatever have been fruitless?—that we have never succeeded in producing a motion veritably perpetual, that is, a motion which will continue forever without alteration in the bodies set to work to accomplish it? The electromotor apparatus (the pile of Volta) has sometimes been regarded as capable of producing perpetual motion; attempts have been made to realize this idea by constructing dry piles said to be unchangeable; but however it has been done, the apparatus has always exhibited sensible deteriorations when its action has been sustained for a time with any energy.

The general and philosophic acceptation of the words *perpetual motion* should include not only a motion susceptible of indefinitely continuing itself after a first impulse received, but the action of an apparatus, of any construction whatever, capable of creating motive power in unlimited quantity, capable of starting from rest all the bodies of nature if they should be found in that condition, of overcoming their inertia; capable, finally, of finding in itself the forces necessary to move the whole universe, to prolong, to accelerate incessantly, its motion. Such would be a veritable creation of motive power. If this were a possibility, it would be useless to seek in currents of air and water or in combustibles this motive power. We should have at our disposal an inexhaustible source upon which we could draw ta will.

ment of equilibrium which shall be accomplished without production of this power should be considered as an actual loss. Now, very little reflection would show that all change of temperature which is not due to a change of volume of the bodies can be only a useless re-establishment of equilibrium in the caloric.* The necessary condition of the maximum is, then, *that in the bodies employed to realize the motive power of heat there should not occur any change of temperature which may not be due to a change of volume.* Reciprocally, every time that this condition is fulfilled the maximum will be attained. This principle should never be lost sight of in the construction of heat-engines; it is its fundamental basis. If it cannot be strictly observed, it should at least be departed from as little as possible.

Every change of temperature which is not due to a change of volume or to chemical action (an action that we provisionally suppose not to occur here) is necessarily due to the direct passage of the caloric from a more or less heated body to a colder body. This passage occurs mainly by the contact of bodies of different temperatures; hence such contact should be avoided as much as possible. It cannot probably be avoided entirely, but it should at least be so managed that the bodies brought in contact with each other differ as little as possible in temperature. When we just now supposed, in our demonstration, the caloric of the body A employed to form steam, this steam was considered as generated at the temperature of the body A; thus the contact took place only between bodies of equal temperatures; the change of temperature occurring afterwards in the steam was due to dilatation, consequently to a change of volume. Finally, condensation took place also without contact of bodies of different temperatures. It occurred while exerting a constant pressure on the steam brought in contact with the body B of the same temperature as itself. The conditions for a maximum are thus found to be fulfilled. In reality the operation cannot proceed exactly as we have assumed. To determine the passage of caloric from one body to another, it is necessary that there should be an excess of temperature in the first, but this excess may be supposed as slight as we please. We can regard it as insensible in theory, without thereby destroying the exactness of the arguments.

* We assume here no chemical action between the bodies employed to realize the motive power of heat. The chemical action which takes place in the furnace is, in some sort, a preliminary action—an operation destined not to produce immediately motive power, but to destroy the equilibrium of the caloric, to produce a difference of temperature which may finally give rise to motion.

A more substantial objection may be made to our demonstration, thus: When we borrow caloric from the body *A* to produce steam, and when this steam is afterwards condensed by its contact with the body *B*, the water used to form it, and which we considered at first as being of the temperature of the body *A*, is found at the close of the operation at the temperature of the body *B*. It has become cool. If we wish to begin again an operation similar to the first, if we wish to develop a new quantity of motive power with the same instrument, with the same steam, it is necessary first to re-establish the original condition—to restore the water to the original temperature. This can undoubtedly be done by at once putting it again in contact with the body *A*; but there is then contact between bodies of different temperatures, and loss of motive power.* It would be impossible to execute the inverse operation, that is, to return to the body *A* the caloric employed to raise the temperature of the liquid.

This difficulty may be removed by supposing the difference of temperature between the body *A* and the body *B* indefinitely small. The quantity of heat necessary to raise the liquid to its former temperature will be also indefinitely small and unimportant relatively to that which is necessary to produce steam—a quantity always limited.

The proposition found elsewhere demonstrated for the case in which the difference between the temperatures of the two bodies is indefinitely small, may be easily extended to the general case. In fact, if it operated to produce motive power by the passage of caloric from the body *A* to the body *Z*, the temperature of this latter body being very different from that of the former, we should imagine a series of bodies *B*, *C*, *D* . . . of temperatures intermediate between those of the bodies *A*, *Z*, and selected so that the differences from *A* to *B*, from *B* to *C*, etc., may all be indefinitely small. The caloric coming from *A* would not arrive at *Z* till after it had passed through

* This kind of loss is found in all steam-engines. In fact, the water destined to feed the boiler is always cooler than the water which it already contains. There occurs between them a useless re-establishment of equilibrium of caloric. We are easily convinced, *à posteriori*, that this re-establishment of equilibrium causes a loss of motive power if we reflect that it would have been possible to previously heat the feed-water by using it as condensing-water in a small accessory engine, when the steam drawn from the large boiler might have been used, and where the condensation might be produced at a temperature intermediate between that of the boiler and that of the principal condenser. The power produced by the small engine would have cost no loss of heat, since all that which had been used would have returned into the boiler with the water of condensation.

the bodies B, C, D, etc., and after having developed in each of these stages maximum motive power. The inverse operations would here be entirely possible, and the reasoning of page 11 would be strictly applicable.

According to established principles at the present time, we can compare with sufficient accuracy the motive power of heat to that of a waterfall. Each has a maximum that we cannot exceed, whatever may be, on the one hand, the machine which is acted upon by the water, and whatever, on the other hand, the substance acted upon by the heat. The motive power of a waterfall depends on its height and on the quantity of the liquid; the motive power of heat depends also on the quantity of caloric used, and on what may be termed, on what in fact we will call, the *height of its fall*,* that is to say, the difference of temperature of the bodies between which the exchange of caloric is made. In the waterfall the motive power is exactly proportional to the difference of level between the higher and lower reservoirs. In the fall of caloric the motive power undoubtedly increases with the difference of temperature between the warm and the cold bodies; but we do not know whether it is proportional to this difference. We do not know, for example, whether the fall of caloric from 100 to 50 degrees furnishes more or less motive power than the fall of this same caloric from 50 to zero. It is a question which we propose to examine hereafter.

We shall give here a second demonstration of the fundamental proposition enunciated on page 12, and present this proposition under a more general form than the one already given.

When a gaseous fluid is rapidly compressed its temperature rises. It falls, on the contrary, when it is rapidly dilated. This is one of the facts best demonstrated by experiment. We will take it for the basis of our demonstration.†

---

* The matter here dealt with being entirely new, we are obliged to employ expressions not in use as yet, and which perhaps are less clear than is desirable.

† The experimental facts which best prove the change of temperature of gases by compression or dilatation are the following:

(1) The fall of the thermometer placed under the receiver of a pneumatic machine in which a vacuum has been produced. This fall is very sensible on the Bréguet thermometer: it may exceed 40° or 50°. The mist which forms in this case seems to be due to the condensation of the watery vapor caused by the cooling of the air.

(2) The igniting of German tinder in the so-called pneumatic tinderboxes; which are, as we know, little pump-chambers in which the air is rapidly compressed.

If, when the temperature of a gas has been raised by compression, we wish to reduce it to its former temperature without subjecting its volume to new changes, some of its caloric must be removed. This caloric might have been removed in proportion as pressure was applied, so that the temperature of the gas would remain constant. Similarly, if the gas is rarefied we can avoid lowering the temperature by supplying it with a certain quantity of caloric. Let us call the caloric employed at such times, when no change of temperature occurs, *caloric due to change of volume*. This denomination does not indicate that the caloric appertains to the volume: it does not appertain to it any more than to pressure, and might as well be called *caloric due to the change of pressure*. We do not know what laws it follows relative to the variations of volume: it is possible that its quantity changes either with the nature of the gas, its density,

---

(3) The fall of a thermometer placed in a space where the air has been first compressed and then allowed to escape by the opening of a cock.

(4) The results of experiments on the velocity of sound. M. de Laplace has shown that, in order to secure results accurately by theory and computation, it is necessary to assume the heating of the air by sudden compression.

The only fact which may be adduced in opposition to the above is an experiment of MM. Gay-Lussac and Welter, described in the *Annales de Chimie et de Physique*. A small opening having been made in a large reservoir of compressed air, and the ball of a thermometer having been introduced into the current of air which passes out through this opening, no sensible fall of the temperature denoted by the thermometer has been observed.

Two explanations of this fact may be given: (1) The striking of the air against the walls of the opening by which it escapes may develop heat in observable quantity. (2) The air which has just touched the ball of the thermometer possibly takes again by its collision with this ball, or rather by the effect of the *détour* which it is forced to make by its rencounter, a density equal to that which it had in the receiver—much as the water of a current rises against a fixed obstacle, above its level.

The change of temperature occasioned in the gas by the change of volume may be regarded as one of the most important facts of physics, because of the numerous consequences which it entails, and at the same time as one of the most difficult to illustrate, and to measure by decisive experiments. It seems to present in some respects singular anomalies.

Is it not to the cooling of the air by dilatation that the cold of the higher regions of the atmosphere must be attributed? The reasons given heretofore as an explanation of this cold are entirely insufficient; it has been said that the air of the elevated regions receiving little reflected heat from the earth, and radiating towards celestial space, would lose caloric, and that this is the cause of its cooling; but this explanation is refuted by the fact that, at an equal height, cold reigns with equal and even more intensity on the elevated plains than on the summit of the mountains, or in those portions of the atmosphere distant from the sun.

or its temperature. Experiment has taught us nothing on this subject. It has only shown us that this caloric is developed in greater or less quantity by the compression of the elastic fluids.

This preliminary idea being established, let us imagine an elastic fluid, atmospheric air for example, shut up in a cylindrical vessel, *abcd* (Fig. 1), provided with a movable diaphragm or piston, *cd*. Let there be also two bodies, *A* and *B*, kept each at a constant temperature, that of *A* being higher than that of *B*. Let us picture to ourselves now the series of operations which are to be described:*

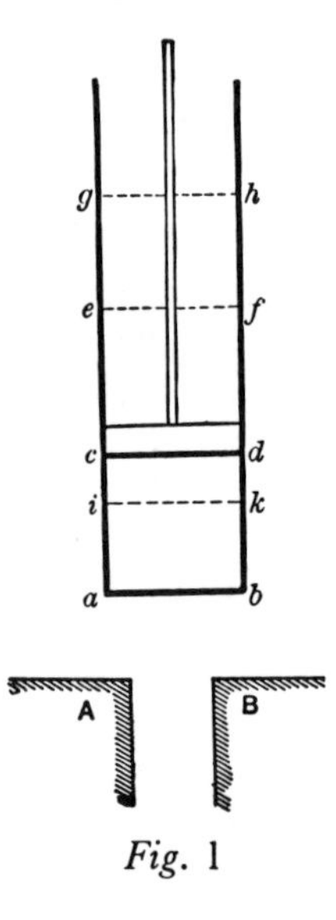

Fig. 1

(1) Contact of the body *A* with the air enclosed in the space *abcd* or with the wall of this space—a wall that we will suppose to transmit the caloric readily. The air becomes by such contact of the same temperature as the body *A*; *cd* is the actual position of the piston.

(2) The piston gradually rises and takes the position *ef*. The body *A* is all the time in contact with the air, which is thus kept at a constant temperature during the rarefaction. The body *A* furnishes the caloric necessary to keep the temperature constant.

(3) The body *A* is removed, and the air is then no longer in contact with any body capable of furnishing it with caloric. The

* ["Caloric" may be taken to mean "entropy." E. M.]

piston meanwhile continues to move, and passes from the position *ef* to the position *gh*. The air is rarefied without receiving caloric, and its temperature falls. Let us imagine that it falls thus till it becomes equal to that of the body *B*; at this instant the piston stops, remaining at the position *gh*.

(4) The air is placed in contact with the body *B*; it is compressed by the return of the piston as it is moved from the position *gh* to the position *cd*. This air remains, however, at a constant temperature because of its contact with the body *B*, to which it yields its caloric.

(5) The body *B* is removed, and the compression of the air is continued, which being then isolated, its temperature rises. The compression is continued till the air acquires the temperature of the body *A*. The piston passes during this time from the position *cd* to the position *ik*.

(6) The air is again placed in contact with the body *A*. The piston returns from the position *ik* to the position *ef*; the temperature remains unchanged.

(7) The step described under number (3) is renewed, then successively the steps (4), (5), (6), (3), (4), (5), (6), (3), (4), (5); and so on.

In these various operations the piston is subject to an effort of greater or less magnitude, exerted by the air enclosed in the cylinder; the elastic force of this air varies as much by reason of the changes in volume as of changes of temperature. But it should be remarked that with equal volumes, that is, for the similar positions of the piston, the temperature is higher during the movements of dilatation than during the movements of compression. During the former the elastic force of the air is found to be greater, and consequently the quantity of motive power produced by the movements of dilatation is more considerable than that consumed to produce the movements of compression. Thus we should obtain an excess of motive power—an excess which we could employ for any purpose whatever. The air, then, has served as a heat-engine; we have, in fact, employed it in the most advantageous manner possible, for no useless re-establishment of equilibrium has been effected in the caloric.

All the above-described operations may be executed in an inverse sense and order. Let us imagine that, after the sixth period, that is to say the piston having arrived at the position *ef*, we cause it to return to the position *ik*, and that at the same time we keep the air in contact with the body *A*. The caloric furnished by this body during the sixth period would return to its source, that is, to the body

A, and the conditions would then become precisely the same as they were at the end of the fifth period. If now we take away the body *A*, and if we cause the piston to move from *ik* to *cd*, the temperature of the air will diminish as many degrees as it increased during the fifth period, and will become that of the body *B*. We may evidently continue a series of operations the inverse of those already described. It is only necessary under the same circumstances to execute for each period a movement of dilatation instead of a movement of compression, and reciprocally.

The result of these first operations has been the production of a certain quantity of motive power and the removal of caloric from the body *A* to the body *B*. The result of the inverse operations is the consumption of the motive power produced and the return of the caloric from the body *B* to the body *A*; so that these two series of operations annul each other, after a fashion, one neutralizing the other.

The impossibility of making the caloric produce a greater quantity of motive power than that which we obtained from it by our first series of operations, is now easily proved. It is demonstrated by reasoning very similar to that employed at page 11; the reasoning will here be even more exact. The air which we have used to develop the motive power is restored at the end of each cycle of operations exactly to the state in which it was at first found, while, as we have already remarked, this would not be precisely the case with the vapor of water.*

We have chosen atmospheric air as the instrument which should develop the motive power of heat, but it is evident that the reasoning would have been the same for all other gaseous substances, and even for all other bodies susceptible of change of temperature through

---

* We tacitly assume in our demonstration, that when a body has experienced any changes, and when after a certain number of transformations it returns to precisely its original state, that is, to that state considered in respect to density, to temperature, to mode of aggregation—let us suppose, I say, that this body is found to contain the same quantitiy of heat that it contained at first, or else that the quantities of heat absorbed or set free in these different transformations are exactly compensated. This fact has never been called in question. It was first admitted without reflection, and verified afterwards in many cases by experiments with the calorimeter. To deny it would be to overthrow the whole theory of heat to which it serves as a basis. For the rest, we may say in passing, the main principles on which the theory of heat rests require the most careful examination. Many experimental facts appear almost inexplicable in the present state of this theory.

successive contractions and dilatations, which comprehends all natural substances, or at least all those which are adapted to realize the motive power of heat. Thus we are led to establish this general proposition:

The motive power of heat is independent of the agents employed to realize it; its quantity is fixed solely by the temperatures of the bodies between which is effected, finally, the transfer of the caloric.

27

Reprinted from *Phil. Trans. Roy. Soc.*, 113, 160–165 (1823)

EVIDENCE AGAINST THE THEORY OF CALORIC

Michael Faraday and Humphry Davy

XIV. *On fluid Chlorine. By M.* FARADAY, *Chemical Assistant in the Royal Institution. Communicated by Sir* H. DAVY, *Bart. Pres. R. S.*

Read March 13, 1823.

It is well known that before the year 1810, the solid substance obtained by exposing chlorine, as usually procured, to a low temperature, was considered as the gas itself reduced into that form; and that SIR HUMPHRY DAVY first showed it to be a hydrate, the pure dry gas not being condensible even at a temperature of —40° F.

I took advantage of the late cold weather to procure crystals of this substance for the purpose of analysis. The results are contained in a short paper in the Quarterly Journal of Science, Vol. XV. Its composition is very nearly 27.7 chlorine, 72.3 water, or 1 proportional of chlorine, and 10 of water.

The President of the Royal Society having honoured me by looking at these conclusions, suggested, that an exposure of the substance to heat under pressure, would probably lead to interesting results; the following experiments were commenced at his request. Some hydrate of chlorine was prepared, and being dried as well as could be by pressure in bibulous paper, was introduced into a sealed glass tube, the upper end of which was then hermetically closed. Being placed in water at 60°, it underwent no change; but when put into water at 100°, the substance fused, the tube became filled with a bright yellow atmosphere, and, on examination, was found to contain two fluid substances: the one, about three-fourths of the whole, was of a faint yellow colour, having very much the appearance of water; the remaining fourth was a heavy bright yellow fluid, lying at the bottom of the former, without any apparent tendency to mix with it. As the tube cooled, the yellow atmosphere condensed into more of the yellow fluid, which floated in a film on the pale fluid, looking very like chloride of nitrogen; and at 70° the pale portion congealed, although even at 32° the yellow portion did not solidify. Heated up to 100° the yellow fluid appeared to boil, and again produced the bright coloured atmosphere.

By putting the hydrate into a bent tube, afterwards hermetically sealed, I found it easy, after decomposing it by a heat of 100°, to distil the yellow fluid to one end of the tube, and so separate it from the remaining portion. In this way a more complete decomposition of the hydrate was effected, and, when the whole was allowed to cool, neither of the fluids solidified at temperatures above 34°, and the yellow portion not even at 0°. When the two were mixed together they gradually combined at temperatures below 60°, and formed the same solid substances as that first introduced. If, when the fluids were separated, the tube was cut in the middle, the parts flew asunder as if with an explosion, the whole of the yellow portion disappeared, and there was a powerful atmosphere of chlorine produced; the pale portion on the contrary remained, and when examined, proved to be a weak solution of chlorine in water, with a little muriatic acid, probably from the impurity of the hydrate used. When that end of the tube in which the yellow fluid lay was broken under a jar of water, there was an immediate production of chlorine gas.

I at first thought that muriatic acid and euchlorine had been formed; then, that two new hydrates of chlorine had been produced; but at last I suspected that the chlorine had been entirely separated from the water by the heat, and condensed into a dry fluid by the mere pressure of its own abundant vapour. If that were true, it followed, that chlorine gas, when compressed, should be condensed into the same fluid, and, as the atmosphere in the tube in which the fluid lay was not very yellow at 50° or 60°, it seemed probable that the pressure required was not beyond what could readily be obtained by a condensing syringe. A long tube was therefore furnished with a cap and stop-cock, then exhausted of air and filled with chlorine, and being held vertically with the syringe upwards, air was forced in, which thrust the chlorine to the bottom of the tube, and gave a pressure of about 4 atmospheres. Being now cooled, there was an immediate deposit in films, which appeared to be hydrate, formed by water contained in the gas and vessels, but some of the yellow fluid was also produced. As this however might also contain a portion of the water present, a perfectly dry tube and apparatus were taken, and the chlorine left for some time over a bath of sulphuric acid before it was introduced. Upon throwing in air and giving pressure, there was now no solid film formed, but the clear yellow fluid was deposited, and more abundantly still upon cooling. After remaining some time it disappeared, having gradually mixed with the atmosphere above it, but every repetition of the experiment produced the same results.

Presuming that I had now a right to consider the yellow fluid as pure chlorine in the liquid state, I proceeded to examine its properties, as well as I could when obtained by héat from the hydrate. However obtained, it always appears very limpid and fluid, and excessively volatile at common pressure. A portion was cooled in its tube to 0°: it remained fluid. The tube was then opened, when a part immediately flew off, leaving the rest so cooled by the evaporation as to remain a fluid under the atmospheric pressure. The temperature could not have been higher than —40° in this case; as Sir HUMPHRY DAVY has shown that dry chlorine does not condense at that temperature under common pressure. Another tube was opened at a temperature of 50°; a part of the chlorine volatilised, and cooled the tube so much as to condense the atmospheric vapour on it as ice.

A tube having the water at one end and the chlorine at the other was weighed, and then cut in two; the chlorine immediately flew off, and the loss being ascertained was found to be 1.6 grains: the water left was examined and found to contain some chlorine: its weight was ascertained to be 5.4 grains. These proportions, however, must not be considered as indicative of the true composition of hydrate of chlorine; for, from the mildness of the weather during the time when these experiments were made, it was impossible to collect the crystals of hydrate, press, and transfer them, without losing much chlorine; and it is also impossible to separate the chlorine and water in the tube perfectly, or keep them separate, as the atmosphere within will combine with the water, and gradually reform the hydrate.

Before cutting the tube, another tube had been prepared

exactly like it in form and size, and a portion of water introduced into it, as near as the eye could judge, of the same bulk as the fluid chlorine: this water was found to weigh 1.2 grains; a result, which, if it may be trusted, would give the specific gravity of fluid chlorine as 1.33; and from its appearance in, and on water, this cannot be far wrong.

Note on the condensation of Muriatic Acid Gas into the liquid form. By Sir H. DAVY, *Bart. Pres. R. S.*

IN desiring Mr. FARADAY to expose the hydrate of chlorine to heat in a closed glass tube, it occurred to me, that one of three things would happen; that it would become fluid as a hydrate; or that a decomposition of water would occur, and euchlorine and muriatic acid be formed; or that the chlorine would separate in a condensed state. This last result having been obtained, it evidently led to other researches of the same kind. I shall hope, on a future occasion, to detail some general views on the subject of these researches. I shall now merely mention, that by sealing muriate of ammonia and sulphuric acid in a strong glass tube, and causing them to act upon each other, I have procured liquid muriatic acid: and by substituting carbonate for muriate of ammonia, I have no doubt that carbonic acid may be obtained, though in the only trial I have made the tube burst. I have requested Mr. FARADAY to pursue these experiments, and to extend them to all the gases which are of considerable density, or to any extent soluble in water; and I hope soon to be able to lay an account of his results, with some applications of them that I propose to make, before the Society.

I cannot conclude this note without observing, that the generation of elastic substances in close vessels, either with or without heat, offers much more powerful means of approximating their molecules than those dependent upon the application of cold, whether natural or artificial: for, as gases diminish only about $\frac{1}{480}$ in volume for every — degree of FAHRENHEIT'S scale, beginning at ordinary temperatures, a very slight condensation only can be produced by the most powerful freezing mixtures, not half as much as would result from the application of a strong flame to one part of a glass tube, the other part being of ordinary temperature: and when attempts are made to condense gases into fluids by sudden mechanical compression, the heat, instantly generated, presents a formidable obstacle to the success of the experiment; whereas, in the compression resulting from their slow generation in close vessels, if the process be conducted with common precautions, there is no source of difficulty or danger; and it may be easily assisted by artificial cold in cases when gases approach near to that point of compression and temperature at which they become vapours.

Part III

ENERGY — THE MID-NINETEENTH-CENTURY BREAKTHROUGH

The previous papers traced the development of the concept we now call energy from classical antiquity up to about 1830. We have noted the gradual emergence of the idea that in a dynamical system there can exist under certain conditions a function of the position and velocity of the various particles composing the system that remains constant, while the particles undergo changes in position and velocity. We have seen that the conservation idea was fostered by the introduction of the concept of *vis viva*, a quantity found by multiplying the mass of each particle in the system by the square of its velocity and summing over all the particles. D'Alembert, Daniel Bernoulli, and Lagrange all realized the importance of this conservation idea, although the name they gave to it, conservation of *vis viva*, was misleading. Even Euler, who evidently did not approve of Leibniz's philosophy and opposed on numerous occasions the adherents of Leibniz's *vis viva* concept, nevertheless in some of his writings on mechanics developed the energy equation for simple collision problems without using the current terminology. Young's introduction of the word "energy" for *vis viva* evidently did not catch on. Theorists like Lagrange evidently did not foresee the future importance of the concept; this is not surprising, since they were chiefly concerned with purely mechanical problems. It is true that Laplace and Poisson in their famous equations made use of Lagrange's potential function V in their attempt to provide a field point of view for mechanics. One might have expected that engineers like Lazare Carnot and Coriolis would have grasped the value of a unified terminology in connection with the behavior of machines.

The key to the whole problem of the generalization of the concept of energy and its extension from a purely mechanical concept to a premier position in science and technology is in the intensive study of the nature of heat. For this reason, we have included in this volume a number of early papers bearing on heat and in particular its relation to mechanical activity (e.g., friction and the production of mechanical energy by engines). It was pointed out that for a long time this connection was not well understood and the opinions expressed about it were purely speculative. The real breakthrough did not come until the 1840s with the work of Mayer, Joule, Colding, and similarly motivated but not so successful researchers and thinkers. The final part of this volume is devoted to this development, commonly called the mechanical theory of heat, which clearly demonstrated the enormous value of the idea of mechanical energy and the possibility of extending its range of application to all physical phenomena.

But first we turn to the work of the Irish mathematician William Rowan Hamilton, who in the 1830s refined and generalized the analytical mechanics of Lagrange and put it in a shape suitable for the more sophisticated applications of the late nineteenth and twentieth centuries in celestial mechanics, statistical mechanics, and quantum mechanics.

Editor's Comments on Papers 28 Through 31

Excerpt from *On a General Method in Dynamics*
28A W. R. HAMILTON

Excerpt from *Second Essay on a General Method in Dynamics*
28B W. R. HAMILTON

On the Forces of Inorganic Nature
29A J. R. MAYER

Excerpt from *The Motions of Organisms and Their Relation to Metabolism*
29B J. R. MAYER

On the Calorific Effects of Magneto-Electricity, and on the Mechanical Value of Heat
30A J. P. JOULE

On the Existence of an Equivalent Relation Between Heat and the Ordinary Forces of Mechanical Power
30B J. P. JOULE

On Matter, Living Force, and Heat
30C J. P. JOULE

Note on the Work of L. A. Colding
31A THE ROYAL DANISH ACADEMY OF SCIENCES

Excerpt from *Investigations of the General Forces of Nature and Their Mutual Dependence*
31B L. A. COLDING

Sir William Rowan Hamilton (1805–1865), Irish mathematical scientist and astronomer, whose name will forever be connected with modern analytical mechanics and quantum mechanics, did research ranging rather widely over the fields of pure and applied mathematics. His chief contributions in physical science were to celestial mechanics, optics, and dynamics. Early in his professional career he developed the famous analogy between light rays and motions of particles, which a century later suggested to Louis de Broglie an early form of quantum mechanics. In the mid-1830s Hamilton recast Lagrangian mechanics in a form that he felt increased its applicability to problems involving the motion of systems of particles. In addition to the quantity T, which he took as one half the *vis viva* of Leibniz (called by him the living force) and which later became known as kinetic energy, Hamilton also introduced what we now recognize as the potential energy ($-U$ in Hamilton's notation). The sum T - U of kinetic and potential energies he calls H, which we now recognize as the total mechanical energy of a conservative system.

In his two memoirs of 1834 and 1835, which are reproduced in part here, Hamilton showed that by introducing what he called the principle of varying action in place of the principle of least action, as used by Lagrange, he was able to derive a set of first-order partial differential equations involving the function H as a function of the generalized coordinates and momenta of the system being described. These equations, now known as Hamilton's canonical equations, have been found to be fundamental in modern analytical mechanics.

In his Royal Institution lectures of the early 1800s, Thomas Young had attached the name "energy" to the living force (see Paper 18), but this terminology was not sufficiently well established in Hamilton's time for the latter to employ it in his memoirs. Nevertheless, he thoroughly appreciated and understood the importance of the T, U, and H functions in the dynamics of systems. He thus provided all the mathematical machinery for the exploitation of the concept of energy. It was still necessary for the more general physical significance of the energy idea to be made clear by the labors of such men as Joule, Mayer, Helmholtz, and Kelvin before the physical advantages of Hamilton's purely analytical work could become evident.

Julius Robert von Mayer (1814–1878) was born a Swabian and spent his life as a practicing physician in his native town of Heilbronn in southern Germany. He had an inquiring mind which made him desire to push his thoughts beyond the confines of medicine. Early stimulated by the curious biological observation that the color of venous blood in healthy patients is more reddish in tropical countries than in the colder climate of northern Europe, as well as by philosophical considerations relating to cause and effect in natural phenomena, he became convinced as a young man that there must be a definite relation between heat and other natural manifestations, which he lumped together under the generic term "force" (German *Kraft*). After a false start due to his inadequate grasp of physics and mechanics in particular, he reached the idea that the numerical relation between heat and what we now call mechanical energy could actually be calculated from the observed differences between the two specific heats of a gas. This he proceeded to do and published his result in Liebig's *Annalen der Chemie und Pharmacie* (**42**, 233, 1842). His value in modern units was 3.58 joules per calorie. As far as we know, this was the first theoretical evaluation of the mechanical equivalent of heat. Unfortunately, Mayer did not provide the details of his calculation, although what he must have done is clear enough from what he said. However, his paper was for a long time completely ignored by those who should have been interested. In this respect he suffered a fate somewhat similar to that of Joule, whose first experimental measurements of the equivalent received little attention. Because of its epoch-making value and its general thread of thought, we reprint Mayer's 1842 paper in English translation in its entirety.

Although somewhat discouraged by the lack of attention to his paper, Mayer was not completely frustrated, and in 1845 brought out at his own expense the privately printed essay "Die organische Bewegung in ihrem Zusammenhang nit dem Stoffwechsel. Ein Beitrag zur Naturkunde" ("The Motions of Organisms and Their Relation to Metabolism. An Essay in Natural Science"). The choice of title was unfortunate, since it practically guaranteed a complete lack of attention to the essay on the part of physical scientists. In the first part of the essay Mayer summarized his train of throught and explained in detail how he had found the mechanical equivalent, as reported in his 1842 paper. He also developed further his ideas on the role of energy (which he still called "force") in all aspects of nature, including the behavior of living things. In his detailed treatment of the relation between nutrition and the thermal and mechanical phenomena connected with organisms he made such strides that he may justly be considered one of the founders of biophysics and biochemistry. We reprint here the first 24 pages of this essay, which deals with the problem of the generalization of energy and omit the more specifically biological part.

Mayer suffered keenly from the neglect of his work during the productive part of his career. He was also affected with serious health and family problems. Fortunately, recognition of his achievement came before his death. For a biographical sketch, a critical analysis of his contributions, particularly to the establishment of the concept of energy in a premier position in science, and the complete text in English translation of his most important writings, see *Julius Robert Mayer—Prophet of Energy* by R. B. Lindsay in the Pergamon Press series "Men of Physics" (Elmsford, N.Y., 1973).

Claims have been made, particularly by P. G. Tait and E. T. Whittaker, on behalf of the priority of C. F. Mohr in the establishment of the mechanical theory of heat. Mohr, a German chemist (1806–1879), in 1837 published in Liebig's *Annalen der Chemie und Pharmacie*

(**24**, 141, 1837) a paper entitled "Über die Natur der Wärme" in which he attacked the caloric theory of heat and developed the idea that heat is due to "vibrations" in bodies. The paper contains no specific mention of the possible relation between heat and mechanical work, although there are some suggestive remarks about the compression and expansion of gases. It is clear that Mohr was on the track of the significance of the two specific heats of a gas, but simply did not go far enough along with this idea to use it as Mayer did in calculating a mechanical equivalent of heat. Mohr did make the statement that the various phenomena of nature are related, but he failed to follow up this notion in the way Mayer did a few years later. Mohr left this field to work in analytical chemistry and pharmacology. It was only in 1874 that he became interested in what by that time had become the well-established field of thermodynamics. He then claimed priority for his earlier work as a basis for the development of thermodynamics. This claim was promoted by P. G. Tait in England. Tait translated Mohr's 1837 paper and published it in the *Philosophical Magazine* (5th series, **2**, 110–114). It is not evident that Mohr's work influenced anyone. Mayer's line of attack was certainly more fundamental, and he followed it up with further significant investigations in the late 1840s. Tait's polemic against Mayer is well known, and this may have had something to do with his support of Mohr.

James Prescott Joule (1818–1889) provides one of the best examples of the amateur scientist who achieves genuine success and renown in his chosen field. Born into a family of brewers at Salford near Manchester, England, he was largely self-taught. During the early part of his life he had independent means and could afford to devote all his time to research on scientific subjects of his own choosing. At the age of 19 he became interested in the relatively new science of electromagnetism and devoted much time to the development of what were then called electromagnetic engines (electric motors), with the idea that such engines might ultimately prove more efficient than steam engines. He was rather soon disabused of this hope. His work was not wasted, however, since it gave him the notion that heat always plays a role in the production of mechanical effects by electricity. He studied (around 1840) the production of heat by electric currents and discovered the famous law known by his name: that the rate of heat production in a current-carrying conductor is proportional to the product of the resistance and the square of the current.

Joule's early experiments made him sceptical of the caloric theory of heat, and he began to suspect (about 1842) that there should be a definite relation between the mechanical effect necessary to produce electric current from an electromagnetic generator and the total heat resulting from the flow of that current. This led to the first of his really great experiments from the standpoint of energy, the results of which were reported to the meeting of the British Association for the Advancement of Science at Cork in 1843, and published in the *Philosophical Magazine* [Ser. 3, **23**, 263, 347, 435 (1843)] under the title "On the Calorific Effects of Magneto-Electricity and on the Mechanical Value of Heat" (*The Scientific Papers of James Prescott Joule*, published by the Physical Society of London, Taylor & Francis Ltd., London, 1884, pp. 123–159). This paper beautifully illustrates the elaborate precautions taken by Joule both to make sure of the meaning of what he was measuring and to achieve the highest possible precision in his measured values. The precision of his temperature measurements was extraordinary for his time and, indeed, scarcely believed by some of his critics. This paper, which is included here in full, led to the first experimental values of the mechanical equivalent of heat that could be taken seriously. Converted to modern Celsius–metric units, Joule's values for the mechanical equivalent varied from 4.5 to 5.46 joules per calorie.

Joule took this work very seriously. To confirm his results, however, he invented his famous paddle box in which the work done in rotating the paddle against the resistance of the water in the box is converted into heat, which can be estimated by measuring the rise in water temperature. Joule reported the results obtained in this way to the British Association Meeting in Cambridge in June 1845, and at the same time published them in the *Philosophical Magazine* [Ser. 3, **27**, 205 (1845)]. In the short paper reprinted here he quotes a result for the mechanical equivalent of heat, which in modern terms is 4.80 joules per calorie. He considered this figure sufficiently close to those obtained in his electromagnetic experiments to justify the conclusion that the mechanical equivalent of heat is a meaningful and definite physical constant. He pursued his frictional experiments with great vigor during the next 2 years and was able to report at the British Association Meeting in Oxford in 1847 a figure equivalent to 4.2 joules per calorie, a result close to the accepted value today. At this meeting he attracted the attention of the young William Thomson (later Lord Kelvin), who saw significance in Joule's work that had been previously overlooked by Joule's contemporaries. From being a neglected amateur and indeed in the eyes of many almost a crank, Joule now became a celebrity.

During this same year (1847), Joule summarized his general views resulting from his experiments about the value of heat and force (what we now call energy) in a lecture entitled "On Matter, Living Force and Heat," which was delivered in Saint Anne's Church Reading Room in Manchester and printed in the Manchester *Courier* newspaper for

May 5 and 12, 1847. This very clear and beautiful summary, from Joule's *Scientific Papers*, is here reprinted in full.

In his lack of mathematical techniques Joule was not a physicist of the school of Helmholtz, Kelvin, Clausius, and Maxwell. But he was one of the most thorough and painstaking experimenters that ever lived. Moreover, he knew what he was after, and so along with Mayer, whose theoretical determination of the mechanical equivalent of heat he experimentally verified, he became one of the founders of thermodynamics and one of the important generalizers of the concept of energy.

Ludwig August Colding (1815–1888), the Danish engineer, was born on the island of Sjaelland, not far from the city of Copenhagen. He early showed great interest in technical matters, and through the acquaintance of his family with the celebrated physicist Hans Christian Oersted, Colding was encouraged to enroll as a student in the Royal Technical University (Polyteknisk Laereanstalt) in 1837. Through further contact with Oersted, he not only developed an insight into engineering problems, but also became interested in their scientific aspects. By 1840 he had begun to think about the possible relation of heat and mechanical work, and his first experiments on heat due to friction date from 1842. His first communication of results was made to the Danish Academy of Sciences in November 1843. This was unfortunately never officially published. However, it led to further investigations with more elaborate equipment, by which Colding was able to measure the mechanical equivalent of heat with fair accuracy, although not with that accuracy which Joule later achieved. Colding was for many years ignorant of the work of Mayer and Joule, and they in turn knew nothing of his. It seems clear, however, that he was one of the first to establish definitively the relation between heat and mechanical energy. This was recognized in England by the publication of papers in English about his work in 1862, 1864, and 1871 [*Philosophical Magazine*, Ser. 4, **25**, 467–472 (1862); **27**, 56–64 (1864); **42**, 1–20 (1871)].

Colding devoted his professional life to hydraulic engineering and became ultimately engineer-in-chief to the city of Copenhagen.

We reproduce here in English translation the comments of the Danish Academy's Committee appointed to examine Colding's preliminary communication of 1843 (unpublished). We also reproduce in English translation extracts from his 1847 memoir, "Investigations of the General Forces of Nature and Their Mutual Dependence," published in the *Proceedings of the Royal Danish Academy of Sciences* (5th Ser., Natural Science and Mathematics Section, vol. 2, Copenhagen, 1851, pp. 121–146).

28 A

Reprinted from *Phil. Trans. Roy. Soc.*, Pt. II, 247–257 (1834)

XV. *On a General Method in Dynamics; by which the Study of the Motions of all free Systems of attracting or repelling Points is reduced to the Search and Differentiation of one central Relation, or characteristic Function.*

By WILLIAM ROWAN HAMILTON,

Member of several scientific Societies in the British Dominions, and of the American Academy of Arts and Sciences, Andrews' Professor of Astronomy in the University of Dublin, and Royal Astronomer of Ireland. Communicated by Captain BEAUFORT, *R.N. F.R.S.*

Received April 1,—Read April 10, 1834.

Introductory Remarks.

THE theoretical development of the laws of motion of bodies is a problem of such interest and importance, that it has engaged the attention of all the most eminent mathematicians, since the invention of dynamics as a mathematical science by GALILEO, and especially since the wonderful extension which was given to that science by NEWTON. Among the successors of those illustrious men, LAGRANGE has perhaps done more than any other analyst, to give extent and harmony to such deductive researches, by showing that the most varied consequences respecting the motions of systems of bodies may be derived from one radical formula; the beauty of the method so suiting the dignity of the results, as to make of his great work a kind of scientific poem. But the science of force, or of power acting by law in space and time, has undergone already another revolution, and has become already more dynamic, by having almost dismissed the conceptions of solidity and cohesion, and those other material ties, or geometrically imaginable conditions, which LAGRANGE so happily reasoned on, and by tending more and more to resolve all connexions and actions of bodies into attractions and repulsions of points: and while the science is advancing thus in one direction by the improvement of physical views, it may advance in another direction also by the invention of mathematical methods. And the method proposed in the present essay, for the deductive study of the motions of attracting or repelling systems, will perhaps be received with indulgence, as an attempt to assist in carrying forward so high an inquiry.

In the methods commonly employed, the determination of the motion of a free point in space, under the influence of accelerating forces, depends on the integration of three equations in ordinary differentials of the second order; and the determination of the motions of a system of free points, attracting or repelling one another, depends on the integration of a system of such equations, in number threefold the number of the attracting or repelling points, unless we previously diminish by unity this latter number, by considering only relative motions. Thus, in the solar system, when we consider only the mutual attractions of the sun and of the ten known planets, the determination of the motions of the latter about the former is reduced, by the usual methods, to the integration of a system of thirty ordinary differential equations of the second order, between the coordinates and the time; or, by a transformation of LAGRANGE, to the integration of a system of sixty ordinary differential equations of the first order, between the time and the elliptic elements: by which integrations, the thirty varying coordinates, or the sixty varying elements, are to be found as functions of the time. In the method of the present essay, this problem is reduced to the search and differentiation of a single function, which satisfies two partial differential equations of the first order and of the second degree: and every other dynamical problem, respecting the motions of any system, however numerous, of attracting or repelling points, (even if we suppose those points restricted by any conditions of connexion consistent with the law of living force,) is reduced, in like manner, to the study of one central function, of which the form marks out and characterizes the properties of the moving system, and is to be determined by a pair of partial differential equations of the first order, combined with some simple considerations. The difficulty is therefore at least transferred from the integration of many equations of one class to the integration of two of another: and even if it should be thought that no practical facility is gained, yet an intellectual pleasure may result from the reduction of the most complex and, probably, of all researches respecting the forces and motions of body, to the study of one characteristic function*, the unfolding of one central relation.

The present essay does not pretend to treat fully of this extensive subject,—a task which may require the labours of many years and many minds; but only to suggest the thought and propose the path to others. Although, therefore, the method may be used in the most varied dynamical researches, it is at present only applied to the orbits and perturbations of a system with any laws of attraction or repulsion, and with one predominant mass or centre of predominant energy; and only so far, even in this one research, as appears sufficient to make the principle itself understood. It may be mentioned here, that this dynamical principle is only another form of that idea which has already been applied to optics in the *Theory of systems of rays*, and that an intention of applying it to the motions of systems of bodies was announced †

* LAGRANGE and, after him, LAPLACE and others, have employed a single function to express the different forces of a system, and so to form in an elegant manner the differential equations of its motion. By this conception, great simplicity has been given to the statement of the problem of dynamics; but the solution of that problem, or the expression of the motions themselves, and of their integrals, depends on a very different and hitherto unimagined function, as it is the purpose of this essay to show.

† Transactions of the Royal Irish Academy, vol. xv. page 80. A notice of this dynamical principle was also lately given in an article "On a general Method of expressing the Paths of Light and of the Planets," published in the Dublin University Review for October 1833.

at the publication of that theory. And besides the idea itself, the manner of calculation also, which has been thus exemplified in the sciences of optics and dynamics, seems not confined to those two sciences, but capable of other applications; and the peculiar combination which it involves, of the principles of variations with those of partial differentials, for the determination and use of an important class of integrals, may constitute, when it shall be matured by the future labours of mathematicians, a separate branch of analysis.

WILLIAM R. HAMILTON.

Observatory, Dublin,
March 1834.

Integration of the Equations of Motion of a System, characteristic Function of such Motion, and Law of varying Action.

1. The known differential equations of motion of a system of free points, repelling or attracting one another according to any functions of their distances, and not disturbed by any foreign force, may be comprised in the following formula:

$$\Sigma \,.\, m\,(x''\,\delta x + y''\,\delta y + z''\,\delta z) = \delta \mathrm{U}. \qquad (1.)$$

In this formula the sign of summation Σ extends to all the points of the system; m is, for any one such point, the constant called its mass; x'', y'', z'', are its component accelerations, or the second differential coefficients of its rectangular coordinates x, y, z, taken with respect to the time; δx, δy, δz, are any arbitrary infinitesimal displacements which the point can be imagined to receive in the same three rectangular directions; and $\delta \mathrm{U}$ is the infinitesimal variation corresponding, of a function U of the masses and mutual distances of the several points of the system, of which the form depends on the laws of their mutual actions, by the equation

$$\mathrm{U} = \Sigma \,.\, m\, m_{\prime} f(r), \qquad (2.)$$

r being the distance between any two points m, $m_{\prime}$, and the function $f(r)$ being such that its derivative or differential coefficient $f'(r)$ expresses the law of their repulsion, being negative in the case of attraction. The function which has been here called U, may be named the *force-function* of a system: it is of great utility in theoretical mechanics, into which it was introduced by LAGRANGE, and it furnishes the following elegant forms for the differential equations of motion, included in the formula (1.):

$$\left.\begin{array}{l} m_1 x''_1 = \dfrac{\delta \mathrm{U}}{\delta x_1}; \; m_2 x''_2 = \dfrac{\delta \mathrm{U}}{\delta x_2}; \; \ldots m_n x''_n = \dfrac{\delta \mathrm{U}}{\delta x_n}; \\ m_1 y''_1 = \dfrac{\delta \mathrm{U}}{\delta y_1}; \; m_2 y''_2 = \dfrac{\delta \mathrm{U}}{\delta y_2}; \; \ldots m_n y''_n = \dfrac{\delta \mathrm{U}}{\delta y_n}; \\ m_1 z''_1 = \dfrac{\delta \mathrm{U}}{\delta z_1}; \; m_2 z''_2 = \dfrac{\delta \mathrm{U}}{\delta z_2}; \; \ldots m_n z''_n = \dfrac{\delta \mathrm{U}}{\delta z_n}; \end{array}\right\} \qquad (3.)$$

the second members of these equations being the partial differential coefficients of the first order of the function U. But notwithstanding the elegance and simplicity of this known manner of stating the principal problem of dynamics, the difficulty of solving that problem, or even of expressing its solution, has hitherto appeared insuperable; so that only seven intermediate integrals, or integrals of the first order, with as many arbitrary constants, have hitherto been found for these general equations of motion of a system of n points, instead of $3n$ intermediate and $3n$ final integrals, involving ultimately $6n$ constants; nor has any integral been found which does not need to be integrated again. No general solution has been obtained assigning (as a complete solution ought to do) $3n$ relations between the n masses $m_1, m_2, \ldots m_n$, the $3n$ varying coordinates $x_1, y_1, z_1, \ldots x_n, y_n, z_n$, the varying time t, and the $6n$ initial data of the problem, namely, the initial coordinates $a_1, b_1, c_1, \ldots a_n, b_n, c_n$, and their initial rates of increase, $a'_1, b'_1, c'_1, \ldots a'_n, b'_n, c'_n$; the quantities called here initial being those which correspond to the arbitrary origin of time. It is, however, possible (as we shall see) to express these long-sought relations by the partial differential coefficients of a new central or radical function, to the search and employment of which the difficulty of mathematical dynamics becomes henceforth reduced.

2. If we put for abridgement

$$T = \tfrac{1}{2} \Sigma . m (x'^2 + y'^2 + z'^2), \qquad (4.)$$

so that 2 T denotes, as in the Mécanique Analytique, the whole living force of the system; (x', y', z', being here, according to the analogy of our foregoing notation, the rectangular components of velocity of the point m, or the first differential coefficients of its coordinates taken with respect to the time;) an easy and well known combination of the differential equations of motion, obtained by changing in the formula (1.) the variations to the differentials of the coordinates, may be expressed in the following manner,

$$d\,T = d\,U, \qquad (5.)$$

and gives, by integration, the celebrated law of living force, under the form

$$T = U + H. \qquad (6.)$$

In this expression, which is one of the seven known integrals already mentioned, the quantity H is independent of the time, and does not alter in the passage of the points of the system from one set of positions to another. We have, for example, an initial equation of the same form, corresponding to the origin of time, which may be written thus,

$$T_0 = U_0 + H \qquad (7.)$$

The quantity H may, however, receive any arbitrary increment whatever, when we pass in thought from a system moving in one way, to the same system moving in another, with the same dynamical relations between the accelerations and positions of its points, but with different initial data; but the increment of H, thus obtained, is evidently connected with the analogous increments of the functions T and U, by the relation

$$\Delta \mathrm{T} = \Delta \mathrm{U} + \Delta \mathrm{H}, \qquad (8.)$$

which, for the case of infinitesimal variations, may conveniently be written thus,

$$\delta \mathrm{T} = \delta \mathrm{U} + \delta \mathrm{H}; \qquad (9.)$$

and this last relation, when multiplied by dt, and integrated, conducts to an important result. For it thus becomes, by (4.) and (1.),

$$\begin{aligned} &\int \Sigma . m\,(d x . \delta x' + d y . \delta y' + d z . \delta z') = \\ &\int \Sigma . m\,(d x' . \delta x + d y' . \delta y + d z' . \delta z) + \int \delta \mathrm{H} . d t, \end{aligned} \qquad (10.)$$

that is, by the principles of the calculus of variations,

$$\delta \mathrm{V} = \Sigma . m\,(x' \delta x + y' \delta y + z' \delta z) - \Sigma . m\,(a' \delta a + b' \delta b + c' \delta c) + t\, \delta \mathrm{H}, \qquad (\mathrm{A}.)$$

if we denote by V the integral

$$\mathrm{V} = \int \Sigma . m\,(x' d x + y' d y + z' d z) = \int_0^t 2\,\mathrm{T}\, d t, \qquad (\mathrm{B}.)$$

namely, the accumulated living force, called often the action of the system, from its initial to its final position.

If, then, we consider (as it is easy to see that we may) the action V as a function of the initial and final coordinates, and of the quantity H, we shall have, by (A.), the following groups of equations; first, the group,

$$\left.\begin{aligned} &\frac{\delta \mathrm{V}}{\delta x_1} = m_1 x'_1; \quad \frac{\delta \mathrm{V}}{\delta x_2} = m_2 x'_2; \ldots \frac{\delta \mathrm{V}}{\delta x_n} = m_n x'_n; \\ &\frac{\delta \mathrm{V}}{\delta y_1} = m_1 y'_1; \quad \frac{\delta \mathrm{V}}{\delta y_2} = m_2 y'_2; \ldots \frac{\delta \mathrm{V}}{\delta y_n} = m_n y'_n; \\ &\frac{\delta \mathrm{V}}{\delta z_1} = m_1 z'_1; \quad \frac{\delta \mathrm{V}}{\delta z_2} = m_2 z'_2; \ldots \frac{\delta \mathrm{V}}{\delta z_n} = m_n z'_n; \end{aligned}\right\} \qquad (\mathrm{C}.)$$

Secondly, the group,

$$\left.\begin{aligned} &\frac{\delta \mathrm{V}}{\delta a_1} = - m_1 a'_1; \quad \frac{\delta \mathrm{V}}{\delta a_2} = - m_2 a'_2; \ldots \frac{\delta \mathrm{V}}{\delta a_n} = - m_n a'_n; \\ &\frac{\delta \mathrm{V}}{\delta b_1} = - m_1 b'_1; \quad \frac{\delta \mathrm{V}}{\delta b_2} = - m_2 b'_2; \ldots \frac{\delta \mathrm{V}}{\delta b_n} = - m_n b'_n; \\ &\frac{\delta \mathrm{V}}{\delta c_1} = - m_1 c'_1; \quad \frac{\delta \mathrm{V}}{\delta c_2} = - m_2 c'_2; \ldots \frac{\delta \mathrm{V}}{\delta c_n} = - m_n c'_n; \end{aligned}\right\} \qquad (\mathrm{D}.)$$

and finally, the equation,

$$\frac{\delta \mathrm{V}}{\delta \mathrm{H}} = t: \qquad (\mathrm{E}.)$$

So that if this function V were known, it would only remain to eliminate H between the $3n + 1$ equations (C.) and (E.), in order to obtain all the $3n$ intermediate integrals, or between (D.) and (E.) to obtain all the $3n$ final integrals of the differential equations of motion; that is, ultimately, to obtain the $3n$ sought relations between the $3n$ varying coordinates and the time, involving also the masses and the $6n$ initial data above mentioned; the discovery of which relations would be (as we have said) the general solution of the general problem of dynamics. We have, therefore, at least reduced that general problem to the search and differentiation of a single function V, which we shall call on this account the CHARACTERISTIC FUNCTION of motion of a system; and the equation (A.), expressing the fundamental law of its variation, we shall call the *equation of the characteristic function*, or the LAW OF VARYING ACTION.

3. To show more clearly that the action or accumulated living force of a system, or in other words, the integral of the product of the living force by the element of the time, may be regarded as a function of the $6n + 1$ quantities already mentioned, namely, of the initial and final coordinates, and of the quantity H, we may observe, that whatever depends on the manner and time of motion of the system may be considered as such a function; because the initial form of the law of living force, when combined with the $3n$ known or unknown relations between the time, the initial data, and the varying coordinates, will always furnish $3n + 1$ relations, known or unknown, to connect the time and the initial components of velocities with the initial and final coordinates, and with H. Yet from not having formed the conception of the action as a *function* of this kind, the consequences that have been here deduced from the formula (A.) for the variation of that definite integral, appear to have escaped the notice of LAGRANGE, and of the other illustrious analysts who have written on theoretical mechanics; although they were in possession of a formula for the variation of this integral not greatly differing from ours. For although LAGRANGE and others, in treating of the motion of a system, have shown that the variation of this definite integral vanishes when the extreme coordinates and the constant H are given, they appear to have deduced from this result only the well known law of *least action*; namely, that if the points or bodies of a system be imagined to move from a given set of initial to a given set of final positions, not as they do nor even as they could move consistently with the general dynamical laws or differential equations of motion, but so as not to violate any supposed geometrical connexions, nor that one dynamical relation between velocities and configurations which constitutes the law of living force; and if, besides, this geometrically imaginable, but dynamically impossible motion, be made to differ infinitely *little* from the actual manner of motion of the system, between the given extreme positions; then the varied value of the definite integral called action, or the accumulated living force of the system in the motion thus imagined, will differ infinitely *less* from the actual value of that integral. But when this well known law of least, or as it might be better called, of *stationary action*, is applied to the determination of the actual motion of a system, it serves only to form, by the rules of the calculus of variations, the differential equations of motion of the second order, which can always be otherwise found. It seems, therefore, to be with reason that LAGRANGE, LAPLACE, and POISSON have spoken lightly of the utility of this principle in the present state of dynamics. A different estimate, perhaps, will be formed of that

other principle which has been introduced in the present paper, under the name of the *law of varying action,* in which we pass from an actual motion to another motion dynamically possible, by varying the extreme positions of the system, and (in general) the quantity H, and which serves to express, by means of a single function, not the mere differential equations of motion, but their intermediate and their final integrals.

Verifications of the foregoing Integrals.

4. A verification, which ought not to be neglected, and at the same time an illustration of this new principle, may be obtained by deducing the known differential equations of motion from our system of intermediate integrals, and by showing the consistence of these again with our final integral system. As preliminary to such verification, it is useful to observe that the final equation (6.) of living force, when combined with the system (C.), takes this new form,

$$\frac{1}{2}\Sigma . \frac{1}{m}\left\{\left(\frac{\delta V}{\delta x}\right)^2 + \left(\frac{\delta V}{\delta y}\right)^2 + \left(\frac{\delta V}{\delta z}\right)^2\right\} = U + H \, ; \quad \text{(F.)}$$

and that the initial equation (7.) of living force becomes by (D.)

$$\frac{1}{2}\Sigma . \frac{1}{m}\left\{\left(\frac{\delta V}{\delta a}\right)^2 + \left(\frac{\delta V}{\delta b}\right)^2 + \left(\frac{\delta V}{\delta c}\right)^2\right\} = U_0 + H. \quad \text{(G.)}$$

These two partial differential equations, initial and final, of the first order and the second degree, must both be identically satisfied by the characteristic function V: they furnish (as we shall find) the principal means of discovering the form of that function, and are of essential importance in its theory. If the form of this function were known, we might eliminate $3n - 1$ of the $3n$ initial coordinates between the $3n$ equations (C.); and although we cannot yet perform the actual process of this elimination, we are entitled to assert that it would remove along with the others the remaining initial coordinate, and would conduct to the equation (6.) of final living force, which might then be transformed into the equation (F.). In like manner we may conclude that all the $3n$ final coordinates could be eliminated together from the $3n$ equations (D.), and that the result would be the initial equation (7.) of living force, or the transformed equation (G.). We may therefore consider the law of living force, which assisted us in discovering the properties of our characteristic function V, as included reciprocally in those properties, and as resulting by elimination, in every particular case, from the systems (C.) and (D.); and in treating of either of these systems, or in conducting any other dynamical investigation by the method of this characteristic function, we are at liberty to employ the partial differential equations (F.) and (G.), which that function must necessarily satisfy.

It will now be easy to deduce, as we proposed, the known equations of motion (3.) of the second order, by differentiation and elimination of constants, from our intermediate integral system (C.), (E.), or even from a part of that system, namely, from the group (C.), when combined with the equation (F.). For we thus obtain

$$\left.\begin{aligned} m_1 x''_1 = \frac{d}{dt}\frac{\delta V}{\delta x_1} &= x'_1 \frac{\delta^2 V}{\delta x^2_1} + x'_2 \frac{\delta^2 V}{\delta x_1 \delta x_2} + \ldots + x'_n \frac{\delta^2 V}{\delta x_1 \delta x_n} \\ &+ y'_1 \frac{\delta^2 V}{\delta x_1 \delta y_1} + y'_2 \frac{\delta^2 V}{\delta x_1 \delta y_2} + \ldots + y'_n \frac{\delta^2 V}{\delta x_1 \delta y_n} \\ &+ z'_1 \frac{\delta^2 V}{\delta x_1 \delta z_1} + z'_2 \frac{\delta^2 V}{\delta x_1 \delta z_2} + \ldots + z'_n \frac{\delta^2 V}{\delta x_1 \delta z_n} \\ = \frac{1}{m_1}\frac{\delta V}{\delta x_1}\frac{\delta^2 V}{\delta x^2_1} &+ \frac{1}{m_2}\frac{\delta V}{\delta x_2}\frac{\delta^2 V}{\delta x_1 \delta x_2} + \ldots + \frac{1}{m_n}\frac{\delta V}{\delta x_n}\frac{\delta^2 V}{\delta x_1 \delta x_n} \\ + \frac{1}{m_1}\frac{\delta V}{\delta y_1}\frac{\delta^2 V}{\delta x_1 \delta y_1} &+ \frac{1}{m_2}\frac{\delta V}{\delta y_2}\frac{\delta^2 V}{\delta x_1 \delta y_2} + \ldots + \frac{1}{m_n}\frac{\delta V}{\delta y_n}\frac{\delta^2 V}{\delta x_1 \delta y_n} \\ + \frac{1}{m_1}\frac{\delta V}{\delta z_1}\frac{\delta^2 V}{\delta x_1 \delta z_1} &+ \frac{1}{m_2}\frac{\delta V}{\delta z_2}\frac{\delta^2 V}{\delta x_1 \delta z_2} + \ldots + \frac{1}{m_n}\frac{\delta V}{\delta z_n}\frac{\delta^2 V}{\delta x_1 \delta z_n} \\ = \frac{\delta}{\delta x_1} \Sigma \,.\, \frac{1}{2m}&\left\{ \left(\frac{\delta V}{\delta x}\right)^2 + \left(\frac{\delta V}{\delta y}\right)^2 + \left(\frac{\delta V}{\delta z}\right)^2 \right\} = \frac{\delta}{\delta x}(U + H); \end{aligned}\right\} \quad \ldots \quad (11.)$$

that is, we obtain

$$m_1 x''_1 = \frac{\delta U}{\delta x_1}: \quad \ldots \ldots \quad (12.)$$

And in like manner we might deduce, by differentiation, from the integrals (C.) and from (F.) all the other known differential equations of motion, of the second order, contained in the set marked (3.); or, more concisely, we may deduce at once the formula (1.), which contains all those known equations, by observing that the intermediate integrals (C.), when combined with the relation (F.), give

$$\left.\begin{aligned} &\Sigma \,.\, m(x''\delta x + y''\delta y + z''\delta z) = \Sigma \left(\frac{d}{dt}\frac{\delta V}{\delta x} \,.\, \delta x + \frac{d}{dt}\frac{\delta V}{\delta y} \,.\, \delta y + \frac{d}{dt}\frac{\delta V}{\delta z} \,.\, \delta z\right) \\ &= \Sigma \,.\, \frac{1}{m}\left(\frac{\delta V}{\delta x}\frac{\delta}{\delta x} + \frac{\delta V}{\delta y}\frac{\delta}{\delta y} + \frac{\delta V}{\delta z}\frac{\delta}{\delta z}\right) \Sigma \left(\frac{\delta V}{\delta x}\delta x + \frac{\delta V}{\delta y}\delta y + \frac{\delta V}{\delta z}\delta z\right) \\ &= \Sigma \left(\delta x \frac{\delta}{\delta x} + \delta y \frac{\delta}{\delta y} + \delta z \frac{\delta}{\delta z}\right) \Sigma \,.\, \frac{1}{2m}\left\{\left(\frac{\delta V}{\delta x}\right)^2 + \left(\frac{\delta V}{\delta y}\right)^2 + \left(\frac{\delta V}{\delta z}\right)^2\right\} \\ &= \Sigma \left(\delta x \frac{\delta}{\delta x} + \delta y \frac{\delta}{\delta y} + \delta z \frac{\delta}{\delta z}\right)(U + H) \\ &= \delta U. \end{aligned}\right\} . \quad (13.)$$

5. Again, we were to show that our intermediate integral system, composed of the equations (C.) and (E.), with the $3n$ arbitrary constants $a_1, b_1, c_1, \ldots a_n; b_n, c_n$, (and involving also the auxiliary constant H,) is consistent with our final integral system of equations (D.) and (E.), which contain $3n$ other arbitrary constants, namely, $a'_1, b'_1, c'_1, \ldots a'_n, b'_n, c'_n$. The immediate differentials of the equations (C.), (D.), (E.), taken with respect to the time, are, for the first group,

$$\left.\begin{array}{l}\frac{d}{dt}\frac{\delta V}{\delta x_1} = m_1 x''_1; \ \frac{d}{dt}\frac{\delta V}{\delta x_2} = m_2 x''_2; \ \ldots \frac{d}{dt}\frac{\delta V}{\delta x_n} = m_n x''_n; \\ \frac{d}{dt}\frac{\delta V}{\delta y_1} = m_1 y''_1; \ \frac{d}{dt}\frac{\delta V}{\delta y_2} = m_2 y''_2; \ \ldots \frac{d}{dt}\frac{\delta V}{\delta y_n} = m_n y''_n; \\ \frac{d}{dt}\frac{\delta V}{\delta z_1} = m_1 z''_1; \ \frac{d}{dt}\frac{\delta V}{\delta z_2} = m_2 z''_2; \ \ldots \frac{d}{dt}\frac{\delta V}{\delta z_n} = m_n z''_n; \end{array}\right\} \quad \ldots \quad \text{(H.)}$$

for the second group,

$$\left.\begin{array}{l}\frac{d}{dt}\frac{\delta V}{\delta a_1} = 0; \ \frac{d}{dt}\frac{\delta V}{\delta a_2} = 0; \ \ldots \frac{d}{dt}\frac{\delta V}{\delta a_n} = 0; \\ \frac{d}{dt}\frac{\delta V}{\delta b_1} = 0; \ \frac{d}{dt}\frac{\delta V}{\delta b_2} = 0; \ \ldots \frac{d}{dt}\frac{\delta V}{\delta b_n} = 0; \\ \frac{d}{dt}\frac{\delta V}{\delta c_1} = 0; \ \frac{d}{dt}\frac{\delta V}{\delta c_2} = 0; \ \ldots \frac{d}{dt}\frac{\delta V}{\delta c_n} = 0; \end{array}\right\} \quad \ldots \quad \text{(I.)}$$

and finally, for the last equation,

$$\frac{d}{dt}\frac{\delta V}{\delta H} = 1. \quad \ldots \quad \text{(K.)}$$

By combining the equations (C.) with their differentials (H.), and with the relation (F.), we deduced, in the foregoing number, the known equations of motion (3.); and we are now to show the consistence of the same intermediate integrals (C.) with the group of differentials (I.), which have been deduced from the final integrals.

The first equation of the group (I.) may be developed thus:

$$\left.\begin{array}{l} 0 = x'_1 \frac{\delta^2 V}{\delta a_1 \delta x_1} + x'_2 \frac{\delta^2 V}{\delta a_1 \delta x_2} + \ldots + x'_n \frac{\delta^2 V}{\delta a_1 \delta x_n} \\ \quad + y'_1 \frac{\delta^2 V}{\delta a_1 \delta y_1} + y'_2 \frac{\delta^2 V}{\delta a_1 \delta y_2} + \ldots + y'_n \frac{\delta^2 V}{\delta a_1 \delta y_n} \\ \quad + z'_1 \frac{\delta^2 V}{\delta a_1 \delta z_1} + z'_2 \frac{\delta^2 V}{\delta a_1 \delta z_2} + \ldots + z'_n \frac{\delta^2 V}{\delta a_1 \delta z_n}; \end{array}\right\} \quad \ldots \quad \text{(14.)}$$

and the others may be similarly developed. In order, therefore, to show that they are satisfied by the group (C.), it is sufficient to prove that the following equations are true,

$$\left.\begin{array}{l} 0 = \frac{\delta}{\delta a_i} \Sigma . \frac{1}{2m} \left\{ \left(\frac{\delta V}{\delta x}\right)^2 + \left(\frac{\delta V}{\delta y}\right)^2 + \left(\frac{\delta V}{\delta z}\right)^2 \right\}, \\ 0 = \frac{\delta}{\delta b_i} \Sigma . \frac{1}{2m} \left\{ \left(\frac{\delta V}{\delta x}\right)^2 + \left(\frac{\delta V}{\delta y}\right)^2 + \left(\frac{\delta V}{\delta z}\right)^2 \right\}, \\ 0 = \frac{\delta}{\delta c_i} \Sigma . \frac{1}{2m} \left\{ \left(\frac{\delta V}{\delta x}\right)^2 + \left(\frac{\delta V}{\delta y}\right)^2 + \left(\frac{\delta V}{\delta z}\right)^2 \right\}, \end{array}\right\} \quad \ldots \quad \text{(L.)}$$

the integer i receiving any value from 1 to n inclusive; which may be shown at once, and the required verification thereby be obtained, if we merely take the variation of the relation (F.) with respect to the initial coordinates, as in the former verification we took its variation with respect to the final coordinates, and so obtained results which agreed with the known equations of motion, and which may be thus collected,

$$\left.\begin{aligned} \frac{\delta}{\delta x_i} \Sigma . \frac{1}{2m} \left\{ \left(\frac{\delta V}{\delta x}\right)^2 + \left(\frac{\delta V}{\delta y}\right)^2 + \left(\frac{\delta V}{\delta z}\right)^2 \right\} = \frac{\delta U}{\delta x_i}; \\ \frac{\delta}{\delta y_i} \Sigma . \frac{1}{2m} \left\{ \left(\frac{\delta V}{\delta x}\right)^2 + \left(\frac{\delta V}{\delta y}\right)^2 + \left(\frac{\delta V}{\delta z}\right)^2 \right\} = \frac{\delta U}{\delta y_i}; \\ \frac{\delta}{\delta z_i} \Sigma . \frac{1}{2m} \left\{ \left(\frac{\delta V}{\delta x}\right)^2 + \left(\frac{\delta V}{\delta y}\right)^2 + \left(\frac{\delta V}{\delta z}\right)^2 \right\} = \frac{\delta U}{\delta z_i}. \end{aligned}\right\} \quad \text{(M.)}$$

The same relation (F.), by being varied with respect to the quantity H, conducts to the expression

$$\frac{\delta}{\delta H} \Sigma . \frac{1}{2m} \left\{ \left(\frac{\delta V}{\delta x}\right)^2 + \left(\frac{\delta V}{\delta y}\right)^2 + \left(\frac{\delta V}{\delta z}\right)^2 \right\} = 1; \quad \text{(N.)}$$

and this, when developed, agrees with the equation (K.), which is a new verification of the consistence of our foregoing results. Nor would it have been much more difficult, by the help of the foregoing principles, to have integrated directly our integrals of the first order, and so to have deduced in a different way our final integral system.

6. It may be considered as still another verification of our own general integral equations, to show that they include not only the known law of living force, or the integral expressing that law, but also the six other known integrals of the first order, which contain the law of motion of the centre of gravity, and the law of description of areas. For this purpose, it is only necessary to observe that it evidently follows from the conception of our characteristic function V, that this function depends on the initial and final positions of the attracting or repelling points of a system, not as referred to any foreign standard, but only as compared with one another; and therefore that this function will not vary, if without making any real change in either initial or final configuration, or in the relation of these to each other, we alter at once all the initial and all the final positions of the points of the system, by any common motion, whether of translation or of rotation. Now by considering three coordinate translations, we obtain the three following partial differential equations of the first order, which the function V must satisfy,

$$\left.\begin{aligned} \Sigma \frac{\delta V}{\delta x} + \Sigma \frac{\delta V}{\delta a} = 0; \\ \Sigma \frac{\delta V}{\delta y} + \Sigma \frac{\delta V}{\delta b} = 0; \\ \Sigma \frac{\delta V}{\delta z} + \Sigma \frac{\delta V}{\delta c} = 0; \end{aligned}\right\} \quad \text{(O.)}$$

and by considering three coordinate rotations, we obtain these three other relations between the partial differential coefficients of the same order of the same characteristic function,

$$\left.\begin{aligned}\Sigma\left(x\frac{\delta V}{\delta y}-y\frac{\delta V}{\delta x}\right)+\Sigma\left(a\frac{\delta V}{\delta b}-b\frac{\delta V}{\delta a}\right)=0;\\ \Sigma\left(y\frac{\delta V}{\delta z}-z\frac{\delta V}{\delta y}\right)+\Sigma\left(b\frac{\delta V}{\delta c}-c\frac{\delta V}{\delta b}\right)=0;\\ \Sigma\left(z\frac{\delta V}{\delta x}-x\frac{\delta V}{\delta z}\right)+\Sigma\left(c\frac{\delta V}{\delta a}-a\frac{\delta V}{\delta c}\right)=0;\end{aligned}\right\}\quad\text{(P.)}$$

and if we change the final coefficients of V to the final components of momentum, and the initial coefficients to the initial components taken negatively, according to the dynamical properties of this function expressed by the integrals (C.) and (D.), we shall change these partial differential equations (O.) (P.), to the following,

$$\Sigma\,.\,m\,x'=\Sigma\,.\,m\,a'\,;\;\Sigma\,.\,m\,y'=\Sigma\,.\,m\,b'\,;\;\Sigma\,.\,m\,z'=\Sigma\,.\,m\,c'\,;\quad\text{(15.)}$$

and

$$\left.\begin{aligned}\Sigma\,.\,m\,(x\,y'-y\,x')=\Sigma\,.\,m\,(a\,b'-b\,a')\,;\\ \Sigma\,.\,m\,(y\,z'-z\,y')=\Sigma\,.\,m\,(b\,c'-c\,b')\,;\\ \Sigma\,.\,m\,(z\,x'-x\,z')=\Sigma\,.\,m\,(c\,a'-a\,c').\end{aligned}\right\}\quad\text{(16.)}$$

In this manner, therefore, we can deduce from the properties of our characteristic function the six other known integrals above mentioned, in addition to that seventh which contains the law of living force, and which assisted in the discovery of our method.

28B

Reprinted from *Phil. Trans. Roy. Soc.*, Pt. II, 95–99 (1835)

VII. *Second Essay on a General Method in Dynamics.*

By WILLIAM ROWAN HAMILTON,
Member of several Scientific Societies in Great Britain and in Foreign Countries, Andrews' Professor of Astronomy in the University of Dublin, and Royal Astronomer of Ireland. Communicated by Captain BEAUFORT, *R.N. F.R.S.*

Received October 29, 1834,—Read January 15, 1835.

Introductory Remarks.

THE former Essay* contained a general method for reducing all the most important problems of dynamics to the study of one characteristic function, one central or radical relation. It was remarked at the close of that Essay, that many eliminations required by this method in its first conception, might be avoided by a general transformation, introducing the time explicitly into a part S of the whole characteristic function V; and it is now proposed to fix the attention chiefly on this part S, and to call it the *Principal Function.* The properties of this part or function S, which were noticed briefly in the former Essay, are now more fully set forth; and especially its uses in questions of perturbation, in which it dispenses with many laborious and circuitous processes, and enables us to express accurately the disturbed configuration of a system by the rules of undisturbed motion, if only the initial components of velocities be changed in a suitable manner. Another manner of extending rigorously to disturbed motion the rules of undisturbed, by the gradual variation of elements, in number double the number of the coordinates or other marks of position of the system, which was first invented by LAGRANGE, and was afterwards improved by POISSON, is considered in this Second Essay under a form perhaps a little more general; and the general method of calculation which has already been applied to other analogous questions in optics and in dynamics by the author of the present Essay, is now applied to the integration of the equations which determine these elements. This general method is founded chiefly on a combination of the principles of variations with those of partial differentials, and may furnish, when it shall be matured by the labours of other analysts, a separate branch of algebra, which may be called perhaps the *Calculus of Principal Functions*; because, in all the chief applications of algebra to physics, and in a very extensive class of purely mathematical questions, it reduces the determination of many mutually connected functions to the search and study of one principal or central relation. When applied to the integration of the equations of varying elements, it suggests, as is now shown, the consideration

* Philosophical Transactions for the year 1834, Second Part.

of a certain *Function of Elements*, which may be variously chosen, and may either be rigorously determined, or at least approached to, with an indefinite accuracy, by a corollary of the general method. And to illustrate all these new general processes, but especially those which are connected with problems of perturbation, they are applied in this Essay to a very simple example, suggested by the motions of projectiles, the parabolic path being treated as the undisturbed. As a more important example, the problem of determining the motions of a ternary or multiple system, with any laws of attraction or repulsion, and with one predominant mass, which was touched upon in the former Essay, is here resumed in a new way, by forming and integrating the differential equations of a new set of varying elements, entirely distinct in theory (though little differing in practice) from the elements conceived by LAGRANGE, and having this advantage, that the differentials of all the new elements for *both* the disturbed and disturbing masses may be expressed by the coefficients of *one* disturbing function.

Transformations of the Differential Equations of Motion of an Attracting or Repelling System.

1. It is well known to mathematicians, that the differential equations of motion of any system of free points, attracting or repelling one another according to any functions of their distances, and not disturbed by any foreign force, may be comprised in the following formula:

$$\Sigma . m\,(x''\delta x + y''\delta y + z''\delta z) = \delta U: \quad \ldots \ldots \ldots \ldots \quad (1.)$$

the sign of summation Σ extending to all the points of the system; m being, for any one such point, the constant called its mass, and $x\, y\, z$ being its rectangular coordinates; while $x''\, y''\, z''$ are the accelerations, or second differential coefficients taken with respect to the time, and δx, δy, δz are any arbitrary infinitesimal variations of those coordinates, and U is a certain *force-function*, introduced into dynamics by LAGRANGE, and involving the masses and mutual distances of the several points of the system. If the number of those points be n, the formula (1.) may be decomposed into $3\,n$ ordinary differential equations of the second order, between the coordinates and the time,

$$m_i\, x''_i = \frac{\delta U}{\delta x_i}; \quad m_i\, y''_i = \frac{\delta U}{\delta y_i}; \quad m_i\, z''_i = \frac{\delta U}{\delta z_i}: \quad \ldots \ldots \ldots \quad (2.)$$

and to integrate these differential equations of motion of an attracting or repelling system, or some transformations of these, is the chief and perhaps ultimately the only problem of mathematical dynamics.

2. To facilitate and generalize the solution of this problem, it is useful to express previously the $3\,n$ rectangular coordinates $x\, y\, z$ as functions of $3\,n$ other and more general marks of position $\eta_1\, \eta_2 \ldots \eta_{3n}$; and then the differential equations of motion take this more general form, discovered by LAGRANGE,

$$\frac{d}{dt}\frac{\delta T}{\delta \eta'_i} - \frac{\delta T}{\delta \eta_i} = \frac{\delta U}{\delta \eta_i}, \quad \text{(3.)}$$

in which

$$T = \tfrac{1}{2}\Sigma . m\,(x'^2 + y'^2 + z'^2). \quad \text{(4.)}$$

For, from the equations (2.) or (1.),

$$\left.\begin{aligned} \frac{\delta U}{\delta \eta_i} &= \Sigma . m\left(x''\frac{\delta x}{\delta \eta_i} + y''\frac{\delta y}{\delta \eta_i} + z''\frac{\delta z}{\delta \eta_i}\right) \\ &= \frac{d}{dt}\Sigma . m\left(x'\frac{\delta x}{\delta \eta_i} + y'\frac{\delta y}{\delta \eta_i} + z'\frac{\delta z}{\delta \eta_i}\right) \\ &\quad - \Sigma . m\left(x'\frac{d}{dt}\frac{\delta x}{\delta \eta_i} + y'\frac{d}{dt}\frac{\delta y}{\delta \eta_i} + z'\frac{d}{dt}\frac{\delta z}{\delta \eta_i}\right); \end{aligned}\right\} \quad \text{(5.)}$$

in which

$$\left.\begin{aligned} &\Sigma . m\left(x'\frac{\delta x}{\delta \eta_i} + y'\frac{\delta y}{\delta \eta_i} + z'\frac{\delta z}{\delta \eta_i}\right) \\ &= \Sigma . m\left(x'\frac{\delta x'}{\delta \eta'_i} + y'\frac{\delta y'}{\delta \eta'_i} + z'\frac{\delta z'}{\delta \eta'_i}\right) = \frac{\delta T}{\delta \eta'_i}, \end{aligned}\right\} \quad \text{(6.)}$$

and

$$\left.\begin{aligned} &\Sigma . m\left(x'\frac{d}{dt}\frac{\delta x}{\delta \eta_i} + y'\frac{d}{dt}\frac{\delta y}{\delta \eta_i} + z'\frac{d}{dt}\frac{\delta z}{\delta \eta_i}\right) \\ &= \Sigma . m\left(x'\frac{\delta x'}{\delta \eta_i} + y'\frac{\delta y'}{\delta \eta_i} + z'\frac{\delta z'}{\delta \eta_i}\right) = \frac{\delta T}{\delta \eta_i}, \end{aligned}\right\} \quad \text{(7.)}$$

T being here considered as a function of the $6n$ quantities of the forms η' and η, obtained by introducing into its definition (4.), the values

$$x' = \eta'_1\frac{\delta x}{\delta \eta_1} + \eta'_2\frac{\delta x}{\delta \eta_2} + \ldots + \eta'_{3n}\frac{\delta x}{\delta \eta_{3n}}, \text{ \&c.} \quad \text{(8.)}$$

A different proof of this important transformation (3.) is given in the Mécanique Analytique.

3. The function T being homogeneous of the second dimension with respect to the quantities η', must satisfy the condition

$$2T = \Sigma . \eta'\frac{\delta T}{\delta \eta'}; \quad \text{(9.)}$$

and since the variation of the same function T may evidently be expressed as follows,

$$\delta T = \Sigma\left(\frac{\delta T}{\delta \eta'}\delta \eta' + \frac{\delta T}{\delta \eta}\delta \eta\right), \quad \text{(10.)}$$

we see that this variation may be expressed in this other way,

$$\delta T = \Sigma\left(\eta'\,\delta\frac{\delta T}{\delta \eta'} - \frac{\delta T}{\delta \eta}\delta \eta\right). \quad \text{(11.)}$$

If then we put, for abridgement,

$$\frac{\delta T}{\delta \eta'_1} = \varpi_1, \ldots \frac{\delta T}{\delta \eta'_{3n}} = \varpi_{3n}, \quad \text{(12.)}$$

and consider T (as we may) as a function of the following form,

$$\mathrm{T} = \mathrm{F}(\varpi_1, \varpi_2, \ldots \varpi_{3n}, \eta_1, \eta_2, \ldots \eta_{3n}), \quad \ldots \quad (13.)$$

we see that

$$\frac{\delta \mathrm{F}}{\delta \varpi_1} = \eta'_1, \ldots \frac{\delta \mathrm{F}}{\delta \varpi_{3n}} = \eta'_{3n}, \quad \ldots \quad (14.)$$

and

$$\frac{\delta \mathrm{F}}{\delta \eta_1} = -\frac{\delta \mathrm{T}}{\delta \eta_1}, \ldots \frac{\delta \mathrm{F}}{\delta \eta_{3n}} = -\frac{\delta \mathrm{T}}{\delta \eta_{3n}}; \quad \ldots \quad (15.)$$

and therefore that the general equation (3.) may receive this new transformation,

$$\frac{d\varpi_i}{dt} = \frac{\delta(\mathrm{U} - \mathrm{F})}{\delta \eta_i}. \quad \ldots \quad (16.)$$

If then we introduce, for abridgement, the following expression H,

$$\mathrm{H} = \mathrm{F} - \mathrm{U} = \mathrm{F}(\varpi_1, \varpi_2, \ldots \varpi_{3n}, \eta_1, \eta_2, \ldots \eta_{3n}) - \mathrm{U}(\eta_1, \eta_2, \ldots \eta_{3n}), \quad . \quad (17.)$$

we are conducted to this new manner of presenting the differential equations of motion of a system of n points, attracting or repelling one another:

$$\left.\begin{array}{l} \frac{d\eta_1}{dt} = \frac{\delta \mathrm{H}}{\delta \varpi_1}; \quad \frac{d\varpi_1}{dt} = -\frac{\delta \mathrm{H}}{\delta \eta_1}; \\ \frac{d\eta_2}{dt} = \frac{\delta \mathrm{H}}{\delta \varpi_2}; \quad \frac{d\varpi_2}{dt} = -\frac{\delta \mathrm{H}}{\delta \eta_2}; \\ \ldots\ldots\ldots \\ \frac{d\eta_{3n}}{dt} = \frac{\delta \mathrm{H}}{\delta \varpi_{3n}}; \quad \frac{d\varpi_{3n}}{dt} = -\frac{\delta \mathrm{H}}{\delta \eta_{3n}}. \end{array}\right\} \quad \ldots \quad (\mathrm{A.})$$

In this view, the problem of mathematical dynamics, for a system of n points, is to integrate a system (A.) of $6n$ ordinary differential equations of the first order, between the $6n$ variables $\eta_i\ \varpi_i$ and the time t; and the solution of the problem must consist in assigning these $6n$ variables as functions of the time, and of their own initial values, which we may call $e_i\ p_i$. And all these $6n$ functions, or $6n$ relations to determine them, may be expressed, with perfect generality and rigour, by the method of the former Essay, or by the following simplified process.

Integration of the Equations of Motion, by means of one Principal Function.

4. If we take the variation of the definite integral

$$\mathrm{S} = \int_0^t \left(\Sigma \,.\, \varpi \frac{\delta \mathrm{H}}{\delta \varpi} - \mathrm{H}\right) dt \quad \ldots \quad (18.)$$

without varying t or dt, we find, by the Calculus of Variations,

$$\delta \mathrm{S} = \int_0^t \delta \mathrm{S}' \,.\, dt, \quad \ldots \quad (19.)$$

in which

$$\mathrm{S}' = \Sigma \,.\, \varpi \frac{\delta \mathrm{H}}{\delta \varpi} - \mathrm{H}, \quad \ldots \quad (20.)$$

and therefore

$$\delta S' = \Sigma \left(\varpi \delta \frac{\delta H}{\delta \varpi} - \frac{\delta H}{\delta \eta} \delta \eta \right), \quad \ldots \ldots \quad (21.)$$

that is, by the equations of motion (A.),

$$\delta S' = \Sigma \left(\varpi \delta \frac{d\eta}{dt} + \frac{d\varpi}{dt} \delta \eta \right) = \frac{d}{dt} \Sigma \,.\, \varpi \delta \eta; \quad \ldots \ldots \quad (22.)$$

the variation of the integral S is therefore

$$\delta S = \Sigma (\varpi \delta \eta - p \delta e), \quad \ldots \ldots \quad (23.)$$

(p and e being still initial values,) and it decomposes itself into the following $6n$ expressions, when S is considered as a function of the $6n$ quantities η_i e_i, (involving also the time,)

$$\left. \begin{array}{l} \varpi_1 = \dfrac{\delta S}{\delta \eta_1}; \;\; p_1 = -\dfrac{\delta S}{\delta e_1}; \\ \varpi_2 = \dfrac{\delta S}{\delta \eta_2}; \;\; p_2 = -\dfrac{\delta S}{\delta e_2}; \\ \ldots\ldots\ldots \\ \varpi_{3n} = \dfrac{\delta S}{\delta \eta_{3n}}; \;\; p_{3n} = -\dfrac{\delta S}{\delta e_{3n}}; \end{array} \right\} \quad \ldots \ldots \quad (B.)$$

which are evidently forms for the sought integrals of the $6n$ differential equations of motion (A.), containing only one unknown function S. The difficulty of mathematical dynamics is therefore reduced to the search and study of this one function S, which may for that reason be called the PRINCIPAL FUNCTION of motion of a system.

This function S was introduced in the first Essay under the form

$$S = \int_0^t (T + U)\, dt,$$

the symbols T and U having in this form their recent meanings; and it is worth observing, that when S is expressed by this definite integral, the conditions for its variation vanishing (if the final and initial coordinates and the time be given) are precisely the differential equations of motion (3.), under the forms assigned by LAGRANGE. The variation of this definite integral S has therefore the double property, of giving the differential equations of motion for any transformed coordinates when the extreme positions are regarded as fixed, and of giving the integrals of those differential equations when the extreme positions are treated as varying.

29A

Reprinted from R. B. Lindsay, *Men of Physics: Julius Robert Mayer, Prophet of Energy,* Pergamon Press, Oxford, 1973, pp. 68–74

ON THE FORCES OF INORGANIC NATURE

Julius Robert Mayer

THE purpose of the following article is to seek for an answer to the question: what are we to understand by *forces* (Kräfte) and how are these related to each other? Through the name *matter* we attribute to objects definite properties such as weight and volume. To the name *force* (Kraft) on the other hand there is joined the idea of the unknown, the inscrutable, the hypothetical. An attempt to make the idea of *force* as precise as that of *matter* and to denote thereby only objects of real inquiry, together with the attempt to understand the consequences flowing from this, might not be unwelcome to friends of a clear, hypothesis-free view of Nature.

Forces are causes. To them there is immediate application of the fundamental principle: *Causa aequat effectum* (the cause equals the effect). If the cause c has the effect e, then $c = e$. If e is in turn the cause of another effect f, $e = f$ and so on: $c = e = f = \ldots = e$. As is clear from the nature of an equation, in a causal chain of this kind, no member nor a part of a member can ever be zero. This property of all causes we call its indestructibility.

If a given cause c has brought about its equal effect e, then c has ceased to exist: c has become e. If after the production of e, the whole or part of c were left over, the part remaining could be the cause of other effects. The total effect of c would therefore exceed e which would contradict the hypothesis $c = e$. Since c is transformed into e, e into f, etc., it follows that we must consider these quantities as different manifestations of one and the same object. The ability to assume different forms is the second essential characteristic of all causes. In summarizing both properties, we say: causes are quantitatively indestructible and qualitatively transformable objects.

In Nature we find two classes of causes between which we learn from experience no transitions take place. The one class is made up of those causes which have the properties of ponderability (weight) and inpenetrability, namely what we commonly call *matter*. The other is made up of those causes which do not possess these properties—these are the forces, which from the designated negative properties are also called *imponderables.* Accordingly forces are indestructible, transformable and imponderable entities.

A cause which brings about the raising of a weight is a force. Its effect, the raised weight, is accordingly likewise a force. Expressed in more general terms this means that the spatial separation of ponderable objects is a force. Since this force brings about the fall of the object, we call it a *fall-force* (Fallkraft), or the force connected with falling. Fall-force and fall, or more generally fall-force and motion are forces which are related to each other as cause and effect. They are forces which can be transformed, one into another; they are two different manifestations, of the same entity. For example, a weight resting on the ground is no force; it is neither the cause of a motion nor the cause of the raising of another weight. It becomes such a cause, however, to the extent to which it is raised above the ground. The cause, the displacement of the weight from the earth and the effect, the amount of motion which is produced, stand in a constant relation to each other, as mechanics shows.

If in considering gravity as the cause of the fall of an object we speak of the force of gravity we confuse the concepts of force and property. Every property must dispense with precisely that which is attached to every force, namely the union of indestructibility and transformability. Between a property and a force (e.g., between gravity and motion) we cannot therefore set up the necessary equation for a correct causal relation: If we call gravity a force we have to think of it as a cause which produces an effect without itself decreasing in magnitude. It therefore involves an incorrect representation of the causal connection of things. In order that a body should be able to fall its elevation from the ground is no less necessary then its weight (gravity). One may not therefore ascribe the falling of the body to gravity alone.

It is the object of mechanics to develop the equations which exist between fall-force and motion, between motion and fall-force and between motions in general. Here we recall only one point. The magnitude of the fall-force (if we take the earth's radius as effectively infinite) is directly proportional to the mass m and the displacement d above the ground. That is, $v = md$. [In modern terminology if m is really the mass of the falling particle, we should write $v = gmd$, where g is the acceleration of gravity. It seems clear that Mayer used mass and weight interchangeably. If his m is weight, his formula is correct. In any case his fall-force v corresponds to potential energy for the case of free fall, in modern terminology—Ed. note.] Suppose with $d = 1$, the final velocity of the mass on reaching the ground is $c = 1$, then we can also express v as equal to mc. In general, however, $v = mc^2$ is the measure of the fall-force. The law of the conservation of *vis visa* is based on the general law of the indestructibility of causes. [Mayer seems to be saying here that we use the product of mass and velocity (mc) as a measure of fall-force only if the velocity c gained by a fall from height d were equal to unity for $d = 1$. He recognizes that $v = mc^2$ is a more valid representation for fall force, with $c = \sqrt{2gd}$, etc. He fails to stress the point that v cannot be measured by both momentum and *vis viva* since they are different dimensionally. Incidentally Mayer consistently uses *vis viva* in place of $mc^2/2$, the modern kinetic energy. He was still following Leibniz—Ed. note.]

In countless cases we see a motion cease without bringing about another motion or the raising of a weight. However, a force once in existence cannot become zero, but must reappear in another form. The question then arises: what other form can be taken on by the force which we designate as fall-force or motion? Experience alone can provide information on this point. In order to experiment profitably in this matter, we must choose instruments which while causing the motion to cease, are themselves changed but little by the objects under investigation. For example, if we rub two metal plates together, we make motion disappear and on the other hand observe the production of heat. The question then arises: is the motion the cause of the heat? In order to be sure about this relation, we must discuss the further question: in the countless cases in which the appearance of heat has been detected on the cessation of motion, has not the motion had some other effect than the appearance of heat and has not the heat another cause than the cessation of motion?

An attempt to demonstrate the results of the cessation of motion has never been seriously carried out. [Mayer was at this time presumably ignorant of the work J. P. Joule had been doing in England as well as that of Colding in Denmark—Ed. note.] Without desiring to dismiss in *a priori* fashion the possible hypothesis that may be set up, we yet call attention to the fact that the effect cannot in general be associated with a change in the state of aggregation of the bodies that are moved and rubbed together, etc. If we assume that a certain quantity of motion v is expended in the transformation of a rubbed material m into n, then $m+v = n$ or $n = m+v$. By the transformation of n back into m, v must reappear in some form. In the very long continued rubbing of two metal plates we can cause repeatedly the cessation of an enormous amount of motion. Can we really expect to find in the metal dust a trace of the force that has disappeared? We repeat that the motion cannot be reduced to *nothing*. Oppositely directed or positive and negative motions cannot become zero any more than oppositely directed motions can arise from nothing or a weight rise by itself alone.

No account can be given of the disappearance of motion without the recognition of a causal connection between motion and heat, any more than we can explain the existence of frictional heat without this same recognition. The production of frictional heat cannot be explained by the decrease in the volume of the bodies being rubbed together. It is well known that one can melt two pieces of ice by rubbing them together in a vacuum, but let anyone try to melt ice by the mere increase in pressure on it. [Some years after the publication of this paper, Sir William Thomson (later Lord Kelvin) showed that the melting point of ice is lowered by the application of pressure. The thermodynamic problem involved is more complicated than Mayer or his contemporaries could have visualized. In any case Kelvin's result applies only to substances which expand on freezing (like ice). For substances which contract on freezing (like bismuth) the effect of pressure is to *raise* the melting point—Ed. note.] Water when shaken violently experiences a rise in temperature as the author found. The heated water (at around 12 °C or 13 °C) experiences an increase in volume. Whence then comes the heat which can be produced in arbitrary amounts by repeated shaking of the water in the same apparatus? The thermal vibration hypothesis leans toward the principle that heat is the effect of motion but does not accept this causal connection at its full value, putting the principal stress on the vibrations themselves.

If it now develops that in many cases no other effect can be found for the vanishing motions save heat *(exceptio confirmat regulam)* [This is a rather strange interpolation. The Latin statement is usually put in the form: *exceptio probat regulam* (The exception tests the rule). Mayer's form may have meant to him that the exception calls attention to the necessity of investigating the rule more carefully!—Ed. note.] and for the heat that appears no other cause can be found than the motion, we prefer the assumption that heat originates from motion to the assumption of a cause without an effect or an effect without a cause, just as the chemist, instead of allowing the disappearance of hydrogen and oxygen and the appearance of water to go without inquiry, postulates a connection between hydrogen and oxygen on the one hand and water on the other.

We can visualize the naturally existing connection between fall-force, motion and heat in the following fashion. We know that heat makes its appearance if the individual particles of a body move closer together; compression produces heat. That which holds for the smallest particles and the smallest spaces separating them must clearly find an application to large masses and measurable spaces. The falling of a weight is a real decrease in the volume of the earth and must therefore stand in some relation to the heat produced. This heat must be exactly proportional to the magnitude of the weight and its original distance from the earth's surface. From this consideration we are led very simply to the relation connecting fall-force, heat and motion as mentioned above.

From the relation connecting fall-force and motion we have no right to draw the conclusion that the essential nature of fall-force *is* motion, nor have we any greater right to draw this conclusion for heat. We should rather take the opposite view that in order to become heat, motion (whether it be simple motion or vibrating motion like light or radiant heat) must cease to be motion.

If fall-force and motion are equivalent to heat, naturally heat must be equivalent to fall-force and motion. Just as heat results as an effect of volume decrease and the cessation of motion, so heat disappears as a cause on the appearance of its effects: motion, volume expansion, elevation of a weight.

In the mills operated by water wheels the motion that is produced and again disappears, resulting indeed from the volume decrease which the earth continually suffers through the fall of the water, produces in turn a significant quantity of heat. Conversely the steam engine serves to transform heat into motion and the raising of weights. The steam locomotive and its accompanying train may be compared to a distillation apparatus; the heat produced in the boiler is transformed into motion, which in turn is changed back into heat (at least in part) in the wheel axles.

We complete our thesis, which necessarily follows from the fundamental principle: *causa aequat effectum* and which stands in complete accord with all natural phenomena, with a practical conclusion. For the solution of the equations connecting fall-force and motion, magnitude of the fall must be determined by experiment for a definite period of time, e.g., for the first second of fall. Similarly, for the solution of the equation existing between fall-force and motion, on the one hand and heat on the other, we must ask ourselves the question: how great is the quantity of heat corresponding to a definite amount of fall-force and motion? For example, we must discover how high a definite weight must be raised above the surface of the earth in order that its fall-force shall be equivalent to the heating of an equal weight of water from 0 °C to 1 °C. That such a relation actually exists in nature can serve as the resumé of the considerations presented in this essay.

By the application of the principles developed here to the heat and volume relations of a gas, we find the decrease in height of a mercury column compressing a gas equivalent to the quantity of heat associated with the compression. If we put the ratio of the specific heat capacities of the gas at constant pressure and constant volume respectively equal to 1.421, it turns out that the fall of a weight from a height of about 365 meters corresponds to the heating of an equal mass of water from 0 °C to 1 °C. Comparing with this result the efficiency of our best steam engines, we see what a small part of the heat transferred to the boiler is really transformed into motion or the raising of a weight. This could justify the attempt to produce motion in other ways than through the use of the chemical reaction between carbon and oxygen. This might be by the transformation of the electricity obtained chemically directly into motion in an efficient fashion.

29B

Reprinted from R. B. Lindsay, *Men of Physics: Julius Robert Mayer, Prophet of Energy,* Pergamon Press, Oxford, 1973, pp. 76–99

THE MOTIONS OF ORGANISMS AND THEIR RELATION TO METABOLISM

Julius Robert Mayer

Introduction

In the course of the last century applied mathematics has attained such a high stage of development and its conclusions have acquired such a high degree of certainty that it has been justified in assuming the first rank among the sciences. It is the beginning and the end for the astronomer, the technologist and the navigator, it is the solid axis of all the natural philosophy of the present time. It is only in biology that the discoveries of Galileo, Newton and Mariotte have borne comparatively little fruit. No formulas have been found for the phenomena of life, for the letter killeth, the spirit alone giveth life!

In the study of motions produced organically the gulf between mathematical physics and physiology, which even the outstanding investigations of people like Schwann and Valentin have not been able to bridge, is vividly perceptible. Therefore the attempt to set up a method by which both sciences can be brought closer together with reference to the matter in question should not be without interest to physiologists.

It must indeed be considered a relapse into the mistakes of ancient natural philosophy or the confusion of modern science if there were in view an attempt to construct a universe *a priori*. When, however, there has been success in tying together countless natural phenomena and from them to deduce a fundamental law of nature one should not be reproached if after careful tests one uses this law as a compass to guide his path with greater assurance over the sea of details.

Proceeding from the laws of inorganic phenomena we take for granted on the one hand the results of mechanics as established truths, while on the other hand we do not feel obliged to accept all the concepts and classifications which mechanics has found it good to set up as binding on our considerations. Mechanics, so to speak, anatomizes or dissects the natural objects with which it deals by abstractions pushed as far as possible until they correspond to numbers in its mathematical analysis, and is content to be able to answer the questions which it raises with admirable sharpness and mathematical accuracy. Mechanics is troubled but little if through its way of looking at things phenomena which are closely associated in nature appear on the boundary of the mechanical domain to be widely separate. Mechanics is concerned just as little about the apparent coincidence, in its domain, of concepts and objects which in the real world have nothing in common.

The concepts which mechanics has constructed for its purposes have been pushed further by other sciences than mechanics itself could tolerate. Suppose the question arises, what is to be understood by a "body"? The geometer will answer: "Without prejudice to the physicist, zoologist, psychologist, etc., a body according to geometrical concepts is a space bounded in three dimensions." The expert in mechanics, who represents the origin, changes in and cessation of motion as brought about by a pressure calls this, *in abstracto*, "force" *(Kraft)*. The ability of a mass to exercise such a pressure, i.e., gravity or weight, he calls a force. However, without sticking to the abstraction force = pressure of the mechanist, other scientists have tended to treat weight as the general type of all forces and thereby have introduced an artificial confusion of the concepts: property, force, cause and effect. This has proved to be a serious handicap in the building of the tower of knowledge.

Before we begin an investigation of physiological laws we may be permitted to make intelligible the concept of force *(Kraft)* and to represent the important inorganic phenomena in their natural connections with one another.

In the composition of the inorganic part of this work the author has taken considerable trouble to set forth the relevant mechanical and physical problems in a generally intelligible way. Should nevertheless individual points arise for the understanding of which a more exact acquaintance with the theorems of mechanics is required, in the nature of the case these could not very well be avoided.

It is to be hoped that physicists for whom the calculus is a tool in their investigations and not an end in itself will not deny an earnest examination to this part of the author's work.

If a mass initially at rest is to be put in motion the expenditure of force is necessary. A motion never arises by itself: it arises from its *cause*, namely force.

Ex nihilo nil fit

We call force an entity which through its expenditure brings about motion. Force as a cause of motion is an indestructible entity. No effect arises without a cause. No cause disappears without a corresponding effect.

Ex nihilo nil fit. Nil fit ad nihilum

The effect is equal to the cause. The effect of force is once again force. The quantitative unchangeability (invariance) of the given is one of the fundamental laws of nature which applies equally to force and matter.

Chemistry teaches us to recognize the qualitative changes which given matter undergoes under different conditions. It provides in every individual case the proof that in chemical processes only the *form* and not the *magnitude* of the given matter is changed.*

What chemistry performs with respect to matter, physics has to perform in the case of force. The only mission of physics is to become acquainted with force in its various forms and to investigate the conditions governing its changes. The creation or destruction of a force, if they have any meaning, lies outside the domain of human thought and action.

Whether in the future it will prove possible to transmute the many chemical elements into one another, or to reduce them to a few simpler elements or even to a single fundamental substance, is more than doubtful. The same situation, however, does not hold for the causes of motion. It can be proved *a priori* and confirmed everywhere by experience that the various forms of forces *can* be transformed into one another.

* A piece of matter A suffers through the addition of another piece of matter B a change in size. Since B as well as A must be considered as given and the sum $A+B$ of their parts taken together is equal, it is clear that the given as a whole suffers no change in size by the composition or separation of the parts.

In truth there exists only a *single* force. In never-ending exchange this circles through all dead as well as living nature. In the latter as well as the former nothing happens without form variation of force! [For *force* here and throughout read *energy* in the modern notation. In the rest of the translation we make this change—Ed. note.]

1

Motion is a form of energy. In the enumeration of forms of energy it merits the first place. Heat warms, motion moves.

When a moving mass meets one at rest, the latter is set in motion whereas the first loses some motion. If in billiards the white ball collides squarely with the red one, the white one loses its velocity and the red one moves on with the velocity the white one has lost. It is the motion of the white ball which when expended brings about the motion of the red one or we may say is transformed into the latter. The motion of the white ball is a form of energy. The motion of the red one is an effect which is equal to cause; it is also a form of energy.

A billiard ball can by collision set many other balls in motion and still remain in motion itself. The magnitude of the *vis viva* (kinetic energy) of the whole system, however, stays the same before and after the collision.

2

A mass at rest at any arbitrary distance above the earth's surface and then released will immediately set itself in motion and will reach the ground with a velocity which is readily calculable. The motion of this mass cannot arise without the expenditure of energy. What is this latter energy?

If we restrict ourselves not to traditional assumptions but to the simple facts of experience, we readily become aware that it is the raising of the weight which is the cause of the motion of the weight. For example, a pound weight [500 grams as Mayer took it] was at rest 15 feet [1 foot = 32.484 centimeters in Mayer's units] above the ground. In falling freely to the ground, the final velocity is 30 feet per second [$v = \sqrt{2\,gh}$, etc.]. The raising of the weight was expended, but the motion of the weight was brought into existence.

Hence raising the weight is the cause of motion and is a form of energy. This energy causes the fall motion. We call it *Fall kraft* [fall energy or in modern terminology potential energy].

If a mass moves along a horizontal plane with a certain velocity it keeps this velocity constant due to the law of inertia, as we are accustomed to say. The same mass, however, with the same initial velocity, if it begins to move vertically upwards loses its motion completely in a few seconds. Suppose a mass of 1 pound starts to move up with a velocity of 30 feet per second. After 1 second the motion has ceased, and the 1 pound mass has been lifted to 15 feet above the ground. The energy which has raised this load is its motion; what was previously effect is now the cause, what was cause has now become effect. Fall energy has been transformed into motion and motion in turn transformed into fall energy.

The magnitude of the fall energy is measured by the product of weight and height (mgh, if g = acceleration of gravity). The magnitude of the motion as energy (motional energy) is given by one-half the product of the mass and the square of the velocity. [Here the author presents some mechanical formulas relating to motion of gravitating masses which are not included here—Ed. note.] Both forms of energy are represented by the collective name of mechanical effect.

If fall energy is transformed into motion or vice versa, the total mechanical effect maintains a constant value. This law, a special case of the axiom of the indestructibility of energy is known in mechanics as the principle of the conservation of *vis viva* [This is bad terminology, as in most cases the *vis viva* actually changes, even when the total mechanical energy = *vis viva*+potential energy stays constant—Ed. note.] As examples consider free fall from any height, fall along prescribed paths, pendulum oscillations, motions of the heavenly bodies.

3

For a thousand years or more, the human race was almost exclusively restricted to the ever-recurring problem of setting resting masses in motion by means of the tools of inorganic nature, in particular the application of given mechanical effects. It was reserved for a later time to add a new type of energy to the energy forms of the old world, i.e., those of streaming wind and flowing water. This third form of energy which our century gazes on with wonder is *heat*.

Heat is a form of energy. It can be transformed into mechanical energy.*

Let us suppose that a wagon train having a mass of 100,000 pounds is given a velocity of 30 feet per second. By the expenditure of an appropriate amount of energy this can be achieved. For example, the wagon train can gain this velocity by rolling down a suitable inclined plane. As a rule, however, the train will be set in motion without the expenditure of "fall energy" (potential energy) and in spite of friction, etc., will maintain this motion. When a rise in elevation of the path of 1 part in 150 is assumed (as equivalent to the friction) then a velocity of 30 feet per second will be enough to raise the train load 720 feet high in 1 hour, which corresponds to an expenditure of 45 horsepower. This enormous quantity of motion originally produced assumes an equally great quantity of expended energy of some kind. The effective energy in the case of a locomotive pulling a train is *heat*.

The expenditure of heat or the transformation of heat into motion rests on the fact that the quantity of heat which is taken up by the steam is continually greater than that which is given up when the steam is exhausted and condensed in the surroundings. The difference is the heat transformed into mechanical activity (work).

Equal quantities of combustible material under the same conditions give equal quantities of heat. However, the coal burned under the boiler provides less free heat when the engine is working then when it is not. The free heat distributes itself to the surroundings and hence is lost for mechanical purposes. The more perfect (efficient?) the apparatus, by

* If here a transformation of heat into mechanical effects is laid down as something that takes place it is stated only as a fact, and in no way is the thing given an explanation. Thus a given amount of ice is transformed into a corresponding amount of water: This fact is simply so and remains independent of the unfruitful speculations on how and why. Real science remains satisfied with positive knowledge and freely leaves to poets and philosophers the solution of such everlasting riddles with the help of fantasy.

so much the less will heat be transferred to the surroundings. The best engines give an efficiency of about 5 per cent. One hundred pounds of hard coal in such an engine provide no greater quantity of free heat than 95 pounds of coal, burning without doing any work.

For the establishment of this important law we must examine the behavior of elastic fluids with respect to heat and mechanical action.

Gay-Lussac has proved by experiment that an elastic fluid (gas) which streams out of a vessel into an equally large evacuated container suffers in the former vessel just as much cooling as the latter vessel warms up. This investigation of outstanding simplicity, which has been confirmed by other observers, shows that a given weight and volume of an elastic fluid can expand two-fold, four-fold or to a volume of any size without experiencing *on the whole* any temperature change. This means that for the expansion of the gas in and for itself no heat expenditure is necessary. At the same time experiment confirms that when a gas expands against pressure it suffers a drop in temperature.

Let us assume that 1 cubic inch of air at 0 °C and a pressure of 27 inches of mercury [standard conditions] is heated by a quantity of heat x at constant volume to 274 °C. When this gas is allowed to expand into an evacuated space of the same volume it will still retain the temperature 274 °C and a medium surrounding the vessels containing the gas will during the expansion experience no change in temperature. Now, however, consider the other case in which 1 cubic inch of air is heated from 0 °C to 274 °C not at constant volume but at constant pressure (namely 27 inches of mercury). In this case a larger quantity of heat is required. Represent this as $x+y$.

In both cases above the air is heated from 0 °C to 274° and in both cases the air expanded from one volume to twice the volume.

In the first case the quantity of heat required was x. In the second it was $x+y$. In the first case the mechanical effect produced was zero, but in the second it was the equivalent of raising 15 pounds 1 inch.

If the air is cooled under the same circumstances under which it was heated, an amount of heat is given back equal to that which was taken up. The given amount of air if it is cooled from 274 °C to 0 °C without the simultaneous expenditure of mechanical work (or with pressure absent) will accordingly give back the quantity of heat = x. However, in cooling under constant pressure with the expenditure of potential energy equivalent to that needed to raise 15 pounds 1 inch, the air will give back the quantity of heat $x+y$.

The steam in the engine when it expands behaves like the air at constant pressure. The quantity of heat needed for the heating and expansion of the steam is $x+y$. In the cooling process the steam experiences no particular pressure and hence the cooling takes place without (or with very small) expenditure of mechanical work. It gives back the heat quantity x. Hence there is associated with every cycle of the piston in the cylinder of the engine a heat loss equal to y. Thus the operation of the engine is inseparately connected with a consumption of heat.*

The quantity of heat which must be expended to produce a definite amount of mechanical work must be evaluated experimentally.

The total expenditure of heat can be calculated from the quantity of combustible material burned in the engine. When the inevitable losses of heat through radiation, conduction and convection are deducted from the above, there remains the heat really available for transformation and this corresponds to the actual performance of the engine. Since, however, by far the greater part of the unused and dissipated heat can be only approximately estimated, even a partially reliable result is hardly to be expected along these lines.

The problem can be solved more simply and precisely by calculation of the quantity of heat which becomes latent† if a gas expands under pressure. If the heat taken up by the gas in heating it by t °C at

* The periodic rise and fall of the engine beam would in and of itself not bring about a continuous expenditure of energy or a consumption of heat. Once set in motion the balance beam of the engine would move by itself [barring friction, etc.]. Like a pendulum in vibration it moves without net energy expenditure.

† The concepts of heat becoming latent and free are equivalent to those of expenditure and production respectively. We can say that motion becomes *latent* when an object moves up from the earth and the motion slows down. It becomes free when the motion is downward. Heat may be thought of as latent motion, just as motion may be thought of as latent heat.

constant volume is x, the heat needed to heat the gas through the same temperature range at constant pressure will be $x+y$. If in the latter case the weight raised is P, then $y = Ph$.

One cubic centimeter of atmospheric air at 0 °C and 0.76 meters barometric pressure weighs (has a mass of) 0.0013 gram. If it is heated through 1 °C the air expands by 1/274 part of its volume and at the same time raises a column of mercury of 1 square centimeter cross-section and 76 centimeters high by 1/274 meter. The weight of this column is 1033 grams. The specific heat of air (that of water taken as unity), from the work of Delaroche and Berard, is 0.267. The quantity of heat which a cubic centimeter of air takes up in order to go from 0 °C to 1 °C at constant pressure is accordingly equal to the heat by which (0.0013)(0.267) = 0.000347 gram of water would have its temperature raised by 1 °C. According to Dulong, whom most physicists follow, the quantity of heat which air takes up to heat itself by 1 °C at constant volume to that for constant pressure is in the ratio 1 : 1.421. If we use this we calculate the heat needed to heat 1 cubic centimeter of air by 1 °C at constant volume as 0.00037/1.421 = 0.000244. [There is an obvious misprint here. Mayer has 1.41 in place of 1.421. His result, however, agrees with the choice of 1.421. Ironically 1.41 is in better agreement with modern values of the ratio of the specific heats—Ed. note.]

The difference $(x+y)-x = y$ is therefore $0.000347-0.000244 = 0.000103$ units of heat [Mayer uses degree (°) of heat. Actually he is using equivalent calories—Ed. note.] By the expenditure of this, 1033 grams of mercury is lifted 1/274 centimeters. Hence 1 unit of heat [1 calorie] is equivalent to 1 gram raised 367 meters [or an energy of 3.59 joules—Ed. note.]*

The same result will be obtained if in place of atmospheric air one

* Mayer inserts a footnote, evidently added in the second edition of his paper, that a more accurate experimental value of the ratio of the specific heat provides a value of the mechanical equivalent more nearly in agreement with Joule's value. He says he wants to leave his original value in the main text, though in subsequent applications, in the revised version of his paper, he uses effectively 4.165 joules per calorie (though not in these units, of course).

takes and applies a similar calculation to any other simple or complex gas.*

The law: "heat equals mechanical energy" is independent of the nature of the elastic fluid in question. The latter acts only as a vehicle for the transformation of the one form of energy into the other. Among the differing (but not too widely separated) data on the heats of combustion of carbon, those of Liebig are probably most reliable. From the direct experimental observations of Dulong which were published by Arago after Dulong's death, Liebig calculated that the quantity of heat developed in the burning of 1 gram of carbon to CO_2 is 8558 calories. [Mayer does not use the term calorie, but this is what he means—Ed. note.] (*Annalen der Chemie von Wöhler und Liebig*, vol. 53, p. 73).

In the combustion of 1 gram of carbon, therefore, work equivalent to the raising of 3.6×10^6 grams of carbon though 1 meter is involved. This result would be achieved if all heat losses were avoided. But we have no more chance of changing a given amount of heat *all* into work in one operation than we have of transforming chlorine, hydrogen

* In other words this means: if the heat capacity of atmospheric air under constant pressure is taken to be equal to *unity*, and if the heat capacity of any other gas under constant pressure is S, and if K is the ratio of the specific heats for this other gas, under the assumption of the same coefficient of expansion [all ideal gases], we must have

$$S\left(\frac{K-1}{K}\right) = \frac{0.421}{1.421}$$

[which is clearly satisfied for $K = 1.421$ and $S = 1$]. This result agrees with the experimental work of Dulong. For CO_2 he gets $S = 1.175$ and $K = 1.388$, so that we should have

$$(1.175)\frac{(0.338)}{1.338} = \frac{0.421}{1.421}$$

and we do. Similarly for olefiant gas Dulong has $S = 1.531$ and $K = 1.240$, so that

$$(1.531)\frac{(0.240)}{1.240} = \frac{0.421}{1.421}$$

checking again. Also Dulong's famous law that all elastic fluids when they are compressed by the same fractional change in volume liberate the same amounts of heat is one of the necessary consequences of the general law: Heat equals mechanical work.

and a metal into a metallic chloride salt without the formation of other products.

It is a technological problem to minimize as much as possible the unwanted effect of combustion, that is the liberation of heat into space (without doing useful work). In the first engines of Watt, according to John Taylor, the quantity of coal burned for a given amount of work done was seventeen times greater than in the engines developed in 1828.

At that time even under the most favorable circumstances engines working with 1 pound of anthracite could lift about a half million pounds 1 foot. While the maximum efficiency attainable was about 5 to 6 per cent, many engines, locomotives in particular, scarcely reached an efficiency of 1 per cent.

Guns achieve a better performance. Let us figure that a 24 pound shot can be given a muzzle velocity of 1500 feet per second by means of the burning of 8 pounds of powder containing 1 pound of carbon. In this case the mechanical efficiency attains a value of 9 per cent. However, it is well known that a gun loaded with ball is heated less with the same charge of powder then the same gun firing a blank.

If we assume that the whole earth's crust could be raised on suitably placed pillars around its surface, the raising of this immeasurable load would require the transformation of an enormous amount of heat.

Since it is clear that such a volume increase in the earth is connected with a corresponding quantity of heat becoming "latent", it is likewise clear that in a volume decrease of the earth a corresponding quantity of heat will be set free. But whatever holds for the earth's crust as a whole must also apply to every fraction thereof. In the raising of the smallest weight, heat (or some equivalent form of energy) must become latent; and by the falling of this weight to the earth's surface, the same quantity of heat must be set free.

It has already been seen that in the raising of 1 kilogram through 425 meters, a unit of heat is necessary. This is equivalent to saying that the raising of one gram through the same height demands 1 calorie of heat. We can also say that a kilogram which drops through 425 meters through collision or friction must develop 1 large calorie of heat (1000 ordinary calories). [Mayer's notation here has been converted into the modern form. He is not always careful to dinstinguish between the ordinary calorie and the kilogram calorie—Ed. note.]

Physicists were prevented by their traditional presuppositions on the energy of motion and motion itself from grasping the above evident fact which is fully confirmed by experience. Newton in his *Principia* (Definition VIII) states expressly that gravity is a *causa mathematica* (mathematical cause) and cautions against treating it as a *causa physica* (physical cause)*. This important distinction was, however, neglected by Newton's successors. Gravity or the cause of acceleration was taken as the cause of motion itself, and thereby the occurrence of motion without expenditure of energy was ordained, in so far as in the falling of a weight, no gravity was expended. As a necessary consequence of its method of origin given motion was allowed under certain circumstances to eventually become nothing.† [Mayer is continually disturbed by the current use of gravity as a force. It is a force in the Newtonian sense, but not in Mayer's sense (i.e. as energy)—Ed. note.]

We find here accordingly two contradictory points of view. Either a given motion by its disappearance is supposed to go to zero, or it will have an indestructible effect equal to itself. If we unconditionally decide for the latter we are appealing to the laws of thought as well as to experience.

If we draw from a reservoir, a lake or even from the ocean a glass of water, we will not be able to detect the corresponding decrease in the large amount of water in question. If, however, one is willing to grant that these bodies of water have actually suffered no loss of substance at all by the withdrawal of a few ounces of water, then necessarily the

* The *causa mathematica* of Newton, in particular the force of gravity, is the cause or measure of acceleration. If v is the force and c the velocity, then $v = dc/dt$ [He leaves out m; strictly $v = mdc/dt$ — Ed. note]. But in the use of force in the energy sense, the *causa physica* is the measure of motion. If the *Kraft* [he means energy] is v, we have $v = mc^2$ [Strictly kinetic energy $\frac{1}{2}mc^2$ — Ed. note].

† The mathematical expression for this second paradox is the so-called Cartesian measure of force, namely quantity of motion or momentum.

conclusion follows that these ounces have been created out of nothing and when given back to the sea revert to nothing.

The same conclusion holds for the forces. We accordingly ask: is the "moving force" which gives a weight falling to the earth from a position 30 feet above its surface a velocity of 30 feet per second, a constant? People are accustomed to answer this as follows: the increase and decrease in gravitational force over such a small distance may be completely neglected and hence "Yes" is the answer to the question. We say "No". If the force were constant, it would have, if acting long enough, to produce an arbitrarily great motion. But this does not happen. The velocity which a weight falling to the earth can attain has a maximum value. It amounts to 11,200 meters per second. For this is the value of the velocity which would be acquired by a weight of mass m falling to the earth of mass T from an infinite distance. The total "fall-force" *(vis viva)* acquired in gaining this maximum velocity G is mG^2 [It should be the kinetic energy $\frac{1}{2}mG^2$, but Mayer always uses *vis viva*—Ed. note.] The "fall-force" associated with a smaller separation distance can be readily calculated as a fraction of this maximum. For heights in the neighborhood of the earth* the numerator of this fraction is the distance of fall, while the denominator is the radius of the earth. In a fall from a height of 15 feet the *vis viva* gained is therefore mG^2 (15/19,600,050) or the velocity attained on reaching the ground is $G\sqrt{15/19\,609\,050}$. If the weight falls from infinite distance to a height of 15 feet above the earth's surface then 1,299,999/1,300,000 of the total *vis viva* has been gained, 1/1,300,000 of this still remains left and it is with the expenditure of this comparatively small amount of *vis viva* that the mass m gains its velocity of 30 feet per second in falling. It is clear that the motion of a falling body provides no exception to the axiomatic rule of the proportionality of motion and energy expenditure. The *vis viva* expenditure is zero only when the weight

* In general the numerator is the height from which fall takes place. The denominator is, however, the product of the original distance of the center of mass of the two bodies (earth and falling body) with the distance left over, treating the earth's radius as unity. A weight which begins to fall from height h has at height h' the velocity $G\sqrt{(h-h')/h\cdot h'}$.

merely presses down but does not move. A constant energy which brings about effects without changing (decreasing) does not exist in physics.

Experience shows on all sides the transformation of mechanical effects into heat. The decisive facts here, the development of heat by collision and friction, are old and have been known for a long time. Are they, however, for this reason less compelling? We observe the heating of the great mill stones and of the flour in the grist mill, the heating of the great driver and the linseed oil in the oil mill, of the wood in the dye mills, the never-ending heating of the axles of all wheels in motion. We remember the famous experiments of Rumford! Everywhere we note the same phenomenon: endless heat production with the expenditure of mechanical activity. [As is well known, Joule and Colding actually used the heat developed by friction to measure the mechanical equivalent of heat—Ed. note.]

The author made some observations on four pulp cylinders in a paper factory. In each cylinder there were about 80 pounds of paper pulp and 1200 pounds of water. The temperature of the pulp went up steadily from the beginning of the motion. The surroundings were at temperature 15 °C: in 32 to 40 minutes the temperature of the pulp went up from 14 °C to 16 °C. The highest temperature observed after several hours of processing was 30 °C. If we calculate that by the exertion of a horsepower in 1 minute 2700 pounds are raised 1 foot, the heating of 1280 pounds of water (not counting that of the apparatus) by 1 °C in 16 minutes is the equivalent of 3.16 horsepower. This agrees sufficiently well with the approximate estimate of the engineers that in the running of a pulp vat about 5 horsepower is needed. Does the mechanical effect of the 5 horsepower in the machinery go to zero? The actual fact is that it is transformed into heat.

The most important physical laws which relate to the transformation of motion into heat can be briefly summarized as follows:

1. Negative motion like negative matter is an imaginary quantity. The destruction of positive motion by negative is a paradox.

2. Just as the quantity of matter is measured by absolute weight [He really means mass—Ed. note.] so the measure of motion is the product

of the mass by the square of the velocity [he really means $\frac{1}{2}mv^2$ or the kinetic energy]. The Cartesian law: force equals the product of mass by velocity is false and was shown to be so by Leibniz.

3. Just as portions of matter of opposite qualities, such as an electropositive base and an electronegative acid, can neutralize each other, so motions in opposite directions can neutralize each other. The continuing entity changed in quality but unchanged in quantity is the neutral salt in the first illustration and heat in the second.

4. The relation in which the quantities of neutralizing matter or motion stand with respect to each other is as a rule not one of equality. This depends rather on the nature of these objects. Acids and bases neutralize each other when the quantities are proportional to their combining weights; motions in opposite directions neutralize each other when the quantities are proportional to their velocities. In this neutralization and in the production of motions the velocity plays the role of combining weight. In mechanics this law is introduced under the name of "principle of virtual velocities" [or virtual work].

4

A fourth way in which physical energy can appear is electricity. Frictional electricity is produced by the application of mechanical action.

We have before us an electrophorus of ideal perfection. The cover has the weight P. The cover is at height h above the region of influence of the under-disk. By the use of an appropriate balancing weight the cover can move up and down freely above the under-disk. Then whether the disk is electrified or not the top cover can move up and down without loss of mechanical energy as long as no electric charge is withdrawn from the under-disk (or base). The situation is otherwise when the electrophorus *works*. If the cover is lifted up from the non-electrified base, the work done against the balancing weight P is Ph. If the base is electrified, the cover is attracted and the corresponding work is greater than Ph, say $Ph+p$. The cover is able when lying on the underdisk to exercise an electrical effect. Let us suppose this has happened, the energy corresponding to it is definitely determined. Let us call it z. Now the attraction is increased even more and to raise the cover requires even greater weight than before. The work of the resultant weight will be greater than $Ph+p$. Let us call it $Ph+p+x$. If the cover is raised to h again, we get a second electrical energy which we may call z', etc. In every descent the work that is won is $Ph+p$, while in every rise to h, the work lost is $Ph+p+x$. Thus while on each trial we apply or expend a mechanical energy x, we win the electrical energy $z+z'$. Hence we must have

$$x = z+z'.$$

The conclusion is simple. Out of nothing we get nothing. The resinous electricity cannot by itself have produced the continual production of electrical energy, since it remains undiminished in amount. The mechanical energy which disappears in each repetition of the charging process cannot have vanished to zero. What then remains to say, if one does not wish to get involved in a double paradox, except that the mechanical energy has been transformed into electricity?* The base of the electrophorus, like the lever or the retort, is nothing more than an

* From the closer inspection of the quantity x in the equation $x = z+z'$, it develops that the quantity of electrical energy is proportional to the square of the electrical potential (Spannung). *Proof:* Let the given negative electrical potential of the hard rubber disc of the electrophorus be denoted by S, the potential of the positive electricity of the cover plate $= S/q$, where q is a constant, which approaches unity. The magnitude of the attraction or repulsion of two electrically charged bodies is, according to the direct measurements of Coulomb, equal to the product of their potentials. The attraction between the positively charged cover plate and the negatively charged hard rubber base is accordingly S^2/q, or proportional to the square of the potential of the base. Since now the attraction between cover and base is at every distance of separation inversely proportional to the square of the distance the mechanical energy $p+x$ which must be expended to separate cover and base to a distance h must be proportional to S^2. We now see that the ratio p/x as well as the ratio z/z' is independent of S and accordingly $p+x$ as well as z and $z+z'$ as well as z' are proportional to the square of S or the square of S/q. Since, however, $(S/q)^2$ is equal to the square of the positive potential of the cover when raised to height h, while z' is the electrical energy obtained by the cover when it is raised to height h, the electrical energy varies as the square of the potential and the magnitude of the total electrical energy is equal to the product of the surface by the square of the potential.

instrument which the experimenter uses to bring about a metamorphosis.

A pendulum set in vibration will continue to move back and forth without loss of amplitude if friction and air resistance can be neglected. However, if a metal pendulum bob (insulated electrically from its surroundings) is permitted to swing near an electrified non-conductor and if periodically sparks are drawn from it while it is in the region of influence of the charged body, the amplitude will be observed to grow smaller: the mechanical energy of the pendulum motion will be successively transformed to electrical energy.

The production of frictional electricity also takes place with the expenditure of mechanical energy. The materials rubbed together are held fast to each other by the attraction of opposite charges. But the necessary separation to produce free electric charges cannot take place without the expenditure of mechanical energy. It is well known that in the production of frictional electricity there is no frictional heat developed. [A doubtful statement—Ed. note.] In the transmission of electricity the attraction relations just discussed are reversed and by the expenditure of electrical energy mechanical energy is produced. In every contact a part of the electricity is neutralized, more or less as motion is in inelastic collisions. The most important laws associated herewith are the same as those mentioned above in connection with the neutralization of motion. Mass and velocity in motion become replaced by surface area and potential respectively in the case of electricity. The dynamics of motion can be reduced to the consideration of two forms of energy, namely motion and heat, and is therefore simpler in its representation than electricity, which must consider three energy forms, motion, heat and specific electrical energy.

In analogy with the production of electricity, magnetism can also be produced by the expenditure of mechanical energy. A given magnet here plays the role of the electrophorus. Through the magnetization of a previously unmagnetized steel bar the same attraction relations ensue as those considered above in the case of the electrophorus. The result is similar: expenditure of mechanical energy leads to the production of electric and/or magnetic potential.

5

The spatial separation of a mass from the earth we have learned above to consider as a form of energy [potential energy]. A gram of mass at infinite* distance from the earth or, as we shall choose to say, at infinite mechanical separation from the earth, represents a form of energy. By the expenditure of this energy, that is through the mechanical joining of the two masses, another form of energy is produced: the motion of 1 gram of mass with the velocity of 34,450 feet per second. By the expenditure of this energy of motion 1 gram of water could be raised 14,987 °C. [This is unrealistic. What Mayer means is that by the expenditure of the above amount of mechanical energy 14,987 calories of heat can be produced—Ed. note.]

Experience now teaches us that the same energy effect can be gained from a chemical combination of certain materials as from a mechanical joining, that is to say, the development of heat. The presence of chemically different substances or rather the chemical differences of various portions of matter constitute a source of energy.

The chemical combination of 1 gram of carbon and 2.6 grams of oxygen is approximately equivalent in order of magnitude to the mechanical joining of a particle of mass 0.5 gram and the earth. In the former case 8500 calories of heat are produced and in the latter case

* The concept of infinite separation is to be taken here in its physical and not in its mathematical sense. It is to be understood in terms of the physical limit of the earth's gravitational field. Such a limit is not mathematically realizable (save at infinity). But physically we may introduce it with the same right with which we introduced a limit to the electrical atmosphere surrounding a charged conductor. For example, if we call a distance of 10,000 earth diameters an effectively infinite distance, this is sufficiently accurate for practical purposes. The mathematical representation of gravitational phenomena must begin with the consideration of either a very large or very small space and must go from these to concrete measurable spaces by slight increases or slight decreases. Neither of these methods can completely replace the other. From the physical standpoint, however, the method which begins with the infinite magnitude has the decided advantage that the true nature of the attraction, that is its decreasing intensity with distance, has already been taken into consideration. It is only in this way that the connection, the inner unity of mechanical, electrical and chemical processes can be completely clarified. The other method begins with the destruction of the physical concept of attraction, since it considers the attraction at infinitely small distances as effectively constant.

7400 calories. The chemical combination of 1 gram of hydrogen with 8 grams of oxygen (it being assumed with Dulong that the heat of combustion of hydrogen is 34,743 calories per gram) is in order of magnitude equivalent to the mechanical combination of a mass of 2 grams with the earth. In the former case the heat developed is about 34,700 calories. In the latter case it is about 30,000 calories.

For relatively small separations and low velocities the energy associated with mechanical effects falls far below that connected with the better known chemical combinations. The situation is otherwise, however, if we look beyond our immediate surroundings into space.

Of all terrestrial substances the detonating mixture of oxygen and hydrogen provides the greatest amount of heat energy when the combination takes place. When 1 gram of the mixture explodes to water, 3850 calories of heat are produced. The combustion of 1 gram of a carbon and oxygen mixture yields 2370 calories. Since, however, 15,000 calories of heat must be expended to separate a mass of 1 gram completely from the earth, it follows that on the earth no chemical difference exists through whose expenditure as much heat can be obtained as by the joining of a mass particle with the earth from a great distance. On the other hand, on the moon only 777 calories of heat correspond to the separation of 1 gram of mass from the moon.

The earth moves in its orbit about the sun with an average velocity of 93,700 feet per second [The German foot used by Mayer is somewhat larger than the English foot—Ed. note.]. In order to produce this motion by the combustion of carbon, an amount of carbon 13 times the mass of the earth would have to be burned. The quantity of heat connected with this would be in calories equal to 110,000 times the mass of the earth in grams. A small fraction of the kinetic energy of the earth's motion in its orbit would be sufficient to disintegrate the earth into the particles composing it. If we assume, however, that a particle equal in mass to the earth were resting on the surface of the sun, in order to remove this load and put it into motion in the present orbit of the earth with the earth's orbital velocity (the average distance of the earth from the sun = 215 times the radius of the sun) it would be necessary to supply 429 times greater energy expenditure or to burn a mass of carbon equal to 5557 times the mass of the earth.

Since the energy of chemical reactions seems to be insufficient to bring about the above-mentioned effects, it may well be asked how we can conceive of an energy expenditure which was sufficient to bring about the planetary motions in the first place? Let us assume that in "the beginning" the earth was at rest at distance from the center of the sun equal to 430 times the radius of the sun and that from here it fell the equivalent of 215 radii of the sun into its present orbit. It would then have attained its present motion. We can make similar statements about all the other planets. The major axes of their orbits provide a measure for the initial distances of the planets at rest from the sun. The major axes are the expression for the magnitude of the mechanical energy of each planet given to it by the creator. They stand as firm in this respect as all past time.

If we ask why the planets which we have assumed to be originally at rest did not fall directly onto the sun's surface and why the planets with almost invariably small eccentricity move about the sun in the same plane and in the same direction, it is clear that one hypothesis after another must be the answer, for:

"That is the curse of a wicked deed,
That it must ever more bring forth more evil."

It is worthy of note that there are compounds whose decomposition takes place with the development of heat and mechanical energy. Such compounds do not originate by themselves alone, but come about only in connection with chemical processes which are accompanied by a release of heat. We must assume that the heat which arises in the one compound, in chemical phraseology, in the nascent state, partially goes into the detonating compound. If an equal mass of chlorine gas combines on the one hand with a solution of ammonium chloride and on the other hand with a solution of ammonia itself, the heat developed in the second case, as the author found, is much greater than in the first. The reason for this must be sought in part in the fact that in the formation of nitrogen chloride heat becomes "latent" and this latent heat is later liberated as free heat and mechanical energy in the ultimate decomposition [by explosion].

The decomposition into chlorine and hydrogen as well as the combination of chlorine and nitrogen to form a compound both correspond to liberation of *energy*. Let us compare with these certain mechanical relations. A raised weight represents an expenditure of energy. A dropped weight which by the expenditure of a further mechanical effect compresses a stiff spring on which it falls and is then propelled upwards again by the expansion of the spring, represents in its deepest significance the expenditure of energy. [It is assumed that external resistance or damping is absent—Ed. note.] However, we find nothing analogous to the action of the spring and the damping in chemical combinations, for the elasticity of chlorine and nitrogen after they have become freed of combination cannot be compared with the compression of the spring, in so far as this elasticity is a consequence and not a cause of the liberated energy in the decomposition.

If one attaches plates of zinc and copper to insulating handles and then brings them into contact and immediately thereafter separates them, the zinc plate will be found to be at a + potential and the copper plate at a — potential. Before contact the metals were neutral (non-electrified), after contact they are charged oppositely. In order to bring about their separation, the expenditure of mechanical energy is necessary, just as in the somewhat analogous case of the electrophorus, previously discussed. Other circumstances prevail if the plates remain in contact. In place of mechanical energy, chemical energy enters. With the expenditure of the chemical separation of metal and oxygen there arises a whole summation of effects, as we have already considered in detail. By means of the lever we can transform a given fall-force [*Fall kraft*] into another. We sacrifice a given spatial displacement in order to bring about another spatial displacement. The wonderful lever of the chemists is the voltaic pile. Reduction phenomena and the development of heat and mechanical energy, which we see arising as effects of the pile, owe their origin to their expenditure of a form of energy, to the given displacement of metal and oxygen, of salt and acid. The equivalence of cause and effect is brought out most effectively by the gas apparatus of Grove.

If we now combine the results of all these investigations into a single general law, we once more obtain the axiom originally set up. This is:

In all physical and chemical processes the energy involved remains constant.

The following scheme provides a summary of the principal forms of energy already considered.

I. Potential energy (due to gravity) (fall-force)
II. Energy of motion
  A. Simple
  B. Vibrational
III. Heat
IV. Magnetism
  Electricity (galvanic current)
V. Chemical separation of certain materials } Chemical
  Chemical combination of certain other materials } energy

On this formulation of the five principal forms of physical energy there follows the task to demonstrate the metamorphoses (changes) in these forms by means of 25 experimental examples. From the most important and simplest facts we assemble here the following:

1. The transformation of one fall-force into another by means of the lever.

2. The transformation of fall-force (potential energy) into motional energy, either by free fall or by falling along a prescribed path.

3. The transformation of one motion into a second motion. This can take place completely through the central collision of elastic particles of the same mass or incompletely through collision and friction.

4. The transformation of energy of motion into fall-force (potential energy) through the motion of a particle upwards from the earth's

surface. Such a transformation of both forms of energy can take place periodically as in the vibration of a pendulum and the central motions of the planets.

5 and 6. Transformation of mechanical energy into heat in the compression of elastic fluids, and by collision and friction. The absorption of light consists in a transformation of vibrational motion into heat.

7 and 8. The transformation of heat into mechanical energy follows from the expansion of gases under pressure, in steam engines, and in the vibrating energy in the radiation from heated bodies.

9. The transformation of one kind of heat into another by means of conduction.

10. The transformation of heat in chemical reactions. If compounds are decomposed by heat they are formed with the development of heat. Examples are the combination of sulphuric acid with water and the combination of lime with water.

11. The transformation of chemical energy into heat as in combustion.

12, 13, 14. The transformation of chemical energy into the galvania current and the further transformation from the current into chemical energy as well as the transformation of the current into chemical energy in the voltaic pile.

15, 16, 17. The transformation of electricity into heat and mechanical energy: in the glowing of a wire conducting current, in the electric spark, the motions of electric and electromagnetic attractions, by electric discharges, especially in the lightning flash.

18. A partial transformation of one electric current into another giving rise to an induced current.

19. The direct transformation of heat into electricity in the phenomenon of thermoelectricity and in the production of cold through the Peltier effect.

20, 21. The transformation of mechanical energy into electricity by friction and induction.

22–25: The transformation of mechanical energy into chemical energy, indirectly through the transformation of the given energy into electricity and heat.

It is prejudices, sanctioned by age and widespread dissemination, as well as primary sense impressions with their ambiguous yet so persuasive evidence, which seem to contradict the principles set forth here, and not the natural phenomena themselves. Against such prejudices we call to witness the history of all sciences.

While we vindicate the right of motion to exist as an entity and to represent substantiality, we must unconditionally deny the material nature of heat and electricity. For would it not be too absurd to look to a fluid for the nature of motion and the displacement of mass or to wish to assign in alternation now a material and now an immaterial nature to the same object?

Let us speak out the great truth: there are no immaterial materials! We realize that we are fighting a struggle with a deep-rooted hypothesis, canonized by high authority; with the imponderables we intend to abolish the last remnants of the gods of Greece from the study of nature. But we also know that nature in its simple truth is greater and more majestic than any structure built with the hand of man and all illusions created by the mind.

In terms of human conceptions the sun is an inexhaustible source of physical energy. The stream of this energy which also pours over our earth is the continually expanding spring that provides the motive power for terrestrial activities. In view of the large amount of energy which our earth is continuously emitting into space in the form of wave motion, its surface without continuous replenishment would soon revert to the cold of death. It is the light of the sun which when transformed into heat brings about the motions in our atmosphere and raises the waters of the earth to the clouds on high and brings about the flow of rivers. The heat which is produced through friction by the wheels of wind and water mills is sent to the earth by the sun in the form of radiation.

30A

Reprinted from *The Scientific Papers on James Prescott Joule*,
Vol. 1, Taylor & Francis Ltd., London, 1884, pp. 123–159

On the Calorific Effects of Magneto-Electricity, and on the Mechanical Value of Heat.

By J. P. JOULE, *Esq.**

[Phil. Mag. ser. 3. vol. xxiii. pp. 263, 347, and 435; read before the Chemical Section of Mathematical and Physical Science of the British-Association meeting at Cork on the 21st of August, 1843.]

IT is pretty generally, I believe, taken for granted that the electric forces which are put into play by the magneto-electrical machine possess, throughout the whole circuit, the same calorific properties as currents arising from other sources. And indeed when we consider heat not as a *substance*, but as a *state of vibration*, there appears to be no reason why it should not be induced by an action of a simply mechanical character, such, for instance, as is presented in the revolution of a coil of wire before the poles of a permanent magnet. At the same time it must be admitted that hitherto no experiments have been made decisive of this very interesting question; for all of them refer to a particular part of the circuit only, leaving it a matter of doubt whether the heat observed was *generated*, or merely *transferred from the coils* in which the magneto-electricity was induced, the coils themselves becoming cold. The latter view did not appear untenable without further experiments, considering the facts which I had already succeeded in proving, viz. that the heat evolved by the voltaic battery is *definite* † for the chemical changes taking place at the same time; and that the heat rendered

* The experiments were made at Broom Hill, near Manchester.

† Phil. Mag. ser. 3. vol. xix. p. 275.

"latent" in the electrolysis of water is at the expense of the heat which would otherwise have been evolved in a free state by the circuit *—facts which, among others, might seem to prove that *arrangement* only, not *generation* of heat, takes place in the voltaic apparatus, the simply conducting parts of the circuit evolving that which was previously latent in the battery. And Peltier, by his discovery that cold is produced by a current passing from bismuth to antimony, had, I conceived, proved to a great extent that the heat evolved by thermo-electricity is transferred † from the heated solder, no heat being *generated.* I resolved therefore to endeavour to clear up the uncertainty with respect to magneto-electrical heat. In this attempt I have met with results which will, I hope, be worthy the attention of the British Association.

Part I.—*On the Calorific Effects of Magneto-Electricity.*

The general plan which I proposed to adopt in my experiments under this head, was to revolve a small compound electro-magnet, immersed in a glass vessel containing water, between the poles of a powerful magnet, to measure the electricity thence arising by an accurate galvanometer, and to ascertain the calorific effect of the coil of the electro-magnet by the change of temperature in the water surrounding it.

The revolving electro-magnet was constructed in the following manner:—Six plates of annealed hoop-iron, each 8 inches long, $1\frac{1}{8}$ inch broad, and $\frac{1}{16}$ inch thick, were insulated from each other by slips of oiled paper, and then bound tightly together by a ribbon of oiled silk. Twenty-one yards of copper wire $\frac{1}{18}$ inch thick, well covered with

* 'Memoirs of the Literary and Philosophical Society of Manchester,' 2nd series, vol. vii. p. 97.

† The quantity of heat thus transferred is, I doubt not, proportional to the square of the difference between the temperatures of the two solders. I have attempted an experimental demonstration of this law, but, owing to the extreme minuteness of the quantities of heat in question, I have not been able to arrive at any satisfactory result.

silk, were wound on the bundle of insulated iron plates, from one end of it to the other and back again, so that both of the terminals were at the same end.

Having next provided a glass tube sealed at one end, the length of which was $8\frac{3}{4}$ inches, the exterior diameter 2·33 inches, and the thickness 0·2 of an inch, I fastened it in a round hole, cut out of the centre of the wooden revolving piece *a*, fig. 44. The glass was then covered with tinfoil,

Fig. 44. Scale $\frac{1}{12}$.

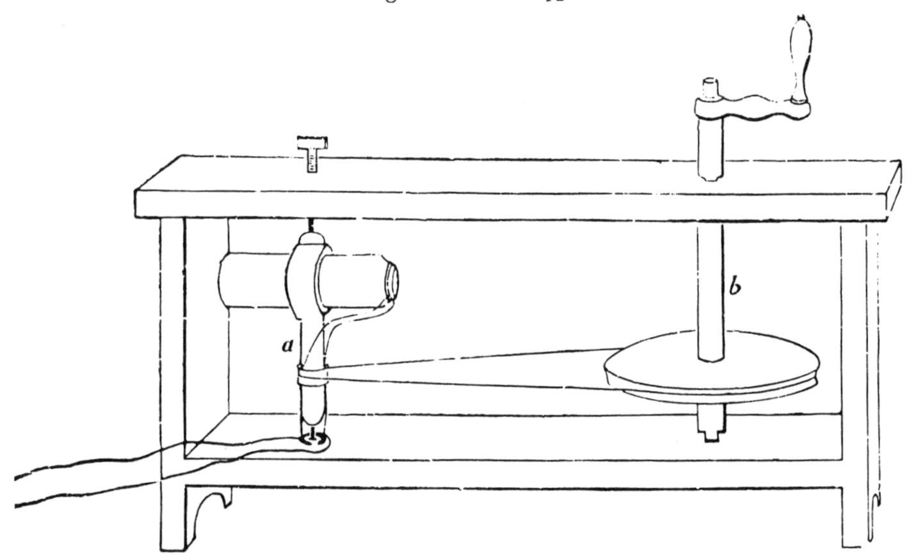

excepting a narrow slip in the direction of its length, which was left in order to interrupt magneto-electrical currents in the tinfoil during the experiments. Over the tinfoil small cylindrical sticks of wood were placed at intervals of about an inch, and over these again a strip of flannel was tightly bound, so as to inclose a stratum of air between it and the tinfoil. Lastly, the flannel was well varnished. By these precautions the injurious effects of radiation, and especially of convection of heat in consequence of the impact of air at great velocities of rotation, were obviated to a great extent.

The small compound electro-magnet was now put into the tube, and the terminals of its wire, tipped with platinum, were arranged so as to dip into the mercury of a commutator *, consisting of two semicircular grooves cut out of the base of the frame, fig. 44. By means of wires connected with the mercury of the commutator, I could connect the revolving electro-magnet with a galvanometer or any other apparatus.

In the first experiments I employed two electro-magnets (formerly belonging to an electro-magnetic engine) for the purpose of inducing the magneto-electricity. They were placed with two of their poles on opposite sides of the revolving electro-magnet, and the other two joining each other beneath the frame. I have drawn fig. 45 representing these

Fig. 45. Scale $\frac{1}{14}$.

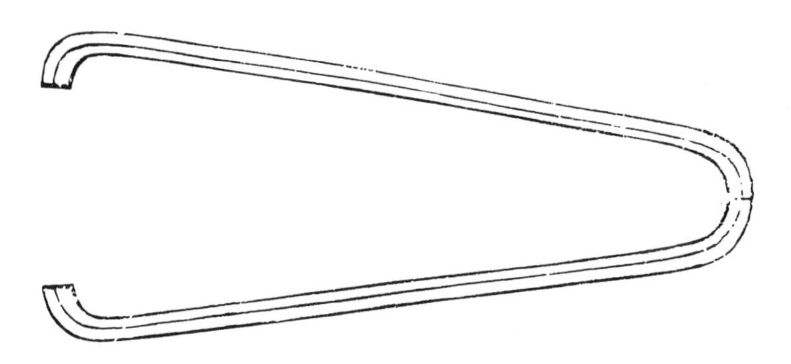

electro-magnets by themselves, to prevent confusing fig. 44. The iron of which they were made was 1 yard 6 inches long, 3 inches broad, and $\frac{1}{2}$ inch thick. The wire which was wound upon them was $\frac{1}{20}$ inch thick; it was arranged so as to form a sixfold conductor a hundred yards long.

The following is the method in which my experiments were made:—Having removed the revolving piece from its place (which is done with great facility by lifting the top of the frame, and with it the brass socket in which the upper steel pivot of the revolving piece works), I filled the

* I had made previous experiments in order to ascertain the best form of commutator, but found none to answer my purpose as well as the above. I found an advantage in covering the mercury with a little water. The steadiness of the needle of the galvanometer during the experiments proved the efficacy of this arrangement.

tube containing the small compound electro-magnet with $9\frac{3}{4}$ oz. of water. After stirring the water until the heat was equally diffused, its temperature was ascertained by a very delicate thermometer, by which I could estimate a change of temperature equal to about $\frac{1}{50}$ of Fahrenheit's degree. A cork covered with several folds of greased paper was then forced into the mouth of the tube, and kept in its place by a wire passing over the whole, and tightened by means of one or two small wooden wedges. The revolving piece was then restored to its place as quickly as possible, and revolved between the poles of the large electro-magnets for a quarter of an hour, during which time the deflections of the galvanometer and the temperature of the room were carefully noted. Finally, another observation with the thermometer detected any change that had taken place in the temperature of the water.

Notwithstanding the precautions taken against the injurious effects of radiation and convection of heat, I was led into error by my first trials: the water had lost heat, even when the temperature of the room was such as led me to anticipate a contrary result. I did not stop to inquire into the cause of the anomaly, but I provided effectually against its interference with the subsequent results by alternating the experiments with others made under the same circumstances, except as regards the communication of the battery with the stationary electro-magnets, which was in these instances broken. And to avoid any objection which might be made with regard to the heat, however trifling, evolved by the wires of the large electro-magnets, the thermometer employed in registering the temperature of the air was situated so as to receive the influence arising from that source equally with the revolving piece.

I will now give a series of experiments in which six Daniell's cells, each 25 inches high and $5\frac{1}{2}$ inches in diameter, were alternately connected and disconnected with the large stationary electro-magnets. The galvanometer, connected through the commutator with the revolving electro-magnet, had a coil a foot in diameter, consisting of five turns of copper wire, and a needle six inches long. Its deflections could be turned into quantities of current by means of a table constructed from previous experiments. The galvanometer was situated so as to be out of the reach of the attractions of the large electro-magnets, and every other precaution was taken to render the experiments worthy of reliance. The rotation was in every instance carried on for exactly a quarter of an hour.

Series No. 1.

		Revolutions of Electro-Magnet per minute.	Deflections of Galvanometer of 5 turns.	Mean Temperature of Room.	Mean Difference.	Temperature of Water.		Loss or Gain.
						Before.	After.	
April 15, P.M.	Battery contact broken.	600	0° 0′	54°·69	0°·19+	54°·90	54°·85	0°·05 loss
	Battery in connexion.	600	21 0	54·67	0·20+	54·85	54·88	0·03 gain
	Battery contact broken.	600	0 0	54·61	0·24+	54·88	54·83	0·05 loss
	Battery in connexion.	600	24 0	54·65	0·23+	54·85	54·92	0·07 gain
	Mean, Battery in connexion.	600	22 30	..	0·21+	..	..	0·05 gain
	Mean, Battery contact broken.	600	0 0	..	0·21+	..	..	0·05 loss
	Corrected Result.	600	22° 30′=0·177* of cur. mag.-elect.					0·10 gain

Having thus detected the evolution of heat from the coil of the magneto-electrical apparatus, my next business was to confirm the fact by exposing the revolving electro-magnet to a more powerful magnetic influence; and to do so with the greater convenience, I determined on the construction of a new stationary electro-magnet, by which I might obtain a

* Throughout the paper I have called that quantity of current *unity*, which, passing equably for an hour of time, can decompose a chemical equivalent expressed in grains.

more advantageous employment of the electricity of the battery. Availing myself of previous experience, I succeeded in producing an electro-magnet possessing greater power of attraction from a distance than any other I believe on record. On this account a description of it in greater detail than is absolutely necessary to the subject of this paper will not, I hope, be deemed superfluous.

A piece of half-inch boiler plate-iron was cut into the shape represented by fig. 46. Its length was 32 inches; its

Fig. 46. Scale $\frac{1}{12}$.

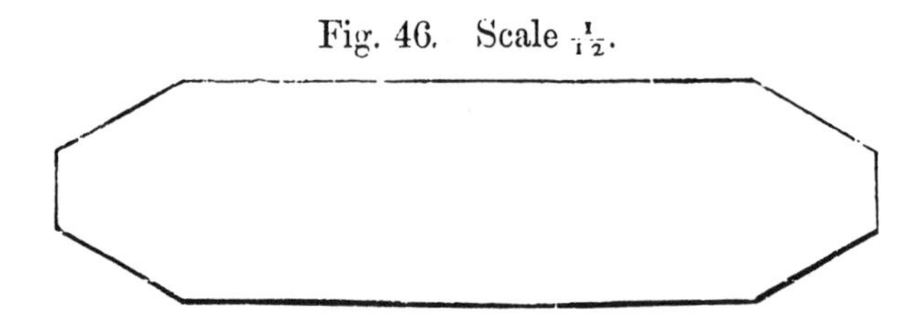

breadth in the middle part 8 inches, at the ends 3 inches. It was bent nearly into the shape of the letter U, so that the shortest distance between the poles was slightly more than 10 inches.

Twenty-two strands of copper wire, each 106 yards long and about one twentieth of an inch in diameter, were now bound tightly together with tape. The insulated bundle of wires, weighing more than sixty pounds, was then wrapped upon the iron, which had itself been previously insulated by a fold of calico. Fig. 47 represents the electro-magnet in its completed state.

Fig. 47. Scale $\frac{1}{12}$.

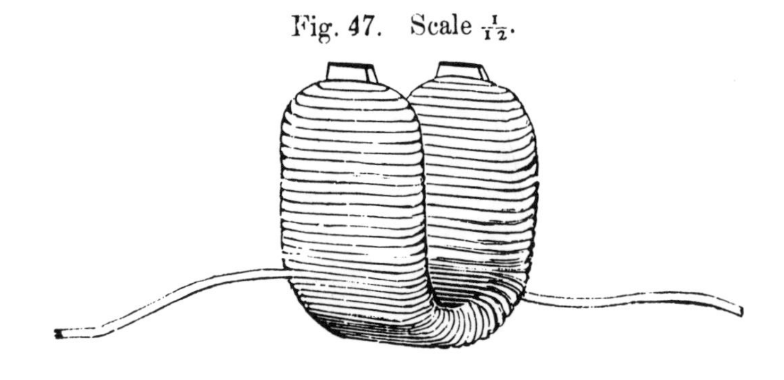

In arranging the voltaic battery for its excitation, care was taken to render the resistance to conduction of the battery equal, as nearly as possible, to that of the coil, Prof. Jacobi having proved that to be the most advantageous arrangement. Ten of my large Daniell's cells, arranged in a series of five double pairs, fulfilled this condition very well, producing a magnetic energy in the iron superior to any thing I had previously witnessed. I will mention the results of a few experiments in order to give some definite idea of it.

1st. The force with which a bar of iron three inches broad and half an inch thick was attracted to the poles was equal, at the distance of $\frac{1}{16}$ of an inch, to 100 lb., at $\frac{1}{4}$ of an inch to 30 lb., at $\frac{1}{2}$ an inch to $10\frac{1}{2}$ lb., and at 1 inch to 4 lb. 13 oz.* 2nd. A small rod of iron three inches long, weighing 148 grains, held vertically under one of the poles, would jump through an interval of $1\frac{3}{4}$ inch: a needle three inches long, weighing 4 grains, would jump from a distance of $3\frac{1}{4}$ inches.

Having fixed the electro-magnet just described with its poles upwards and on opposite sides of the revolving electro-magnet, I arranged to it the battery of ten cells, in a series of five double pairs, and, experimenting as before, I obtained a second series of results. The galvanometer used in the present instance was in every respect similar to that previously described, with the exception of the coil, which now consisted of a single turn of thick copper wire. Great care was taken to prevent, by its distance from and relative position with the electro-magnet, any interference of the latter with its indications.

* The above electro-magnet being constructed for a specific purpose, was not adapted for displaying itself to the best advantage in one respect. On account of the extension of its poles (three inches by half an inch) many of the lines of magnetic attraction were necessarily in very oblique directions. Theoretically, circular poles should give the greatest attraction from small distances.

No. 2.

		Revolutions of Electro-magnet per minute.	Deflections of Galvanometer of one turn.		Mean Temperature of Room.	Mean Difference.	Temperature of Water.		Loss or Gain.
							Before.	After.	
May 6, A.M.	Battery in connexion.	600	22°	0′	58°·93	0°·17+	58°·20	60°·00	1°·80 gain
	Battery contact broken.	600	0	0	59·60	0·40+	60·02	59·98	0·04 loss
	Battery in connexion.	600	24	0	59·55	1·23+	59·90	61·67	1·77 gain
May 6, P.M.	Battery contact broken.	600	0	0	59·45	0·19+	59·78	59·50	0·28 loss
	Battery in connexion.	600	24	45	58·30	0·05+	57·35	59·35	2·00 gain
May 8, A.M.	Battery in connexion.	600	22	0	57·74	0·32+	57·28	58·83	1·55 gain
	Battery contact broken.	600	0	0	58·35	0·49+	58·83	58·85	0·02 gain
	Battery in connexion.	600	21	20	58·73	0·78+	58·83	60·20	1·37 gain
	Mean, Battery in connexion.	600	22	49	..	0·51+	..	..	1·70 gain
	Mean, Battery contact broken.	600	0	0	..	0·36+	..	..	0·10 loss
	Corrected Result.	600	22° 49′=0·902 of cur. mag.-elect.						1·84 gain

The corrected result is obtained, as before, by adding the loss sustained when contact with the battery was broken to the heat gained when the battery was in connexion. I have in the present instance, however, made a further correction of 0°·04 on account of the difference between the *mean differences* 0°·51 and 0°·36. The ground of this correction is the result of a previous experiment, in which, by revolving the apparatus at 94° in an atmosphere of 60°, the water sustained a loss of 7°·6, or about one quarter of the difference between the temperature of the atmosphere and the mean temperature of the water.

With the same electro-magnet, but using a battery of only four cells arranged in a series of two double pairs, by which I expected to obtain about half as much magnetism in the iron, the following results were obtained :—

No. 3.

		Revolutions of Electro-magnet per minute.	Deflections of Galvanometer of 5 turns.	Mean Temperature of Room.	Mean Difference.	Temperature of Water.		Loss or Gain.
						Before.	After.	
May 8, P.M.	Battery in connexion.	600	38° 0	57°·00	0°·02 –	56°·73	57°·23	0°·50 gain
	Battery contact broken.	600	0 0	57·25	0·0	57·23	57·27	0·04 gain
	Battery in connexion.	600	38 30	57·53	0·09 +	57·35	57·90	0·55 gain
May 9, P.M.	Battery in connexion.	600	39 45	56·37	0·45 –	55·60	56·25	0·65 gain
	Battery contact broken.	600	0 0	56·75	0·39 –	56·27	56·45	0·18 gain
	Battery in connexion.	600	38 45	57·14	0·37 –	56·50	57·05	0·55 gain
	Mean, Battery in connexion.	600	38 45	..	0·19 –	..	..	0·56 gain
	Mean, Battery contact broken.	600	0 0	..	0·19 –	..	..	0·11 gain
	Corrected Result.	600	38° 45′ = 0·418 of cur. mag.-elect.					0·45 gain

In the next experiments a battery of ten cells in a series of five double pairs was used for the purpose of exciting the large stationary electro-magnet. But, dismissing the galvanometer and the other extra parts of the circuit, I connected the terminal wires of the revolving electro-magnet together, so as to obtain the whole effect of the magneto-electricity. The resistance of the coil of the revolving electro-magnet being to that of the whole circuit employed in the experiments No. 2 as 1 : 1·13, and 0·902 of current being obtained in those experiments, I expected to obtain the calorific effect of 1·019 in the new series.

No. 4.

		Revolutions of Electro-magnet per minute.	Mean Temperature of Room.	Mean Difference.	Temperature of Water.		Loss or Gain.
					Before.	After.	
May 10, P.M.	Battery in connexion.	600	56·85°	0·61° −	54·98°	57·50°	2·52° gain
	Battery contact broken.	600	57·37	0·12+	57·48	57·50	0·02 gain
	Battery in connexion.	600	57·52	1·08+	57·48	59·73	2·25 gain
	Mean, Battery in connexion.	600	..	0·23+	..	..	2·38 gain
	Mean, Battery contact broken.	600	..	0·12+	..	..	0·02 gain
	Corrected Result.	600	1·019 of cur. mag.-elect.				2·39 gain

It seemed to me very desirable to repeat the experiments, substituting *steel* magnets for the stationary electro-magnets hitherto used. With this intention I constructed two magnets, each consisting of a number of thin plates of hard steel, —an arrangement which we owe to Dr. Scoresby. My metal was, unfortunately, not of very good quality; but nevertheless an attractive force was obtained sufficiently powerful to overcome the gravity of a small key weighing 47 grains, placed at the distance of three eighths of an inch. The following results were obtained by revolving the electro-magnet acted on by the poles of the steel magnets.

No. 5.

		Revolutions of Electro-magnet per minute.	Deflections of Galvanometer of 5 turns.	Mean Temperature of Room.	Mean Difference.	Temperature of Water. Before.	Temperature of Water. After.	Loss or Gain.
May 16, A.M.	Circuit complete.	600	26° 0′	59°·72	0°·0	59°·73	59°·70	0°·03 loss
	Circuit broken.	600	0 0	59·82	0·20 –	59·70	59·55	0·15 loss
	Circuit complete.	600	29 0	59·95	0·41 –	59·55	59·53	0·02 loss
	Circuit broken.	600	0 0	59·58	0·12 –	59·52	59·40	0·12 loss
	Circuit complete.	600	27 0	59·65	0·25 –	59·40	59·40	0
	Mean, Circuit complete.	600	27 20	..	0·22 –	..	..	0·016 loss
	Mean, Circuit broken.	600	0 0	..	0·16 –	..	..	0·135 loss
	Corrected Result.	600	27° 20′ = 0·236 of cur. mag.-elect.					0·10 gain

In order to obtain the whole calorific effect, I now, as in Series No. 4, connected the terminal wires of the revolving electro-magnet, and alternated the experiments with others in which that connexion was broken. The resistance of the coil of the revolving electro-magnet being to the resistance of the whole circuit used in the experiments marked No. 5 as 1 : 1·44, and 0·236 of current electricity being obtained in those experiments, I expected to obtain in the present series the calorific effect of 0·34 of current magneto-electricity.

No. 6.

		Revolutions of Electro-magnet per minute.	Mean Temperature of Room.	Mean Difference.	Temperature of Water.		Loss or Gain.
					Before.	After.	
May 17, A.M.	Terminals joined.	600	59°·07	0°·20—	58°·82	58°·92	0°·10 gain
	Terminals separated.	600	59·07	0·22—	58·92	58·78	0·14 loss
	Terminals joined.	600	58·96	0·20—	58·75	58·78	0·03 gain
	Terminals separated.	600	58·88	0·18—	58·78	58·63	0·15 loss
	Mean, Terminals joined.	600	..	0·20—	..	..	0·065 gain
	Mean, Terminals separated.	600	..	0·20—	..	..	0·145 loss
	Corrected Result.	600	0·34 of cur. mag.-elect.				0·21 gain

Although any considerable development of electrical currents in the iron of the revolving electro-magnet was prevented by its disposition in a number of thin plates insulated from each other, I apprehended that they might, under a powerful inductive influence, exist separately in each plate to such an extent as to produce an appreciable quantity of heat. To ascertain the fact, the terminals of the wire of the revolving electro-magnet were insulated from each other, while the latter was revolved between the poles of the large electro-magnet excited by ten cells in a series of five double pairs. The experiments were alternated with others in which contact with the battery was broken. As we shall hereafter give in detail experiments of the same class, it will not be necessary to do more at present than to state that the mean result of the present series, consisting of eight trials, gave 0°·28 as the quantity of heat evolved by the iron alone.

We are now able to collect the results of the preceding

experiments, so as to discover the laws by which the development of the heat is regulated. The fourth column of the following table, containing the heat due to the currents circulating in the iron alone, is constructed on the basis of a law which we shall subsequently prove, viz. *the heat evolved by a bar of iron revolving between the poles of a magnet is proportional to the square of the inductive force.* Column 5 gives the heat evolved by the coils of the electro-magnet alone. No elimination is required for the results of Series Nos. 5 and 6, because in them the iron of the revolving electro-magnet was subject to the influence of the steel magnets in the alternating as well as in the other experiments.

TABLE I.

Series of Experiments.	Current Magneto-electricity.	Heat actually evolved.	Correction for Currents in the Iron.	Corrected Heat.	Squares of Numbers proportional * to those in column 2.	Heat due to Voltaic Current of the intensities given in col. 2.	The Numbers of column 7 multiplied by $\frac{4}{3}$.
No. 1.	0·177	°0·10	°0·02	°0·08	°0·062	°0·040	°0·053
No. 2.	0·902	1·84	0·28	1·56	1·614	1·040	1·386
No. 3.	0·418	0·45	0·09	0·36	0·346	0·224	0·299
No. 4.	1·019	2·39	0·28	2·11	2·060	1·327	1·769
No. 5.	0·236	0·10	0	0·10	0·109	0·071	0·091
No. 6.	0·340	0·21	0	0·21	0·229	0·148	0·197
1.	2.	3.	4.	5.	6.	7.	8.

On comparing the corrected results in column 5 with the squares of magneto-electricity given in column 6, it will be

* This proportion is arbitrary, and greater than that in Table 2. Column 6 is superfluous. It will be found that in both tables the numbers in column 7 are to those in column 5 nearly as 2 to 3, which difference, as explained in the text, is owing to the uniform intensity of the current in the former case and its pulsatory character in the latter.—*Note*, 1881.

abundantly manifest that *the heat evolved by the coil of the magneto-electrical machine is proportional* (*cæteris paribus*) *to the square of the current.*

Column 7, containing the heat due to voltaic currents of the quantities stated in column 2, is constructed on the basis of three very careful experiments on the heat evolved by passing currents through the coil of the small compound electro-magnet. I observed an increase in the temperature of the water equal to 5°·3, 5°·46, and 5°·9 respectively, when 2·028, 2·078, and 2·145 of current voltaic electricity were passed, each during a quarter of an hour, through the coil. Reducing the first and second experiments to the electricity of the third according to the squares of the current, we have 5°·93, 5°·82, and 5°·9 for 2·145 of current. The mean of these is 5°·88, a datum from which the theoretical results of the preceding and subsequent tables are calculated.

But in comparing the heat evolved by magneto- with that evolved by voltaic electricity, we must remember that the former is propagated by pulsations, the latter uniformly. Now, since the square of the mean of unequal numbers is always less than the mean of their squares, it is obvious that the magnetic effect at the galvanometer will bear a greater proportion to the heat evolved by the voltaic, than the magneto-electricity; so that it is impossible to institute a strict comparison without ascertaining previously the intensity of the magneto-electricity at every instant of the revolution of the revolving electro-magnet. I have not been able to devise any very accurate means for attaining this object; but judging from the comparative brilliancy of the sparks when the commutator was arranged so as to break contact with the mercury at different positions of the revolving electro-magnet with respect to the poles of the stationary electro-magnet, there appeared to be but little variation in the intensity of the magneto-electricity during $\frac{3}{4}$ of each revolution. The remaining $\frac{1}{4}$ (during which the revolving electro-magnet passes the poles of the stationary electro-magnet) is occupied in the reversal of the direction of the electricity. In the experiments all flow of electricity during this $\frac{1}{4}$ is cut off by the divisions of the commutator. In illustration of this I have drawn fig. 48, in which the direction and intensity of the magneto-electricity are represented by ordinates A x, &c., perpendicular to the straight line A B C D E; the intermediate spaces B C, D E, &c. represent the time during which the electricity is wholly cut off by the divisions of the commutator. Were A x x' B &c. perfect rectangles, it is obvious that the heat due to a given deflection of the galvanometer would be

Fig. 48.

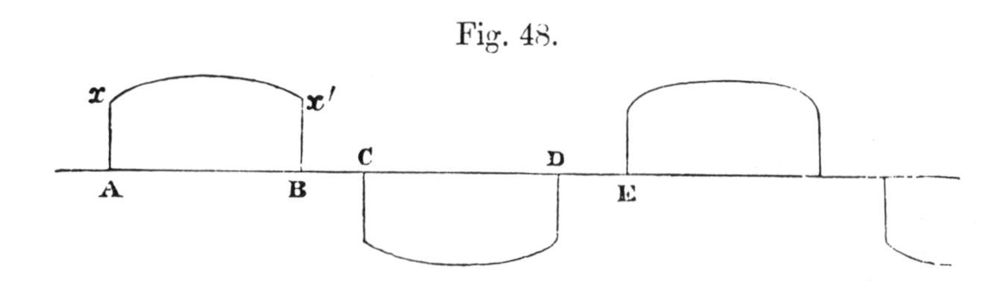

$\frac{4}{3}$ of that due to the same deflection and a uniform current, and column 8 of the table would contain exact theoretical results. But as this is not precisely the case, the numbers in that column are somewhat under the truth.

Bearing this in mind in the comparison of columns 5 and 8, it will, I think, be admitted that the experiments afford decisive evidence that *the heat evolved by the coil of the magneto-electrical machine is governed by the same laws as those which regulate the heat evolved by the voltaic apparatus, and exists also in the same quantity under comparable circumstances.*

Although very little doubt could exist with regard to the heating power of magneto-electricity *beyond* the coil, I thought it would nevertheless be well to follow it there, in order to render the investigation more complete: I am not aware of any previous experiments of the kind.

I immersed five or six yards of insulated copper wire of $\frac{1}{40}$ inch diameter in a flask holding about 12 oz. of water. The terminals of the wire were connected on the one hand with the galvanometer of five turns and on the other with the commutator, and the circuit was completed by a wire extending from the galvanometer to the other compartment of the commutator. The revolving electro-magnet, being now subjected to the inductive influence of the large electro-magnet excited by ten cells in a series of five double pairs, was rotated at the rate of 600 revolutions per minute during a quarter of an hour. The needle of the galvanometer, which remained, as usual, pretty steady, indicated a mean deflection of 32° 40′=0·31 of current; and the heat evolved was found to be 0°·46, after the correction on account of the temperature of the surrounding air had been applied. Another experiment gave me 0°·4 for 0·286. The mean of the two is 0°·43 for 0·298 current magneto-electricity.

By passing a voltaic current from four cells in series through the wire, I found that 2·02 of current flowing uniformly evolved 12°·0 in a quarter of an hour. Reducing this to 0·298 of current we have $\left(\frac{0 \cdot 298}{2 \cdot 02}\right)^2 \times 12° = 0° \cdot 261$. The product of this by $\frac{4}{3}$ (on account of the pulsatory character of the magneto-current) gives 0°·348, which, as theory demands, is somewhat less than the quantity found by experiment.

On the Calorific Effects of Magneto- with Voltaic Electricity.

I now proceeded to consider the heat evolved by voltaic currents when they are counteracted or assisted by magnetic induction. For this purpose it was only necessary to introduce a battery into the magneto-electrical circuit: then, by turning the wheel in one direction, I could oppose the voltaic current; or, by turning in the other direction, I could increase the intensity of the voltaic by the assistance of the magneto-electricity. In the former case the apparatus possessed all the properties of the electro-magnetic engine; in the latter it presented the reverse, viz. the *expenditure* of mechanical power.

No. 7.

		Revolutions of Electro-magnet per minute.	Deflections of Galvanometer of 1 turn.	Mean Temperature of Room.	Mean Difference.	Temperature of Water.		Loss or Gain.
						Before.	After.	
May 19.	Circuits complete.	600	22° 40′	57°·43	1°·03 –	55°·62	57°·18	1°·56 gain
May 19.	Circuits broken.	600	0 0	57·15	0·41 –	57·08	57·00	0·08 loss
May 20.	Circuits complete.	600	20 45	59·40	0·08 –	58·65	60·00	1·35 gain
May 20.	Circuits broken.	600	0 0	59·40	0·51 +	60·00	59·83	0·17 loss
May 20.	Circuits complete.	600	23 0	59·10	1·29 +	59·78	61·00	1·22 gain
	Mean, Circuits complete.	600	22 8	....	0·06 +	....	....	1·38 gain
	Mean, Circuits broken.	600	....	....	0·05 +	....	....	0·12 loss
	Corrected Result.	600	22° 8′=0·864 of current.					1·50 gain

In the preceding series I used the steel magnets previously described, as the inductive force; and I had two of the large Daniell's cells in series, arranged so as to pass a current of electricity through the revolving electro-magnet and galvanometer. The wheel was turned in the direction which it would have taken had the friction been sufficiently reduced to allow of the motion of the apparatus without assistance.

I give another series, in which every thing else remaining the same, the direction of revolution was reverse, so as to *increase* the intensity of the voltaic electricity by superadding that of the magneto-electricity.

No. 8.

		Revolutions of Electro-magnet per minute.	Deflections of Galvanometer of 1 turn.	Mean Temperature of Room.	Mean Difference.	Temperature of Water. Before.	Temperature of Water. After.	Loss or Gain.
May 23.	Circuits complete.	600	30° 15′	60°·55	0°·19+	59°·30	62°·18	2°·88 gain
May 23.	Circuits broken.	600	0 0	62·28	0·48+	62·92	62·60	0·32 loss
May 24.	Circuits complete.	600	29 40	60·90	0·03–	59·50	62·25	2·75 gain
May 24.	Circuits broken.	600	0 0	59·50	0·0	59·50	59·50	0
May 26.	Circuits complete.	600	29 50	61·85	0·18–	60·33	63·02	2·69 gain
May 26.	Circuits broken.	600	0 0	60·90	0·49–	60·50	60·33	0·17 loss
	Mean, Circuits complete.	600	29 55	....	0·02–	....	....	2·77 gain
	Mean, Circuits broken.	600	....	....	0·01 –	....	....	0·16 loss
	Corrected Result.	600	29° 55′=1·346 of current.					2·93 gain

Dismissing the steel magnets, which did not appear to have lost any of the magnetic virtue which they possessed at first, I now substituted for them the large stationary electro-magnet, excited by eight of the Daniell's cells arranged in a series of four double pairs. The revolving electro-magnet completed, as before, a circuit containing the galvanometer and two of Daniell's cells in series. The motive power of the apparatus was now so great that it would revolve rapidly in spite of very considerable friction. In order to give the requisite velocity it was necessary, however, to assist the motion by the hand.

No. 9.

		Revolutions of Electro-magnet per minute.	Deflections of Galvanometer of 1 turn.	Mean Temperature of Room.	Mean Difference.	Temperature of Water.		Loss or Gain.
						Before.	After.	
May 31, P.M.	Circuits complete.	600	16°	62°·50	0°·11 –	62°·00	62°·78	0°·78 gain
May 31, P.M.	Circuits broken.	600	0	63·00	0·23 –	62·73	62·82	0·09 gain
June 1, A.M.	Circuits complete.	600	14	62·65	0·11 –	62·18	62·90	0·72 gain
June 1, A.M.	Circuits broken.	600	0	63·15	0·20 –	62·90	63·00	0·10 gain
	Mean, Circuits complete.	600	15	....	0·11 –	....	....	0·75 gain
	Mean, Circuits broken.	600	....	....	0·21 –	....	....	0·095 ga.
	Corrected Result.	600	15° = 0·543 of current.					0·68 gain

The following series of results was obtained with the same apparatus, by turning the wheel in the opposite direction.

No. 10.

		Revolutions of Electro-magnet per minute.	Deflections of Galvanometer of 1 turn.	Mean Temperature of Room.	Mean Difference.	Temperature of Water.		Loss or Gain.
						Before.	After.	
June 2, A.M.	Circuits complete.	600	35° 10′	63°·38	0°·62 +	63°·25	68°·75	5°·50 gain
June 2, A.M.	Circuits broken.	600	0 0	64·73	0·75 +	65·51	65·45	0·06 loss
June 3, A.M.	Circuits complete.	600	37 10	65·10	1·40 +	63·33	69·66	6·33 gain
June 3, A.M.	Circuits broken.	600	0 0	64·93	1·23 +	66·28	66·05	0·23 loss
	Mean, Circuits complete.	600	36 10	....	1·01 +	....	....	5·915 ga.
	Mean, Circuits broken.	600	....	....	0·99 +	....	....	0·145 loss
	Corrected Result.	600	36° 10′ = 1·845 of current.					6·06 gain

I give two series more, in which only one cell was connected with the revolving electro-magnet, and the revolution was in the direction of the attractive forces. The magneto-electricity

No. 11.

		Revolutions of Electro-magnet per minute.	Deflections of Galvanometer of 1 turn.	Mean Temperature of Room.	Mean Difference.	Temperature of Water. Before.	Temperature of Water. After.	Loss or Gain.
June 3, P.M.	Circuits complete.	350	0	64·02°	0·38°–	63·57°	63·72°	0·15° gain
	Circuits broken.	350	0	63·75	0·02–	63·73	63·73	0
	Circuits complete.	400	0	63·80	0·02–	63·70	63·86	0·16 gain
	Circuits broken.	400	0	64·35	0·08–	64·27	64·27	0
	Mean, Circuits complete.	375	0	....	0·20–	....	....	0·155 ga.
	Mean, Circuits broken.	375	0	....	0·05–	....	....	0
	Corrected Result.	375	0					0·12 gain

No. 12.

		Revolutions of Electro-magnet per minute.	Deflections of Galvanometer of 1 turn.	Mean Temperature of Room.	Mean Difference.	Temperature of Water. Before.	Temperature of Water. After.	Loss or Gain.
June 5, A.M.	Circuits complete.	600	9° 40′	60·40°	0·16°–	60·02°	60·47°	0·45° gain
	Circuits broken.	600	0	60·64	0·17–	60·47	60·47	0
	Corrected Result.	600	9° 40′=0·34	current in the opposite direction.				0·45 gain

was so intense, at a velocity of 600 per minute, as to overpower the intensity of the single cell, causing the needle to be permanently and steadily deflected to between 9° and 10° in the opposite direction. The command of the magneto-electricity over the voltaic current arising from one cell was beautifully illustrated by the sparks at the commutator. Turning slowly, they were bright and snapping*: increasing the rapidity of revolution, they decreased in brightness; until at a velocity of about 370 per minute they ceased altogether. They were plainly visible again when the velocity reached 600 per minute.

The results of the preceding series of experiments are collected in the following table along with theoretical results calculated in precisely the same manner as those of Table I. The correction for heat evolved by the iron of the revolving electro-magnet is estimated at 0°·18, the product of 0°·28 by $\left(\frac{4}{5}\right)^2$; because in the above experiments the large electro-magnet was excited by $\frac{4}{5}$ of the battery used when 0°·28 was obtained. No correction is needed for the series in which the steel magnets were used, because they remained in their places during the alternating experiments.

TABLE II.

Series of Experiments.	Current Magneto-electricity.	Heat actually evolved.	Correction for Currents in the Iron.	Corrected Heat.	Squares of numbers proportional† to those in col. 2.	Heat due to Voltaic Currents of the intensities given in col. 2.	The Numbers of column 7 multiplied by $\frac{4}{3}$.
No. 7.	0·864	1°·50	0°	1°·50	1°·291	0°·954	1°·272
No. 8.	1·346	2·93	0	2·93	3·133	2·316	3·088
No. 9.	0·543	0·68	0·18	0·50	0·510	0·377	0·503
No. 10.	1·845	6·06	0·18	5·88	5·886	4·351	5·801
No. 11.	0	0·12	0·18	−0·06	0	0	0
No. 12.	0·340	0·45	0·18	0·27	0·200	0·148	0·197
1.	2.	3.	4.	5.	6.	7.	8.

* The most splendid sparks are obtained, when the voltaic is assisted by the magneto-electricity, by turning an electro-magnetic engine in a direction contrary to the attractive forces.

† See note to Table I. (p. 136).

In all these experiments, as well as in those collected in Table I., the time occupied by the platinum wires in crossing the divisions of the commutator was found to be exactly $\frac{1}{4}$ of that occupied by an entire revolution; hence the multiplication by $\frac{4}{3}$ in order to obtain true theoretical results on the supposition that the current flows uniformly during $\frac{3}{4}$ of a revolution. It will be observed that these theoretical results are not so much inferior to the experimental results of column 5 as they were in Table I. The principal reason of this arises from the mixture of the constant effect of the battery with the variable magneto-electrical current, as will be readily seen on inspecting figs. 49 and 50, the former of which represents the currents in series No. 9; the latter, those in series No. 10. The dotted rectangles *a b c d*, &c., represent the constant effect of the battery of two cells, which is in one instance diminished, in the other increased by the magneto-electricity.

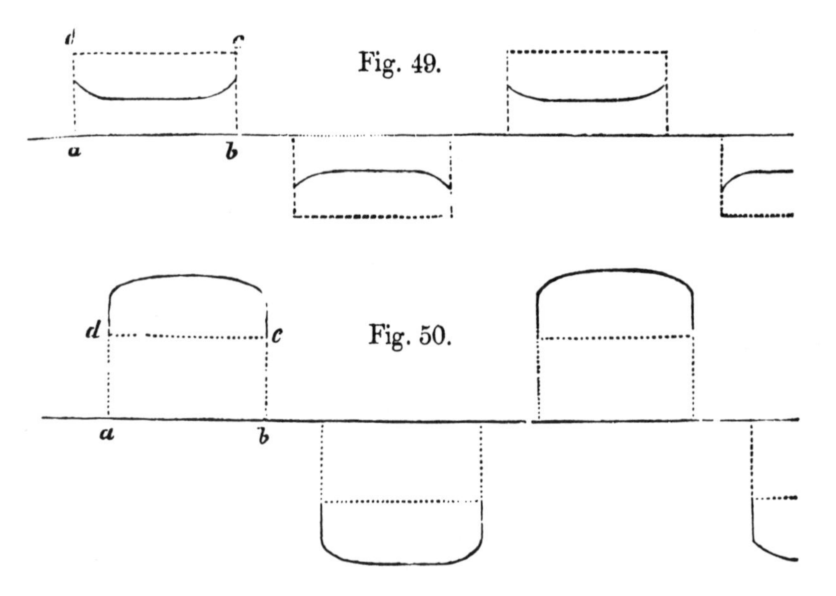

Fig. 49.

Fig. 50.

On comparing columns 6 and 8 with column 5, it is manifest that the law of the square of the electric current still obtains, and is not affected either by the assistance or resistance which the magneto-electricity presents to the voltaic current. Now the increase or diminution of the chemical effects occurring in the battery during a given time is proportional to the magneto-electrical effect, and the heat evolved is always proportional to the square of the current; therefore the heat due to a given chemical action is subject to an increase or to a diminution directly proportional to the intensity of the magneto-electricity assisting or opposing the voltaic current.

We have therefore in magneto-electricity an agent capable by simple mechanical means of destroying or generating heat. In a subsequent part of this paper I shall make an attempt to connect heat with mechanical power in absolute numerical relations. At present we shall turn to a question intimately connected with the previous investigations, and which indeed has already been partly developed.

On the Heat evolved by a Bar of Iron rotating under Magnetic Influence.

Having removed the small electro-magnet from the tube of the revolving piece, I fixed in its stead, in the centre

No. 13.

		Revolutions of the Bar per minute.	Deflections of Galvanometer of 1 turn.	Mean Temperature of Room.	Mean Difference.	Temperature of Water. Before.	Temperature of Water. After.	Gain or Loss.
June 17, P.M.	Electro-magnet in action.	600	70° 55′	67°·38	0°·35−	66°·27	67·80	1°·53 gain
	Battery contact broken.	600	..	67·60	0·23+	67·77	67·90	0·13 gain
	Electro-magnet in action.	600	70 45	67·85	0·67+	67·85	69·20	1·35 gain
	Battery contact broken.	600	..	68·92	0·30+	69·18	69·27	0·09 gain
	Mean, Electro-magnet in action.	600	70 50	..	0·16+	..	..	1·44 gain
	Mean, Battery contact broken.	600	..	..	0·26+	..	..	0·11 gain
	Corrected Result.	600	70° 50′=9·85 current.					1·31 gain

of the tube, a solid cylinder of iron 8 inches long and $\frac{3}{4}$ inch in diameter. The tube was then, as before, filled with water and rotated for a quarter of an hour between the poles of the large electro-magnet. In the first experiments the electro-magnet was excited by ten cells in a series of five double pairs, a galvanometer being included in the circuit in order to indicate the electric force applied. It was of course placed, as before, so as not to be affected by the powerful attraction of the electro-magnet. The precaution of alternating the experiments was adopted as usual. The results are recorded in No. 13.

Every thing else remaining the same, I now used a battery of six cells arranged in a series of three double pairs to excite the electro-magnet in the experiments of No. 14.

No. 14.

		Revolutions of the Bar per minute.	Deflections of Galvanometer of 1 turn.	Mean Temperature of Room.	Mean Difference.	Temperature of Water.		Gain or Loss.
						Before.	After.	
June 19, A.M.	Electro-magnet in action.	600	64° 10′	65°·42	0°·17 –	65°·00	65°·50	0°·50 gain
	Battery contact broken.	600	..	65·55	0·10 –	65·48	65·42	0·06 loss
	Electro-magnet in action.	600	64 10	65·42	0·17 +	65·33	65·86	0·53 gain
	Battery contact broken.	600	..	65·75	0·03 +	65·80	65·77	0·03 loss
	Mean, Electro-magnet in action.	600	64 10	..	0	..	..	0·515 gain
	Mean, Battery contact broken.	600	..	..	0·03 –	..	..	0·045 loss
	Corrected Result.	600	64° 10′ = 6·67 current.					0·56 gain

I give another series obtained with a battery of two cells in series.

No. 15.

		Revolutions of the Bar per minute.	Deflections of Galvanometer of 1 turn.	Mean Temperature of Room.	Mean Difference.	Temperature of Water.		Gain or Loss.
						Before.	After.	
June 19, P.M.	Electro-magnet in action.	600	54°	63°·65	0°·43 –	63°·12	63·32	0°·20 gain
	Battery contact broken.	600	..	63·80	0·38 –	63·40	63·45	0·05 gain
	Electro-magnet in action.	600	54	63·75	0·20 –	63·45	63·65	0·20 gain
	Battery contact broken.	600	..	64·07	0·37 –	63·68	63·73	0·05 gain
	Mean, Electro-magnet in action.	600	54	..	0·32 –	..	..	0·20 gain
	Mean, Battery contact broken.	600	..	..	0·38 –	..	..	0·05 gain
	Corrected Result.	600	54° = 4·17 current.					0·16 gain

The results of the preceding experiments are collected in the following table:—

TABLE III.

Series of Experiments.	Current employed in exciting the Electro-Magnet.	Heat evolved.	Square of Numbers proportional to those of Column 2.
No. 13.	9·85	1°·31	1·2290
No. 14.	6·77	0·56	0 5807
No. 15.	4·17	0·16	0·2203
1	2	3	4

It was discovered by Prof. Jacobi, and by myself also one or two months afterwards*, that the attraction of electro-magnets, either towards one another or for their armatures, is (below the point of saturation †) proportional to the square of the electric force. The *magnetism* in an electro-magnet is therefore simply as the electric force. Consequently the numbers in column 2 are proportional to the magnetic virtue of the electro-magnet. But on comparing columns 3 and 4 together, it will be seen that the heat evolved is as the square of the electricity. Therefore *the heat evolved by a revolving bar of iron is proportional to the square of the magnetic influence to which it is exposed.*

After the preceding experiments there can be no doubt that heat would be evolved by the rotation of non-magnetic substances in proportion to their conducting power, Dr. Faraday having proved the existence of currents in such circumstances, and that their quantity is proportional, *cæteris paribus,* to the conducting power of the body in which they are excited. I have not made any experiments on this subject; but in the next part we shall have occasion to avail ourselves of the good conducting power of copper, in conjunction with the magnetic virtue of the bar of iron, in order to obtain a maximum result from the revolution of a metallic bar.

Part II.—*On the Mechanical Value of Heat.*

Having proved that *heat* is *generated* by the magneto-electrical machine, and that by means of the inductive power of magnetism we can *diminish* or *increase* at pleasure the *heat* due to chemical changes, it became an object of great interest to inquire whether a constant ratio existed between it and the mechanical power gained or lost. For this purpose it was only necessary to repeat some of the previous experiments, and to ascertain, at the same time, the mechanical force necessary in order to turn the apparatus.

* Annals of Electricity, vol. iv. p. 131. Jacobi and Lenz communicated their report to the Petersburg Academy in March 1839—two months previous to the date of my paper.

† I am not aware that Jacobi and Lenz made any limitation to the law.—*Note*, 1881.

To accomplish the latter purpose, I resorted to a very simple device, yet one peculiarly free from error. The axle *b*, fig. 44 (p. 125), was wound with a double strand of fine twine, and the strings (as represented in fig. 51) were carried over very easily-working pulleys, placed on opposite sides of the axle, at a distance from each other of about 30 yards. By means of weights placed in the scales attached to the ends of the strings, I could easily ascertain the force necessary to move the apparatus at any given velocity; for, having given in the first instance the required velocity with the hand, it was easily observed, in the course of about 40 revolutions of the axle, corresponding to

Fig. 51.

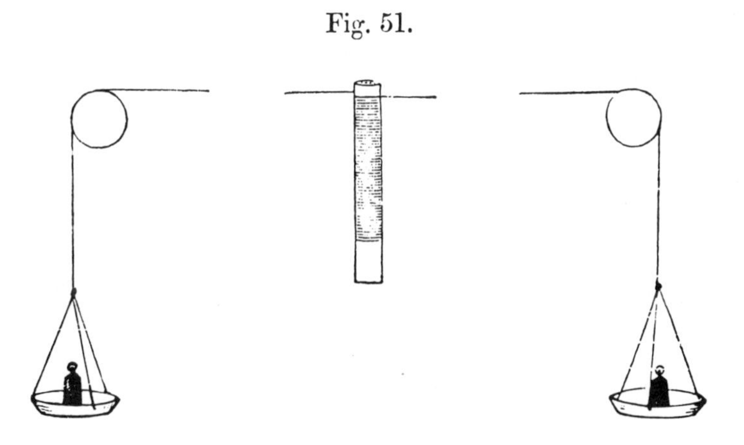

about 270 revolutions of the revolving piece, whether the weights placed in the scales were just able to maintain that velocity.

The experiments selected for repetition first were those of series No. 2. Ten cells, in a series of five double pairs, were connected with the large electro-magnet; and the small compound electro-magnet (restored to its place in the centre of the revolving tube) was connected, through the commutator, with the galvanometer. Under these circumstances a velocity of 600 revolutions per minute was found to produce a steady deflection of the needle to 24° 15′, indicating 0·983 of current magneto-electricity.

To maintain the velocity of 600 per minute, 5 lb. 3 oz. had to be placed in each scale; but when the battery was thrown out of communication with the electro-magnet, and the motion was opposed solely by friction and the resistance of the air, only 2 lb. 13 oz. were required for the same purpose. The difference, 2 lb. 6 oz., represents the force spent during the connexion of the battery with the electro-magnet in overcoming magnetic attractions and repulsions. The perpendicular descent of the weights was at the rate of 517 feet per 15 minutes.

According to series No. 2, Table I., the heat due to 0·983 of current magneto-electricity is $\left(\frac{983}{902}\right)^2 \times 1^\circ{\cdot}56 = 1^\circ{\cdot}85$. But as the resistance of the coil of the revolving electro-magnet was to that of the whole circuit as 1 : 1·13, the heat evolved by the whole conducting circuit was $1^\circ{\cdot}85 \times 1{\cdot}13 = 2^\circ{\cdot}09$. Adding to this, $0^\circ{\cdot}33$ on account of the heat evolved by the iron of the revolving electro-magnet, and $0^\circ{\cdot}04$ on account of the sparks* at the commutator, we have a total of $2^\circ{\cdot}46$. Now in order to refer this to the capacity of a lb. of water, I found:—

	lb.		lb.	
Weight of glass tube	= 1·65 =	capacity for heat of	0·300	of water.
Weight of water........	= 0·61 =	.. ..	0·610	..
Weight of electro-magnet.	= 1·67 =	.. ..	0·204	..
Total weight..	= 3·93 =	.. ..	1·114	..

$2^\circ{\cdot}46 \times 1{\cdot}114 = 2^\circ{\cdot}74$ †; and this has been obtained by the power which can raise 4 lb. 12 oz. to the perpendicular height of 517 feet.

1° of heat per lb. of water is therefore equivalent to a mechanical force capable of raising a weight of 896 lb. to the perpendicular height of one foot.

Two other experiments, conducted precisely in the same manner, gave a degree of heat to mechanical forces represented respectively by 1001 lb. and 1040 lb.

* The heat evolved by sparks in the above and subsequent instances had been determined by previous experiments.

† The thermal effects on the large stationary electro-magnet appear to have been neglected in this summary as insignificant But they were nevertheless in all probability sufficiently great to account for the difference between 838, the equivalent deduced from these experiments, and 772 which was ultimately arrived at.—*Note*, 1881.

I now made an experiment similar to those of series No. 10. Eight cells in a series of four double pairs were connected with the large electro-magnet, and two in series with the small revolving electro-magnet. The velocity of revolution was at the rate of 640 per minute, contrary to the direction of the attractive forces, causing the needle to be deflected to 37° 20′, which indicates a current of 1·955.

A weight of 6 lb. 4 oz. placed in each scale was just able to maintain the above velocity when the circuits were complete; but when they were broken, and friction alone opposed the motion, a weight of 2 lb. 8 oz. only was required, which is less than the former by 3 lb. 12 oz. The fall of the weights was in this instance 551 feet per 15 minutes.

According to series 10, Table II., the heat due to the current observed in the present instance is $\left(\frac{1\cdot955}{1\cdot845}\right)^2 \times 5^\circ\cdot88 = 6^\circ\cdot6$. But I had found by calculations, based as usual upon the laws of Ohm, that in the present experiment the resistance of the coil of the revolving electro-magnet was to that of the whole circuit, including the two cells, as 1 : 1·303. Therefore the heat evolved by the whole circuit, including 0°·18 on account of the iron of the revolving electro-magnet, and 0°·12 on account of sparks at the commutator, was 8°·9, or 9°·92 per capacity of a lb. of water.

Now, when the revolving electro-magnet was stationary, the two cells could pass through it a uniform current of 1·483. The heat evolved from the whole circuit by such a current is $\left(\frac{1\cdot483}{2\cdot145}\right)^2 \times 5^\circ\cdot88 \times 1\cdot303 \times 1\cdot114 = 4^\circ\cdot08$ per lb. of water per 15 minutes, according to data previously given. Hence the quantity of heat due to the chemical reactions in the experiment is $\frac{1\cdot955}{1\cdot483} \times 4^\circ\cdot08 = 5^\circ\cdot38$, instead of 9°·92, the quantity actually evolved.

Hence 4°·54 were evolved in the experiment over and above the heat due to the chemical changes taking place in the battery, by the agency of a mechanical power capable of raising 7 lb. 8 oz. to the height of 551 feet. In other words, one degree is equivalent to 910 lb. raised to the height of one foot.

An experiment was now made, using the same apparatus as an electro-magnetic engine. The power of the magnetic attractions and repulsions alone, without the assistance of any weights, was able to maintain a velocity of 320 revolutions per minute. But when the circuits were broken, a weight of 1 lb. 2 oz. had to be placed in each scale in order to obtain the same velocity. The deflection of the needle was in this instance $17^\circ\ 15' = 0{\cdot}63$ of current electricity. The perpendicular descent of the weights was 275 feet per 15 minutes.

Now, calculating in a similar manner to that adopted in the last experiment, we have, from series 9, Table II., and other data previously given, $\left(\frac{630}{543}\right)^2 \times 0^\circ{\cdot}50 \times 1{\cdot}303 = 0^\circ{\cdot}877$, which, on applying a correction of $0^\circ{\cdot}012$ on account of sparks at the commutator, and $0^\circ{\cdot}18$ on account of the iron of the revolving electro-magnet, and then reducing to the capacity of a pound of water, gives $1^\circ{\cdot}191$ as the quantity of heat evolved by the whole circuit in 15 minutes.

The quantity of current which the two cells could pass through the revolving electro-magnet when the latter was stationary was in this instance $1{\cdot}538$; and $\left(\frac{1{\cdot}538}{2{\cdot}115}\right)^2 \times 5^\circ{\cdot}88 \times 1{\cdot}303 \times 1{\cdot}114 = 4^\circ{\cdot}38$. Hence, as before, the quantity of heat due to the chemical reactions during the experiment is $\frac{0{\cdot}63}{1{\cdot}538} \times 4^\circ{\cdot}38 = 1^\circ{\cdot}794$, which is $0^\circ{\cdot}603$ more than was obtained during the revolution of the electro-magnet.

Hence $0^\circ{\cdot}603$ has been converted into a mechanical power equal to raise 2 lb. 4 oz. to the height of 275 feet. In other words, one degree per lb. of water may be converted into the mechanical power which can raise 1026 lb. to the height of one foot.

Another experiment, conducted in precisely the same manner as the above, gave, per degree of heat, a mechanical power capable of raising 587 lb. to the height of one foot.

As the preceding experiments are somewhat complicated,

and therefore subject to the accumulation of small errors of observation, I thought it would be desirable to execute some of a more simple character. For this purpose I determined upon an arrangement in which the whole * of the heat would be evolved in the revolving tube.

The iron cylinder used in previous experiments was placed in an electrotype apparatus constructed in such a manner as to render every part of it equally exposed to the voltaic action. In four days 11 oz. of copper were deposited in a hard compact stratum. The ends of the cylinder were then filed until the iron just appeared. Thus I had a cylinder of iron immediately surrounded by a hollow cylinder of pure copper nearly one eighth of an inch thick. This was placed in the centre of a new revolving tube fitted up in precisely the same manner as the former one (which had been accidentally broken), and surrounded with $11\frac{1}{4}$ oz. of water. I give the following series of experiments in which the above was rotated between

No. 16.

		Revolutions of the Bar per minute.	Deflections of Galvanometer of 1 turn.	Mean Temperature of Room.	Mean Difference.	Temperature of Water.		Gain or Loss.
						Before.	After.	
July 5, A.M.	Battery contact broken.	600	° ′ ..	67°·50	0°·15 −	67°·37	67°·33	0°·04 loss
July 5, A.M.	Electro-magnet in action.	600	72 35	69·32	0·42 −	67·50	70·30	2·80 gain
July 4, P.M.	Battery contact broken.	600	..	68·80	0·16 +	69·00	68·93	0·07 loss
July 4, P.M.	Electro-magnet in action.	600	72 25	69·70	0·56 +	69·00	71·52	2·52 gain
	Mean, Electro-magnet in action.	600	72 30	..	0·07 +	..	..	2·66 gain
	Mean, Battery contact broken.	600	..	..	..	..	..	0·05 loss
	Corrected Result.	600	72° 30′ = 10·93 current.					2·73 gain

* See note p. 151.

the poles of the large electro-magnet excited by ten cells arranged in a series of five double pairs, a galvanometer being included in the circuit to indicate the electric force to which the electro-magnet was exposed.

I now proceeded to ascertain, by means already described, the mechanical power by which the above effects were produced. First, I ascertained the current passing through the coil of the electro-magnet; then the weights necessary to maintain the velocity of 600 revolutions per minute, both when the magnet was in action and when contact with the battery was broken. I have collected the results of my experiments on this subject in the following table. The first five

TABLE IV.

	Deflections of the Galvanometer of one turn completing the circuit of the Electro-magnet.	Weight in each scale, the Electro-magnet being in action.	Weight in each scale, the Electro-magnet being not in action.	Difference.
		lb. oz.	lb. oz.	lb. oz.
	72° 30′	4 4	2 5	1 15
	72 30	4 4	2 3	2 1
	72 25	4 2	2 0	2 2
	72 15	5 0	2 10	2 6
	72 5	4 0	2 0	2 0
	68 0	3 14	2 8	1 6
	66 10	3 0	2 0	1 0
Mean of the first 5 experiments	72° 21′=10·82 current.			2·1 lb.
Mean of the last 2 experiments	67° 5′=7·91 current.			1·19 lb.

were obtained with a battery of ten cells in a series of five; the last two with a battery of five pairs in series.

Referring to Series 16, we see that 2°·73 were obtained when the bar was revolved between the poles of the electro-magnet excited by a current of 10·93. Therefore the quantity of heat due to the mean current in the first five experiments of the above table is $\left(\frac{10\cdot82}{10\cdot93}\right)^2 \times 2^\circ\cdot73 = 2^\circ\cdot675$.

To reduce this to the capacity of a pound of water, I had in the present instance the following data :—

	lb.		lb.	
Weight of glass tube .	= 1·125 =	capacity for heat of	0·205	of water.
Weight of water	= 0·687 =	.. ..	0·687	..
Weight of metallic bar	= 1·688 =	.. ..	0·202	..
Total weight	= 3·500 =	.. ..	1·094	..

2°·926, the product of 1·094 and 2°·675, is therefore the heat generated by a mechanical force capable of raising 4·2 lb. to the height of 517 feet.

In other words, one degree of heat per lb. of water may be generated by the expenditure of a mechanical power capable of raising 742 lb. to the height of one foot.

By a similar calculation, I find the result of the last two experiments of the table to be 860 lb.

The foregoing are all the experiments I have hitherto made on the mechanical value of heat. I admit that there is a considerable difference between some of the results, but not, I think, greater than may be referred with propriety to mere errors of experiment. I intend to repeat the experiments with a more powerful and more delicate apparatus. At present we shall adopt the mean result of the thirteen experiments given in this paper, and state generally that,

The quantity of heat capable of increasing the temperature of a pound of water by one degree of Fahrenheit's scale is equal to, and may be converted into, a mechanical force capable of raising 838 lb. to the perpendicular height of one foot.

Among the practical conclusions which may be drawn from the convertibility of heat and mechanical power into one another, according to the above absolute numerical relations, I will content myself with selecting two of the more important. The former of these is in reference to the duty of steam-engines; the latter, to the practicability of employing electro-magnetism as an economical motive force.

1. In his excellent treatise on the Steam-engine, Mr. Russell has given a statistical table *, containing the number of

* Encycl. Brit., 7th Edition, vol. xx. part 2, p. 685.

pounds of fuel evaporating one cubic foot of water, from the initial temperature of the water, and likewise from the temperature of 212°. From these facts it appears that in the Cornish boilers at Huel Towan, and the United Mines, the combustion of a lb. of Welsh coal gives 183° to a cubic foot of water, or otherwise 11,437° to a lb. of water. But we have shown that one degree is equal to 838 lb. raised to the height of one foot. Therefore the heat evolved by the combustion of a lb. of coal is equivalent to the mechanical force capable of raising 9,584,206 lb. to the height of one foot, or to about ten times the duty of the best Cornish engines.

2. From my own experiments, I find that a lb. of zinc consumed in Daniell's battery produces a current evolving about 1320°; in Grove's battery, about 2200° per lb. of water. Therefore *the mechanical forces of the chemical affinities which produce the voltaic currents in these arrangements are, per lb. of zinc, equal respectively to* 1,106,160 *lb. and* 1,843,600 *lb. raised to the height of one foot.* But since it will be practically impossible to convert more than about one half of the heat of the voltaic circuit into useful mechanical power, it is evident that the electro-magnetic engine, worked by the voltaic batteries at present used, will never supersede steam in an economical point of view.

Broom Hill, Pendlebury,
near Manchester, July 1843.

P.S.—We shall be obliged to admit that Count Rumford was right in attributing the heat evolved by boring cannon to friction, and not (in any considerable degree) to any change in the capacity of the metal. I have lately proved experimentally that *heat is evolved by the passage of water through narrow tubes.* My apparatus consisted of a piston perforated by a number of small holes, working in a cylindrical glass jar containing about 7 lb. of water. I thus obtained one degree of heat per lb. of water from a mechanical force capable of raising about 770 lb. to the height of one foot, a result which will be allowed to be very strongly confirmatory of our previous deductions. I shall lose no time in repeating and extending these experiments, being satisfied that the grand agents of nature are, by the Creator's fiat, *indestructible*; and that wherever mechanical force is expended, an exact equivalent of heat is *always* obtained.

On conversing a few days ago with my friend Mr. John Davies, he told me that he had himself, a few years ago, attempted to account for that part of animal heat which Crawford's theory had left unexplained, by the friction of the blood in the veins and arteries, but that, finding a similar hypothesis in Haller's 'Physiology'*, he had not pursued the subject further. It is unquestionble that heat is produced by such friction, but it must be understood that the mechanical force expended in the friction is a part of the force of affinity which causes the venous blood to unite with oxygen; so that the whole heat of the system must still be referred to the chemical changes. But if the animal were engaged in turning a piece of machinery, or in ascending a mountain, I apprehend that in proportion to the muscular effort put forth for the purpose, a *diminution* of the heat evolved in the system by a given chemical action would be experienced.

I will observe, in conclusion, that the experiments detailed in the present paper do not militate against, though they certainly somewhat modify, the views I had previously entertained with respect to the electrical origin of chemical heat. I had before endeavoured to prove that when two atoms combine together, the heat evolved is exactly that which would have been evolved by the electrical current due to the chemical action taking place, and is therefore proportional to the intensity of the chemical force causing the atoms to combine. I now venture to state more explicitly, that it is not precisely the attraction of affinity, but rather the mechanical force expended by the atoms in falling towards one another, which determines the intensity of the current, and consequently the quantity of heat evolved; so that we have a simple hypothesis by which we may explain why heat is evolved so freely in the combination of gases, and by

* Haller's 'Physiology,' vol. ii. p. 304.

which, indeed, we may account "latent heat" as a mechanical power prepared for action as a watch-spring is when wound up. Suppose, for the sake of illustration, that 8 lb. of oxygen and 1 lb. of hydrogen were presented to one another in the gaseous state, and then exploded, the heat evolved would be about one degree Fahr. in 60,000 lb. of water, indicating a mechanical force expended in the combination equal to a weight of about 50,000,000 lb. raised to the height of one foot. Now if the oxygen and hydrogen could be presented to each other in a liquid state, the heat of combination would be less than before, because the atoms, in combining, would fall through less space. The hypothesis is, I confess, sufficiently crude at present; but I conceive that ultimately we shall be able to represent the whole phenomena of chemistry by exact numerical expressions, so as to be enabled to predict the existence and properties of new compounds.

August, 1843. J. P. J.

30B

Reprinted from *The Scientific Papers on James Prescott Joule,* Vol. 1, Taylor & Francis Ltd., London, 1884, pp. 202–205

On the Existence of an Equivalent Relation between Heat and the ordinary Forms of Mechanical Power.

By JAMES P. JOULE, *Esq.*

[In a letter to the Editors of the 'Philosophical Magazine.']

['Philosophical Magazine,' ser. 3. vol. xxvii. p. 205.]

GENTLEMEN,

The principal part of this letter was brought under the notice of the British Association at its last meeting at Cambridge. I have hitherto hesitated to give it further publication, not because I was in any degree doubtful of the conclusions at which I had arrived, but because I intended

* The experiments were made at Oak Field, Whalley Range.

to make a slight alteration in the apparatus calculated to give still greater precision to the experiments. Being unable, however, just at present to spare the time necessary to fulfil this design, and being at the same time most anxious to convince the scientific world of the truth of the positions I have maintained, I hope you will do me the favour of publishing this letter in your excellent Magazine.

The apparatus exhibited before the Association consisted of a brass paddle-wheel working *horizontally* in a can of water. Motion could be communicated to this paddle by means of weights, pulleys, &c., exactly in the manner described in a previous paper*.

The paddle moved with great resistance in the can of water, so that the weights (each of four pounds) descended at the slow rate of about one foot per second. The height of the pulleys from the ground was twelve yards, and consequently, when the weights had descended through that distance, they had to be wound up again in order to renew the motion of the paddle. After this operation had been repeated sixteen times, the increase of the temperature of the water was ascertained by means of a very sensible and accurate thermometer.

A series of nine experiments was performed in the above manner, and nine experiments were made in order to eliminate the cooling or heating effects of the atmosphere. After reducing the result to the capacity for heat of a pound of water, it appeared that for each degree of heat evolved by the friction of water a mechanical power equal to that which can raise a weight of 890 lb. to the height of one foot had been expended.

The equivalents I have already obtained are:—1st, 823 lb., derived from magneto-electrical experiments†; 2nd, 795 lb., deduced from the cold produced by the rarefaction of air‡;

* Phil. Mag. ser. 3. vol. xxiii. p. 436. The paddle-wheel used by Rennie in his experiments on the friction of water (Phil. Trans. 1831, plate xi. fig. 1) was somewhat similar to mine. I employed, however, a greater number of "floats," and also a corresponding number of stationary floats, in order to prevent the rotatory motion of the water in the can.

† Phil. Mag. ser. 3. vol. xxiii. pp. 263, 347. ‡ Ibid. May 1845, p. 369.

and 3rd, 774 lb. from experiments (hitherto unpublished) on the motion of water through narrow tubes. This last class of experiments being similar to that with the paddle-wheel, we may take the mean of 774 and 890, or 832 lb., as the equivalent derived from the friction of water. In such delicate experiments, where one hardly ever collects more than half a degree of heat, greater accordance of the results with one another than that above exhibited could hardly have been expected. I may therefore conclude that the existence of an equivalent relation between heat and the ordinary forms of mechanical power is proved; and assume 817 lb., the mean of the results of three distinct classes of experiments, as the equivalent, until more accurate experiments shall have been made.

Any of your readers who are so fortunate as to reside amid the romantic scenery of Wales or Scotland could, I doubt not, confirm my experiments by trying the temperature of the water at the top and at the bottom of a cascade. If my views be correct, a fall of 817 feet will of course generate one degree of heat, and the temperature of the river Niagara will be raised about one fifth of a degree by its fall of 160 feet.

Admitting the correctness of the equivalent I have named, it is obvious that the *vis viva* of the particles of a pound of water at (say) 51° is equal to the *vis viva* possessed by a pound of water at 50° plus the *vis viva* which would be acquired by a weight of 817 lb. after falling through the perpendicular height of one foot.

Assuming that the expansion of elastic fluids on the removal of pressure is owing to the centrifugal force of revolving atmospheres of electricity, we can easily estimate the absolute quantity of heat in matter. For in an elastic fluid the pressure will be proportional to the square of the velocity of the revolving atmospheres, and the *vis viva* of the atmospheres will also be proportional to the square of their velocity; consequently the pressure will be proportional to the *vis viva*. Now the ratio of the pressures of elastic fluids at the temperatures 32° and 33° is 480 : 481; consequently the zero of temperature must be 480° below the freezing-point of water.

We see then what an enormous quantity of *vis viva* exists in matter. A single pound of water at 60° must possess 480° + 28° = 508° of heat; in other words, it must possess a *vis viva* equal to that acquired by a weight of 415036 lb. after falling through the perpendicular height of one foot. The velocity with which the atmospheres of electricity must revolve in order to present this enormous amount of *vis viva* must of course be prodigious, and equal probably to the velocity of light in the planetary space, or to that of an electric discharge as determined by the experiments of Wheatstone.

I remain, Gentlemen,

Yours respectfully,

JAMES P. JOULE.

Oak Field, near Manchester,
August 6, 1845.

30c

Reprinted from *The Scientific Papers of James Prescott Joule*, Vol. 1, Taylor & Francis Ltd., London, 1884, pp. 265-276

On Matter, Living Force, and Heat.

By J. P. Joule,
Secretary of the Manchester Literary and Philosophical Society.

[A Lecture at St. Ann's Church Reading-Room; and published in the Manchester 'Courier' newspaper, May 5 and 12, 1847.]

In our notion of matter two ideas are generally included, namely those of *impenetrability* and *extension*. By the extension of matter we mean the space which it occupies; by its impenetrability we mean that two bodies cannot exist at the same time in the same place. Impenetrability and extension cannot with much propriety be reckoned among the *properties* of matter, but deserve rather to be called its *definitions*, because nothing that does not possess the two qualities bears the name of matter. If we conceive of impenetrability and extension we have the idea of matter, and of matter only.

Matter is endowed with an exceedingly great variety of wonderful properties, some of which are common to all matter, while others are present variously, so as to constitute a difference between one body and another. Of the first of these classes, the attraction of gravitation is one of the most important. We observe its presence readily in all solid bodies, the component parts of which are, in the opinion of Majocci, held together by this force. If we break the body in pieces, and remove the separate pieces to a distance from each other, they will still be found to attract each other, though in a very slight degree, owing to the force being one which diminishes very rapidly as the bodies are removed further from one another. The larger the bodies are the more powerful is the force of attraction subsisting between them. Hence, although the force of attraction between small bodies can only be appreciated by the most delicate apparatus except in the case of contact, that which is occasioned by a body of immense magnitude, such as the earth, becomes very considerable. This attraction of bodies towards the earth constitutes what is called their *weight* or *gravity*, and is always exactly proportional to the quantity of matter. Hence, if any body be found to weigh 2 lb., while another only weighs 1 lb., the former will contain exactly twice as much matter as the latter; and this is the case, whatever the bulk of the bodies may be: 2-lb. weight of air contains exactly twice the quantity of matter that 1 lb. of lead does.

Matter is sometimes endowed with other kinds of attraction besides the attraction of gravitation; sometimes also it possesses the faculty of *repulsion,* by which force the particles tend to separate further from each other. Wherever these forces exist, they do not supersede the attraction of gravitation. Thus the weight of a piece of iron or steel is in no way affected by imparting to it the magnetic virtue.

Besides the force of gravitation, there is another very remarkable property displayed in an equal degree by every kind of matter—its perseverance in any condition, whether of rest or motion, in which it may have been placed. This faculty has received the name of *inertia,* signifying passiveness, or the inability of any thing to change its own state. It is in consequence of this property that a body at rest cannot be set in motion without the application of a certain amount of force to it, and also that when once the body has been set in motion it will never stop of itself, but continue to move straight forwards with a uniform velocity until acted upon by another force, which, if applied contrary to the direction of motion, will retard it, if in the same direction will accelerate it, and if sideways will cause it to move in a curved direction. In the case in which the force is applied contrary in direction, but equal in degree to that which set the body first in motion, it will be entirely deprived of motion whatever time may have elapsed since the first impulse, and to whatever distance the body may have travelled.

From these facts it is obvious that the force expended in setting a body in motion is carried by the body itself, and exists with it and in it, throughout the whole course of its motion. This force possessed by moving bodies is termed by mechanical philosophers *vis viva,* or *living force.* The term may be deemed by some inappropriate, inasmuch as there is no life, properly speaking, in question; but it is *useful*, in order to distinguish the moving force from that which is stationary in its character, as the force of gravity. When, therefore, in the subsequent parts of this lecture I employ the term *living force*, you will understand that I simply mean the force of bodies in motion. The living force of bodies is regulated by their weight and by the velocity of their motion. You will readily understand that if a body of a certain weight possess a certain quantity of living force, twice as much living force will be possessed by a body of twice the weight, provided both bodies move with equal velocity. But the law by which the *velocity* of a body regulates its living force is not so obvious. At first sight one would imagine that the living force would be simply proportional to the velocity, so that if a body moved twice as fast as another, it would have twice the impetus or living force. Such, however, is not the case; for if three bodies of equal weight move with the respective velocities of 1, 2, and 3 miles per hour, their living forces will be found to be proportional to those numbers multiplied by themselves, viz. to 1×1, 2×2, 3×3, or 1, 4, and 9, the squares of 1, 2, and 3. This remarkable law may be proved in several ways. A bullet fired from a gun at a certain velocity will pierce a block of wood to only one quarter of the depth it would if propelled at twice the velocity. Again, if a cannon-ball were found to fly at a certain velocity when propelled by a given charge of gunpowder, and it were required to load the cannon so as to propel the ball with twice that velocity, it would be found necessary to employ four times the weight of powder previously used. Thus, also, it will be found that a railway-train going at 70 miles per hour possesses 100 times the impetus, or living force, that it does when travelling at 7 miles per hour.

A body may be endowed with living force in several ways. It may receive it by the impact of another body. Thus, if a perfectly elastic ball be made to strike another similar ball of equal weight at rest, the striking ball will communicate the whole of its living force to the ball struck, and, remaining at rest itself, will cause the other ball to move in the same direction and with the same velocity that it did itself before the collision. Here we see an instance of the facility with which living force may be transferred from one body to another. A body may also be endowed with living force by means of the action of gravitation upon it through a certain distance. If I hold a ball at a certain height and drop it, it will have acquired when it arrives at the ground a degree of living force proportional to its weight and the height from which it has fallen. We see, then, that living force may be produced by the action of gravity through a given distance or space. We may therefore say that the former is of equal value, or *equivalent*, to the latter. Hence, if I raise a weight of 1 lb. to the height of one foot, so that gravity may act on it through that distance, I shall communicate to it that which is of equal value or equivalent to a certain amount of living force; if I raise the weight to twice the height, I shall communicate to it the equivalent of twice the quantity of living force. Hence, also, when we compress a spring, we communicate to it the equivalent to a certain amount of living force; for in that case we produce molecular attraction between the particles of the spring through the distance they are forced asunder, which is strictly analogous to the production of the attraction of gravitation through a certain distance.

You will at once perceive that the living force of which we have been speaking is one of the most important qualities with which matter can be endowed, and, as such, that it would be absurd to suppose that it can be destroyed, or even lessened, without producing the equivalent of attraction through a given distance of which we have been speaking. You will therefore be surprised to hear that until very recently the universal opinion has been that living force could be absolutely and irrevocably destroyed at any one's option. Thus, when a weight falls to the ground, it has been generally supposed that its living force is absolutely annihilated, and that the labour which may have been expended in raising it to the elevation from which it fell has been entirely thrown away and wasted, without the production of any permanent effect whatever. We might reason, *à priori,* that such absolute destruction of living force cannot possibly take place, because it is manifestly absurd to suppose that the powers with which God has endowed matter can be destroyed any more than that they can be created by man's agency; but we are not left with this argument alone, decisive as it must be to every unprejudiced mind. The common experience of every one teaches him that living force is not *destroyed* by the friction or collision of bodies. We have reason to believe that the manifestations of living force on our globe are, at the present time, as extensive as those which have existed at any time since its creation, or, at any rate, since the deluge—that the winds blow as strongly, and the torrents flow with equal impetuosity now, as at the remote period of 4000 or even 6000 years ago; and yet we are certain that, through that vast interval of time, the motions of the air and of the water have been incessantly obstructed and hindered by friction. We may conclude, then, with certainty, that these motions of air and water, constituting living force, are not *annihilated* by friction. We lose sight of them, indeed, for a time; but we find them again reproduced. Were it not so, it is perfectly obvious that long ere this all nature would have come to a dead standstill. What, then, may we inquire, is the cause of this apparent anomaly? How comes it to pass that, though in almost all natural phenomena we witness the arrest of motion and the apparent destruction of living force, we find that no waste or loss of living force has actually occurred? Experiment has enabled us to answer these questions in a satisfactory manner; for it has shown that, wherever living force is *apparently* destroyed, an equivalent is produced which in process of time may be reconverted into living force. This equivalent is *heat.* Experiment has shown that wherever living force is apparently destroyed or absorbed, heat is produced. The most frequent way in which living force is thus converted into heat is by means of friction. Wood rubbed against wood or against any hard body, metal rubbed against metal or against any other body—in short, all bodies, solid or even liquid, rubbed against each other are invariably heated, sometimes even so far as to become red-hot. In all these instances the quantity of heat produced is invariably in proportion to the exertion employed in rubbing the bodies together—that is, to the living force absorbed. By fifteen or twenty smart and quick strokes of a hammer on the end of an iron rod of about a quarter of an inch in diameter placed upon an anvil an expert blacksmith will render that end of the iron visibly red-hot. Here heat is produced by the absorption of the living force of the descending hammer in the soft iron; which is proved to be the case from the fact that the iron cannot be heated if it be rendered hard and elastic, so as to transfer the living force of the hammer to the anvil.

The general rule, then, is, that wherever living force is *apparently* destroyed, whether by percussion, friction, or any similar means, an exact equivalent of heat is restored. The converse of this proposition is also true, namely, that heat cannot be lessened or absorbed without the production of living force, or its equivalent attraction through space. Thus, for instance, in the steam-engine it will be found that the power gained is at the expense of the heat of the fire,—that is, that the heat occasioned by the combustion of the coal would have been greater had a part of it not been absorbed in producing and maintaining the living force of the machinery. It is right, however, to observe that this has not as yet been demonstrated by experiment. But there is no room to doubt that experiment would prove the correctness of what I have said; for I have myself proved that a conversion of heat into living force takes place in the expansion of air, which is analogous to the expansion of steam in the cylinder of the steam-engine. But the most convincing proof of the conversion of heat into living force has been derived from my experiments with the electro-magnetic engine, a machine composed of magnets and bars of iron set in motion by an electrical battery. I have proved by actual experiment that, in exact proportion to the force with which this machine works, heat is abstracted from the electrical battery. You see, therefore, that living force may be converted into heat, and that heat may be converted into living force, or its equivalent attraction through space. All, three, therefore—namely, heat, living force, and attraction through space (to which I might also add *light*, were it consistent with the scope of the present lecture)—are mutually convertible into one another. In these conversions nothing is ever lost. The same quantity of heat will always be converted into the same quantity of living force. We can therefore express the equivalency in definite language applicable at all times and under all circumstances. Thus the attraction of 817 lb. through the space of one foot is equivalent to, and convertible into, the living force possessed by a body of the same weight of 817 lb. when moving with the velocity of eight feet per second, and this living force is again convertible into the quantity of heat which can increase the temperature of one pound of water by one degree Fahrenheit. The knowledge of the equivalency of heat to mechanical power is of great value in solving a great number of interesting and important questions. In the case of the steam-engine, by ascertaining the quantity of heat produced by the combustion of coal, we can find out how much of it is converted into mechanical power, and thus come to a conclusion how far the steam-engine is susceptible of further improvements. Calculations made upon this principle have shown that at least ten times as much power might be produced as is now obtained by the combustion of coal. Another interesting conclusion is, that the animal frame, though destined to fulfil so many other ends, is as a machine more perfect than the best contrived steam-engine—that is, is capable of more work with the same expenditure of fuel.

Behold, then, the wonderful arrangements of creation. The earth in its rapid motion round the sun possesses a degree of living force so vast that, if turned into the equivalent of heat, its temperature would be rendered at least 1000 times greater than that of red-hot iron, and the globe on which we tread would in all probability be rendered equal in brightness to the sun itself. And it cannot be doubted that if the course of the earth were changed so that it might fall into the sun, that body, so far from being cooled down by the contact of a comparatively cold body, would actually blaze more brightly than before in consequence of the living force with which the earth struck the sun being converted into its equivalent of heat. Here we see that our existence depends upon the *maintenance* of the living force of the earth. On the other hand, our safety equally depends in some instances upon the *conversion* of living force into heat. You have, no doubt, frequently observed what are called *shooting-stars*, as they appear to emerge from the dark sky of night, pursue a short and rapid course, burst, and are dissipated in shining fragments. From the velocity with which these bodies travel, there can be little doubt that they are small planets which, in the course of their revolution round the sun, are attracted and drawn to the earth. Reflect for a moment on the consequences which would ensue, if a hard meteoric stone were to strike the room in which we are assembled with a velocity sixty times as great as that of a cannon-ball. The dire effects of such a collision are effectually prevented by the atmosphere surrounding our globe, by which the velocity of the meteoric stone is checked and its living force converted into heat, which at last becomes so intense as to melt the body and dissipate it into fragments too small probably to be noticed in their fall to the ground. Hence it is that, although multitudes of shooting-stars appear every night, few meteoric stones have been found, those few corroborating the truth of our hypothesis by the marks of intense heat which they bear on their surfaces.

Descending from the planetary space and firmament to the surface of our earth, we find a vast variety of phenomena connected with the conversion of living force and heat into one another, which speak in language which cannot be misunderstood of the wisdom and beneficence of the Great Architect of nature. The motion of air which we call *wind* arises chiefly from the intense heat of the torrid zone compared with the temperature of the temperate and frigid zones. Here we have an instance of heat being converted into the living force of currents of air. These currents of air, in their progress across the sea, lift up its waves and propel the ships; whilst in passing across the land they shake the trees and disturb every blade of grass. The waves by their violent motion, the ships by their passage through a resisting medium, and the trees by the rubbing of their branches together and the friction of their leaves against themselves and the air, each and all of them generate heat equivalent to the diminution of the living force of the air which they occasion. The heat thus restored may again contribute to raise fresh currents of air; and thus the phenomena may be repeated in endless succession and variety.

When we consider our own animal frames, "fearfully and wonderfully made," we observe in the motion of our limbs a continual conversion of heat into living force, which may be either converted back again into heat or employed in producing an attraction through space, as when a man ascends a mountain. Indeed the phenomena of nature, whether mechanical, chemical, or vital, consist almost entirely in a continual conversion of attraction through space, living force, and heat into one another. Thus it is that order is maintained in the universe—nothing is deranged, nothing ever lost, but the entire machinery, complicated as it is, works smoothly and harmoniously. And though, as in the awful vision of Ezekiel, "wheel may be in the middle of wheel," and every thing may appear complicated and involved in the apparent confusion and intricacy of an almost endless variety of causes, effects, conversions, and arrangements, yet is the most perfect regularity preserved—the whole being governed by the sovereign will of God.

A few words may be said, in conclusion, with respect to the real nature of heat. The most prevalent opinion, until of late, has been that it is a *substance* possessing, like all other matter, impenetrability and extension. We have, however, shown that heat can be converted into living force and into attraction through space. It is perfectly clear, therefore, that unless matter can be converted into attraction through space, which is too absurd an idea to be entertained for a moment, the hypothesis of heat being a substance must fall to the ground. Heat must therefore consist of either living force or of attraction through space. In the former case we can conceive the constituent particles of heated bodies to be, either in whole or in part, in a state of motion. In the latter we may suppose the particles to be removed by the process of heating, so as to exert attraction through greater space. I am inclined to believe that both of these hypotheses will be found to hold good,--that in some instances, particularly in the case of *sensible* heat, or such as is indicated by the thermometer, heat will be found to consist in the living force of the particles of the bodies in which it is induced; whilst in others, particularly in the case of *latent* heat, the phenomena are produced by the separation of particle from particle, so as to cause them to attract one another through a greater space. We may conceive, then, that the communication of heat to a body consists, in fact, in the communication of impetus, or living force, to its particles. It will perhaps appear to some of you something strange that a body apparently quiescent should in reality be the seat of motions of great rapidity; but you will observe that the bodies themselves, considered as wholes, are not supposed to be in motion. The constituent particles, or atoms of the bodies, are supposed to be in motion, without producing a gross motion of the whole mass. These particles, or atoms, being far too small to be seen even by the help of the most powerful microscopes, it is no wonder that we cannot observe their motion. There is therefore reason to suppose that the particles of all bodies, their constituent atoms, are in a state of motion almost too rapid for us to conceive, for the phenomena cannot be otherwise explained. The velocity of the atoms of water, for instance, is at least equal to a mile per second of time. If, as there is reason to think, some particles are at rest while others are in motion, the velocity of the latter will be proportionally greater. An increase of the velocity of revolution of the particles will constitute an increase of temperature, which may be distributed among the neighbouring bodies by what is called *conduction*—that is, on the present hypothesis, by the communication of the increased motion from the particles of one body to those of another. The velocity of the particles being further increased, they will tend to fly from each other in consequence of the centrifugal force overcoming the attraction subsisting between them. This removal of the particles from each other will constitute a new condition of the body—it will enter into the state of fusion, or become melted. But, from what we have already stated, you will perceive that, in order to remove the particles violently attracting one another asunder, the expenditure of a certain amount of living force or heat will be required. Hence it is that heat is always absorbed when the state of a body is changed from solid to liquid, or from liquid to gas. Take, for example, a block of ice cooled down to zero; apply heat to it, and it will gradually arrive at 32°, which is the number conventionally employed to represent the temperature at which ice begins to melt. If, when the ice has arrived at this temperature, you continue to apply heat to it, it will become melted; but its temperature will not increase beyond 32° until the whole has been converted into water. The explanation of these facts is clear on our hypothesis. Until the ice has arrived at the temperature of 32° the application of heat increases the velocity of rotation of its constituent particles; but the instant it arrives at that point, the velocity produces such an increase of the centrifugal force of the particles that they are compelled to separate from each other. It is in effecting this separation of particles strongly attracting one another that the heat applied is *then* spent; not in increasing the velocity of the particles. As soon, however, as the separation has been effected, and the fluid water produced, a further application of heat will cause a further increase of the velocity of the particles, constituting an increase of temperature, on which the thermometer will immediately rise above 32°. When the water has been raised to the temperature of 212°, or the boiling-point, a similar phenomenon will be repeated; for it will be found impossible to increase the temperature beyond that point, because the heat then applied is employed in separating the particles of water so as to form steam, and not in increasing their velocity and living force. When, again, by the application of cold we condense the steam into water, and by a further abstraction of heat we bring the water to the solid condition of ice, we witness the repetition of similar phenomena in the reverse order. The particles of steam, in assuming the condition of water, fall together through a certain space. The living force thus produced becomes converted into heat, which must be removed before any more steam can be converted into water. Hence it is always necessary to abstract a great quantity of heat in order to convert steam into water, although the temperature will all the while remain exactly at 212°; but the instant that all the steam has been condensed, the further abstraction of heat will cause a diminution of temperature, since it can only be employed in diminishing the velocity of revolution of the atoms of water. What has been said with regard to the condensation of steam will apply equally well to the congelation of water.

I might proceed to apply the theory to the phenomena of combustion, the heat of which consists in the living force occasioned by the powerful attraction through space of the combustible for the oxygen, and to a variety of other thermo-chemical phenomena; but you will doubtless be able to pursue the subject further at your leisure.

I do assure you that the principles which I have very imperfectly advocated this evening may be applied very extensively in elucidating many of the abstruse as well as the simple points of science, and that patient inquiry on these grounds can hardly fail to be amply rewarded.

31A

This article was translated expressly for this Benchmark volume by R. Bruce Lindsay, Brown University, from "Summary Report" of the Proceedings of the Royal Danish Academy of Sciences for 1844, *p. 3 (1845)*

NOTE ON THE WORK OF L. A. COLDING

The Royal Danish Academy of Sciences

[*Ed. note:* At the end of the report of the Academy session of January 5, 1844, there is the following note about the work of L. A. Colding.]

A committee established to examine an essay communicated by polytechnic candidate Colding renders the following judgment:

"The principal idea in the communication of polytechnic candidate Colding, on which the Academy has asked our opinion, is that the forces which appear to be lost in the output of machines, owing to frictional resistance, pressure, and the like, produce internal effects in bodies, as, for example, heat, electricity, and the like, and that these effects stand in a causal relation with the lost forces. To confirm his point he has carried out a series of experiments on the heat produced by friction.

We find that the leading idea is fully worthy of further experimental explorations, and that his researches are as satisfactory as could be desired, considering the means he has had at his disposal. We therefore desire to encourage him to continue this research and to supply him with financial assistance not to exceed the sum of 200 Rbdlra."

Copenhagen, January 4, 1844
H. C. Oersted, Ramos, Hoffmann

31B

This article was translated expressly for this Benchmark volume by R. Bruce Lindsay, Brown University, from "Undersøgelse om de almindelige Naturkraefte og deres gjnsige Afhaenighed," in Proceedings of the Royal Danish Academy of Sciences, 2, *123–131 (1851)*

INVESTIGATIONS OF THE GENERAL FORCES OF NATURE AND THEIR MUTUAL DEPENDENCE

Ludwig August Colding

I am convinced that experience has shown without any doubt that the various forces of nature are closely connected with each other. Experience indicates that each force through its activity is able to produce other forces and to set these free to become active in their turn.

Some relevant phenomena are so obvious that they have, so to speak, always been known. A good example is that vigorous rubbing of solid bodies against each other can produce heat or even light. It was early observed that amber can be electrified by rubbing, that is, acquire the property of attracting other bodies. It was long ago observed that a vapor or gas enclosed in a space of constant volume, when heated, would produce a force of expansion that could bring about mechanical action and motion. It was also observed that gases become warm when they are compressed and that one can indeed bring about the production of light in this way.

Experience shows that heat is produced when electric charges of opposite sign combine and that the chemical forces involved in the combination of substances produce heat, which when the forces are strong enough can lead to fire. It also turns out that chemical reactions can produce strong electric currents, which in turn can produce heat and light.

All these and many other results relating to the natural effects of forces on each other are provided by experience. At different times different investigators, impressed by the extent of this harmony between the forces of nature, concerned themselves with different subjects arising from them. No matter how isolated from each other these appeared to be, they at length, each in its turn, showed significant connections. The first publication of this kind that has come to my attention is the article by Clapeyron "Über die bewegende Kraft der Wärme" (On the Motive Power of Heat, *Poggendorfs Annalen* B, Vol. 59, p. 446). This article was based on the fundamental principle set forth by Sadi Carnot, to wit, that it is an absurdity to assume that one can develop moving force or heat from nothing. After the author had expressed his belief in Carnot's principle and stated his feeling that there is a definite connection between the quantity of activity in a certain quantity of heat and the mechanical activity that one can get out of it, he went on to develop mathematically the relation between such exchanges of activity, and arrived through the integration of a certain partial differential equation at an algebraic equation from which, among other things, he deduced the result obtained previously by Dulong through direct experiment (see *Memoirs of the Royal Academy of Sciences and the Institute of France,* Vol. 10, p. 188): that equal volumes of all elastic fluids at the same temperature and pressure give up or receive the same quantity of heat when they are respectively compressed or expanded by the same fraction of their volume. Since Clapeyron's formulas contain many arbitrary functions, they are not immediately useful. Later Suerman (*Poggendorfs Annalen* B., Vol. 41, p. 474) worked on these formulas and was able to achieve results agreeing exactly with experiment.

[*Ed. note:* There follows a rather long discussion of problems connected with forces in chemical reactions and in electromagnetic phenomena, clearly indicating Colding's wide and thorough grasp of the literature dealing with the sorts of fundamental relations for which he was seeking. He also introduces references to the work of Laplace, Poisson, and Clapeyron in connection with the heat developed in gases. He mentions the experiments of Colladon and Sturm on the velocity of sound in water and its connection with the compressibility of water, and the investigations of H. C. Oersted on the heat developed by the compression of water. There is no reference to the researches of Mayer and Joule, of which at that time he had heard nothing. Libraries in Copenhagen did not maintain a file of the *Philosophical Magazine.* At that time scientific relations were evidently closer with Germany than with England. We proceed now to reproduce Colding's remarks when he finally said something about his own work on the heat developed by friction.]

In the earlier communication that I presented to the Academy on November 1, 1843, I gave a summary of the results of my first investigations and showed how these agreed with the ideas about "lost" forces and the general notion of the relations connecting different kinds of natural forces. I there presented results of several series of experiments on the heat produced by friction. Each experiment was performed with apparatus constructed and described by me. The investigations described were carried out with brass, zinc, lead, iron, and lime wood, in every case rubbed on brass. The bar of brass on which the various other solids were made to slide expanded owing to the heat produced by the friction. This expansion was measured for each case.

[*Ed. note:* Although Colding does not state it explicitly here, his more detailed description later in this article makes clear that he used the expansion of the rubbed metal bar, measured by a delicate spherometer, to obtain (with the knowledge of the coefficient of thermal expansion of the solids in question) the rise in temperature accompanying the frictional heat and hence the total heat produced. He measured the work done by means of a dynamometer attached to the sliding weight.]

The results of a series of trials are presented in the accompanying table. For convenience the friction of brass rubbed on brass with a load of 31 pounds is listed as one unit of friction, and the associated quantity of heat produced is listed as one unit of heat. In all these trials the path length of slide was the same [*Ed. note:* One trial has been omitted here as being of little relevance.]

Trial no.	*Sliding material*	*Material rubbed*	*Load in pounds*	*Friction in terms of unit friction*	*Heat developed in terms of unit heat*
1	Brass	Brass	88.75	2.75	2.77
2	Brass	Brass	53.5	.179	1.83
3	Brass	Brass	31	1.00	1.00
4	Zinc	Brass	53.5	1.84	2.08
5	Zinc	Brass	31	1.24	1.20
6	Lead	Brass	31	1.77	1.76
7	Iron	Brass	53.5	1.74	1.80
8	Wood	Brass	53.5	1.68	1.66

From these results I have felt myself justified in concluding that the quantities of heat produced are proportional to the vanished moving forces.

[*Ed. note:* In this report of his early researches Colding does not report any value of the mechanical equivalent of heat resulting from them. In the latter part of the paper reproduced in part here, after introducing the details of 13 different trials, he does finally calculate a value of the equivalent from a summary of his results and gives the result in the form that the heat needed to raise the temperature of 1 pound of water 1°C is equivalent to the work of raising 1,185.4 pounds of water 1 foot. Assuming that the Danish pound and foot were the same as their contemporary English equivalents, the result works out to be about 3.54 joules/calorie in standard modern notation. This value agrees rather well with the value calculated by Mayer in 1842, although both values are some 20 percent lower than those ultimately obtained by Joule, and indeed by Colding and others much later. It is unfortunately not clear from Colding's published work whether he actually calculated a value for the equivalent in his earlier researches around 1843, or whether at that time he merely showed, as the table indicates, that his results verified his belief in a precise numerical relationship between frictional heat and work done in overcoming the friction. As far as the equivalent is concerned, it would seem that from the standpoint of priority he will have to stand on his 1851 paper, although his early work as reported was certainly of fundamental significance in the light of the generalization of the concept of energy. It is unfortunate that no formal publication of his earlier work appears to exist in the recognized scientific literature.

It is very interesting that instead of running his sliding apparatus in water and measuring the change in temperature of the water, more or less as Joule did with his paddle box, Colding preferred to measure the heat through the temperature rise corresponding to the observed thermal expansion. This required a degree of precision he was evidently not quite able to achieve, although in the light of the equipment available at his time it must be admitted that he did rather well.]

AUTHOR CITATION INDEX

SUBJECT INDEX